AF454028

Urban and Buildings Regeneration Strategy to Climatic Change Mitigation, Energy, and Social Poverty after a World Health and Economic Global Crisis

Urban and Buildings Regeneration Strategy to Climatic Change Mitigation, Energy, and Social Poverty after a World Health and Economic Global Crisis

Editor

Pilar Mercader-Moyano

MDPI • Basel • Beijing • Wuhan • Barcelona • Belgrade • Manchester • Tokyo • Cluj • Tianjin

Editor
Pilar Mercader-Moyano
Escuela Técnica Superior de
Arquitectura, Universidad de
Sevilla

Editorial Office
MDPI
St. Alban-Anlage 66
4052 Basel, Switzerland

This is a reprint of articles from the Special Issue published online in the open access journal *Sustainability* (ISSN 2071-1050) (available at: https://www.mdpi.com/journal/sustainability/special_issues/Urban_Buildings_Regeneration).

For citation purposes, cite each article independently as indicated on the article page online and as indicated below:

LastName, A.A.; LastName, B.B.; LastName, C.C. Article Title. *Journal Name* **Year**, *Article Number*, Page Range.

ISBN 978-3-0365-2424-5 (Pbk)
ISBN 978-3-0365-2425-2 (PDF)

© 2021 by the authors. Articles in this book are Open Access and distributed under the Creative Commons Attribution (CC BY) license, which allows users to download, copy and build upon published articles, as long as the author and publisher are properly credited, which ensures maximum dissemination and a wider impact of our publications.

The book as a whole is distributed by MDPI under the terms and conditions of the Creative Commons license CC BY-NC-ND.

Contents

About the Editor

Pilar Mercader-Moyano has a PhD in Architecture, and a Master's in Architecture and Historical Heritage and in Expert Assessment and Repair of Buildings. She is a lecturer and vice dean of Quality and Sustainable Habitat at the Higher Technical School of Architecture at the University of Seville, Spain. She is also the president of the Organising Committee of the five editions of the International Congress on Sustainable Construction and Ecoefficient Solutions and the director of the five editions of the International and Interuniversitary Master on Ecoefficient Refurbishment of Buildings and Neighbourhoods at the University of Seville. Her professional activity working as an architect for public administration and also for private clients in the field of architectural rehabilitation and sustainability in buildings has also been correlated with her research activity in the field of Sustainable Construction and Environmental Impact of Buildings.

 *sustainability*

Editorial

Special Issue "Urban and Buildings Regeneration Strategy to Climatic Change Mitigation, Energy, and Social Poverty after a World Health and Economic Global Crisis"

Pilar Mercader-Moyano * and Antonio Serrano-Jiménez

Departamento de Construcciones Arquitectónicas I, University of Seville, 41012 Seville, Spain; aserrano5@us.es
* Correspondence: pmm@us.es; Tel.: +34-618-305-559 (P.M.-M.); +34-646-580-435 (A.S.-J.)

 check for
updates

Citation: Mercader-Moyano, P.;
Serrano-Jiménez, A. Special Issue
"Urban and Buildings Regeneration
Strategy to Climatic Change
Mitigation, Energy, and Social
Poverty after a World Health and
Economic Global Crisis".
Sustainability **2021**, *13*, 11850.
https://doi.org/10.3390/su132111850

Received: 22 October 2021
Accepted: 25 October 2021
Published: 27 October 2021

Publisher's Note: MDPI stays neutral
with regard to jurisdictional claims in
published maps and institutional affiliations.

Copyright: © 2021 by the authors.
Licensee MDPI, Basel, Switzerland.
This article is an open access article
distributed under the terms and
conditions of the Creative Commons
Attribution (CC BY) license (https://
creativecommons.org/licenses/by/
4.0/).

1. Introduction

Throughout the 21st century, urban reports demand solutions to the obsolescence and aging process suffered by the existing buildings, due to the growth and expansion of cities that took place in the second half of the 20th century [1]. This context has meant that today, more than 40% of the European building stock is over 50 years old [2]. This factor together with continued global warming has led to world strategies on the adaptation and renovation of the built environment, that allow one to comply with current comfort, energy, and climatic requirements, and to achieve a sustainable urban transition by opting for bioclimatic conditioning techniques in different regions [3]. Additionally, the global crisis due to COVID-19 has supposed an adverse impact on urban regeneration of the built environment, with a recent world GDP drop of around 4.9%, and an estimated 20–30% decrease in economic funds intended for building renovation, with new decision support systems required, in order to carry out feasible and appropriate action strategies [4].

In this context, global guidelines on urban renovation aim to support low-carbon energy transition through sustainable and viable renovation strategies [5]. The European Energy Performance of Buildings Directive [6] underlines the need to implement new procedures for retrofitting the building stock, ensuring viable and efficient operations to achieve energy targets, as well as to improve the comfort conditions of users, also in families with low income levels that represent an important barrier for carrying out the building renovation or in other vulnerable social groups in which the ageing population predominates and specific accessibility conditions are required for the adaptation of buildings [7].

Thus, the topic of this Special Issue is focused on "Urban and Buildings Regeneration Strategy to Climatic Change Mitigation, Energy, and Social Poverty after a World Health and Economic Global Crisis", in response to the main urban challenges of this 21st century, and taking into account the new socioeconomic scenario that emerged after the global health crisis. This Special Issue of *Sustainability* mainly focuses on:

Thus, it is important to identify and discuss architectural, environmental, and economic problems in the activity of the construction sector, putting special emphasis on alternatives that provide solutions to the main works of building rehabilitation, with special attention to the residential sector, as well as guaranteeing technical, social and economic in feasibility in the action proposals.

- Providing new tools and methods to support the diagnosis and an adequate decision-making in building renovation, also considering different scenarios of vulnerability in the architectural, social, and economic fields.
- Introducing sustainable design patterns in urban planning, along with establishing users' awareness guidelines to guarantee a sustainable urban transition under this new socioeconomic context in emerging countries.
- Optimising the use of resources, energy consumption, materials, and providing an adequate indoor and outdoor environmental quality in the building sector, ensuring

1

an efficient management of all the construction phases, from the early design and decision-making, to guaranteeing a feasible building maintenance.

- Reactivating the construction sector, in the context of this world economic crisis, through solutions that strengthen the circular economy in eco-efficient building renovation and urban regeneration.
- Establishing an action proposal already carried out that serve as a reference for future actions, as an example of a sustainable, efficient pattern and towards an eco-efficient urban transition.

2. Background and Contents

Global targets on urban and social policies have recently established the urgency of adjusting technical and holistic decision support systems to effectively adapt and renovate the built environment, in their most vulnerable areas, in order to comply with current regulatory requirements and follow the main aims of the Sustainable Development Agenda for Sustainable Buildings [8].

Nowadays, a priority issue to deal with is the increase in obsolescence and deterioration in existing neighbourhoods, due to the expansion of cities carried out in the second half of the 20th century, which increasingly affects the guarantee of well-being and quality of life in society. European reports state that more than 40% of existing buildings were built over 50 years ago, and around 85% of residential environments are over 25 years old, demonstrating the need to design appropriate assessment and valuation models that incorporate technical, social and economic disciplines to decide the most appropriate actions to promote towards an effective and sustainable integrated urban regeneration in the coming years.

The global health crisis due to COVID-19 will have an impact on economic, social, and urban policies in the near future, with a world GDP drop of around 4.9% in 2020, and an expected decrease in the distribution of economic funds and priorities towards the urban regeneration of cities, which will make a higher challenge to adjust actions and optimise public resources in the urban, social, and housing fields [9]. This situation highlights the need to help governments with indicators that allow one to objectively detect certain parameters and urgencies that must be addressed with public investments, considering the real needs of users and meeting sustainable development goals [10], as an economic promotion for recovering the declining GDP in the post-COVID-19 crisis from the sector of construction, with this impact being more pronounced in vulnerable countries where extreme poverty is estimated to return to percentages of 20 years ago [11]. These alarming data raise the creation of new mechanisms that evaluate the vulnerability of urban spaces and buildings, involving important regulatory non-compliances and daily inconveniences for users, that must be quantified from multiple approaches to identify action priorities in the built environment [6].

Faced with this technical, social, and economic scenario, the circular economy is presented as an effective and sustainable solution for an integrated urban regeneration, at the same time that it can stimulate the vulnerability of emerging countries with effective action strategies. The incorporation of circular economy patterns allows generating more jobs, more industries, and the design of tools can guarantee economic viability with an optimization of resources. Thus, the current inefficiency of the building stock, together with the social precariousness, must be faced by an effective stimulus through research and the dissemination of good practices in the area of urban regeneration, ensuring, at the same time, the basic conditions of comfort, habitability, and safety.

3. Publications and Insights of the Special Issue

Starting with the introduction of sustainable criteria, circular economy and eco-efficient patterns in the urban regeneration process of the built environment under the current socioeconomic context, the research developed by Soto et al. [12] incorporates a detailed review and assessment of multiple environmental parameters and criteria required in different guides and regulations of the Spanish public administration, also compared with other international guidelines, carrying out a discussion of their usefulness and replicability as well as concluding with different specifications required to be developed and incorporate into new future environmental policies and strategies in the construction sector, especially in those promoted by the public administration.

Following this scope of application, the paper presented by Herrera-Limones et al. [13] is already making progress on this issue, and incorporates a new methodological proposal to satisfy the Sustainable Development Goals (SDGs) in the design of healthy and sustainable actions in the urban environment; for this, it contributes two new tools: the aura method and the aura matrix, both of them having been applied, tested, and compared in two architectural proposals presented to the Solar Decathlon Latin America of the Aura Team of the University of Seville, as a two-way learning between teachers and students. The results of the application allow one to demonstrate that both tools provide two useful concepts to incorporate multidisciplinary approaches to health, economic viability, or well-being requirements in the action proposals of the urban design of our cities, with the ultimate aim of improving the quality of life of citizens, from a transversal and sustainable perspective, fulfilling the established 2030 Agenda.

With a particular focus on the socioeconomic instability generated by the COVID-19 health crisis, the research developed by Mercader-Moyano et al. [14] has contributed in this issue with a novel assessment methodology for existing housing environments occupied by population groups that are in a situation of social vulnerability. This model has introduced the design of 51 different indicators of satisfaction, related to terms of habitability, grouped into different technical and social disciplines. The novelty of this study is that a tool is generated that combines social questionnaires with technical inspections to identify the main deficiencies and precarious conditions of housing environments, being this assessment methodology applied and corroborated in a representative neighbourhood of Mexico, which has allowed us to demonstrate the usefulness of these methodological mechanisms to prioritise highly effective actions in a context of different emergency scenarios.

In line with this previous study, on the situation of economic vulnerability generated by the pandemic, the study carried out by Alba-Rodríguez et al. [15] incorporates a deep reflection on the problem of energy poverty and its increase due to global warming, presenting an assessment model for evaluating retrofitting projects that considers real energy consumption, comfort demands, tenant's health, and the economic situation of users, therefore taking into account the occupant behaviour due to the limitation of their economic resources. This model incorporates, as a novelty, an Index of Vulnerable Homes to quantify the suitability of the actions and the adaptability to climate change of certain proposals, generating a new approach to energy poverty and providing more complete information for the adaptation capacity of the owners according to the climate variability and the different economic scenarios of intervention costs and energy costs that they can afford.

Another key aspect to address in the social vulnerability that exists in the built environment is the aging of the population, since there are increasingly elderly people living in obsolete and inadequate housing. This issue is addressed in the Special Issue in the research published by Agost-Felip et al. [16], which presents a study that brings together the specific needs of the elderly in the public open space, and designs a model to assess the vulnerability of public space through indicators from this particular approach. As a main contribution, this model and this approach implies recognizing the importance of the elderly in making decisions about the design of urban planning actions, in order to create age-friendly environments and even cities. Additionally, Gómez-Jiménez et al. [17] identify key aspects to take into account to satisfy the needs of the elderly in social housing or in specific social services, in order to be introduced in upcoming public policies. In order to have obtained important insights, the procedure carried out is to establish the main characteristics for an "ideal" adequate housing environment for elderly users, taking into account the existing characteristics and the regulations in force in Spain, launching new opportunities to propose a new Spanish regulation on elderly needs in the built environment.

Carrying out a special focus on ideas and new contributions regarding action proposals in housing renovation, different publications have incorporated new findings and important insights. The study published by Pitarch et al. [18] analyses the important impact of rehabilitating the roofs towards a sustainable regeneration of the existing building. The study introduces theoretical benefits and highlights the potential of gratings with regard to energy saving, the opportunity to incorporate green areas and other social uses. The different rehabilitation solutions were proposed and applied in a case study in Castellón, and their usefulness was discussed from different sustainable, economic, social and architectural criteria.

Following this line of research, the paper published by Domínguez-Torres et al. [19] introduces a new optimized and innovative construction system applicable to roofs that combines thermal insulation techniques with different thicknesses and solar reflective coatings, carrying out a dynamic analysis that quantifies savings and performance throughout the life cycle. The results show that the ideal balance between combining cool roof emissivity and insulation layer thickness produces an ideal solution for social housing renovation, with an average cost reduction of 63%, as well as good performance in its life cycle analysis, from an energy and economic point of view.

This Special Issue also incorporates two published papers that include innovative experiences of real application in different buildings and architectural elements. This real application is a confirmation of good, sustainable, and efficient practices for transforming the built environment. First, the publication by Ramos-Carranza et al. [20] describes the research methodology carried out, based on scientific observation, for the design of a prefabricated construction solution for rehabilitation for the application in water mills, as a family of singular and representative buildings that today require specific solutions for their renovation. The paper confirms the usefulness of this research methodology, and the environmental and economic benefits of the intervention strategy designed, carried out in "El Rodezno" as a case study, promoting the acceptance of prefabrication and its advantages for a circular economy.

Finally, the study presented by Calvo-Serrano et al. [21] presents a practical experimental case that shows that it is possible to carry out an eco-sustainable rehabilitation in buildings of special protection using the church of "Santiago Apostol" in Montilla, in the province of Córdoba, which is included in the list of Assets of Cultural Interests from 2001. The study details the eco-efficient criteria taken into account to carry out a design and decision-making methodology that has made it possible to quantify some indicators in different intervention strategies, which has made it possible to select a feasible proposal that has guaranteed environmental, economic and social sustainability for the inhabitants of this municipality.

As a final reflection, the contributions made in the 10 papers that have been selected for publication in this Special Issue introduce important findings towards a sustainable, satisfactory and efficient urban regeneration in the face of a new socioeconomic scenario caused by the pandemic. The results, ideas, approaches, and conclusions are expected to be of great significance for different researchers of these lines of action, and to be a reference for future studies of great impact, allowing the generation of important synergies that lead to a reconversion of the built environment to the current needs of society.

Author Contributions: Conceptualization and methodology, P.M.-M. and A.S.-J.; investigation and writing—original draft preparation, A.S.-J.; writing—review and editing and supervision, P.M.-M. All authors have read and agreed to the published version of the manuscript.

Funding: This special issue has been possible thanks to the funding received by the research funds of the University of Seville "Ayuda del VI Plan Propio de Investigación y Transferencia de la Universidad de Sevilla 2021, para Proyectos de Investigación Precompetitivos (IV.4)" through the research project "REGEBIM" (Gestión integral BIM de soluciones de rehabilitación de envolventes, con criterios de economía circular y sostenibilidad regenerativa, como impulso a la recuperación económica). This work was also developed thanks to the collaboration of the Andalusian Government (Junta de Andalucía—Consejería de Economía, Innovación y Ciencia) through a postdoctoral contract DOC-00950 granted to Antonio Serrano-Jiménez.

Acknowledgments: We would like to express our gratitude to all the authors and reviewers for their contributions and constructive support and comments. Honestly, they have provided important insights to increase the scientific quality of this Special Issue, on such an important approach for the built environment. We also thank the editorial assistance office of MDPI, which demonstrated interest and effort to support during the process of reviewing and publishing this volume.

Conflicts of Interest: The authors declare no conflict of interest.

References

1. Santangelo, A.; Tondelli, S. Occupant behaviour and building renovation of the social housing stock: Current and future challenge. *Energy Build.* **2017**, *145*, 276–283. [CrossRef]
2. Eurostats, Newsrealease-Euroindicators February 2019. Available online: https://ec.europa.eu/eurostat/documents/2995521/9967985/3-10072019-BP-EN.pdf/e152399b-cb9e-4a42-a155-c5de6dfe25d1 (accessed on 5 October 2021).
3. WHO. Supporting Older People during the COVID-19 Pandemic Is Everyone's Business. 2020. Available online: http://www.euro.who.int/en/health-topics/health-emergencies/coronaviruscovid19/news/news/2020/4/supporting-olderpeople-during-the-covid-19-pandemic-is-everyones-business. (accessed on 7 October 2021).
4. Mercader-Moyano, P.; Morat, O.; Serrano-Jiménez, A. Urban and social vulnerability assessment in the built environment: An interdisciplinary index-methodology towards feasible planning and policy-making under a crisis context. *Sustain. Cities Soc.* **2021**, *73*, 103082. [CrossRef]
5. European Commission. *An EU Strategy on Heating and Cooling (COM(2016) 51 Final)*; Brussels, Belgium, 16 February 2016. Available online: https://ec.europa.eu/transparency/documents-register/detail?ref=COM(2016)51&lang=en (accessed on 24 October 2021).
6. European Commission. Directive (EU) 2018/844 of the European Parliament on the energy performance of buildings and energy efficiency. *Off. J. Eur. Union* **2018**, *156*, 75–91.
7. Serrano-Jiménez, A.; Femenías, P.; Thuvander, L.; Barrios-Padura, Á. A multi-criteria decision support method towards selecting feasible and sustainable housing renovation strategies. *J. Clean. Prod.* **2020**, *278*, 123588. [CrossRef]
8. GBCe, European Agenda for Sustainable Buildings. 2020. Available online: https://gbce.es/documentos/Agenda-de-la-UE-para-la-edificacion-sostenible.pdf. (accessed on 20 July 2021).
9. OECD, GDP. 2021. Available online: https://data.oecd.org/gdp/gross-domestic-product-gdp.htm. (accessed on 18 August 2021).
10. ONU. *Transforming Our World: The 2030 Agenda for Sustainable Development*; United Nations: New York, NY, USA, 2015.
11. The World Bank. Supporting Countries in Unprecedented Times. *World Bank Annu. Rep.* **2020**, 1–106. Available online: https://www.worldbank.org/en/about/annual-report/world-bank-group-downloads. (accessed on 18 March 2021).
12. Soto, T.; Escrig, T.; Serrano-Lanzarote, B.; Matarredona, N. An Approach to Environmental Criteria in Public Procurement for the Renovation of Buildings in Spain. *Sustainability* **2020**, *12*, 7590. [CrossRef]
13. Herrera-Limones, R.; LopezDeAsiain, M.; Borrallo-Jiménez, M.; García, M.T. Tools for the Implementation of the Sustainable Development Goals in the Design of an Urban Environmental and Healthy Proposal. A Case Study. *Sustainability* **2021**, *13*, 6431. [CrossRef]
14. Mercader-Moyano, P.; Morat-Pérez, O.; Muñoz-González, C. Housing Evaluation Methodology in a Situation of Social Poverty to Guarantee Sustainable Cities: The Satisfaction Dimension for the Case of Mexico. *Sustainability* **2021**, *13*, 11199. [CrossRef]
15. Alba-Rodríguez, M.; Rubio-Bellido, C.; Tristancho-Carvajal, M.; Castaño-Rosa, R.; Marrero, M. Present and Future Energy Poverty, a Holistic Approach: A Case Study in Seville, Spain. *Sustainability* **2021**, *13*, 7866. [CrossRef]
16. Agost-Felip, R.; Ruá, M.; Kouidmi, F. An Inclusive Model for Assessing Age-Friendly Urban Environments in Vulnerable Areas. *Sustainability* **2021**, *13*, 8352. [CrossRef]
17. Gómez-Jiménez, M.; Antonio, V.-Y. Key Elements for a New Spanish Legal and Architectural Design of Adequate Housing for Seniors in a Pandemic Time. *Sustainability* **2021**, *13*, 7838. [CrossRef]

18. Pitarch, Á.; Ruá, M.; Reig, L.; Arín, I. Contribution of Roof Refurbishment to Urban Sustainability. *Sustainability* **2020**, *12*, 8111. [CrossRef]
19. Domínguez-Torres, C.-A.; Domínguez-Torres, H.; Domínguez-Delgado, A. Optimization of a Combination of Thermal Insulation and Cool Roof for the Refurbishment of Social Housing in Southern Spain. *Sustainability* **2021**, *13*, 10738. [CrossRef]
20. Ramos-Carranza, A.; Añón-Abajas, R.; Rivero-Lamela, G. A Research Methodology for Mitigating Climate Change in the Restoration of Buildings: Rehabilitation Strategies and Low-Impact Prefabrication in the "El Rodezno" Water Mill. *Sustainability* **2021**, *13*, 8869. [CrossRef]
21. Calvo-Serrano, M.; Castillejo-González, I.; Montes-Tubío, F.; Mercader-Moyano, P. The Church Tower of Santiago Apóstol in Montilla: An Eco-Sustainable Rehabilitation Proposal. *Sustainability* **2020**, *12*, 7104. [CrossRef]

sustainability

Article

Housing Evaluation Methodology in a Situation of Social Poverty to Guarantee Sustainable Cities: The Satisfaction Dimension for the Case of Mexico

Pilar Mercader-Moyano [1,*], Oswaldo Morat-Pérez [2] and Carmen Muñoz-González [3]

[1] Escuela Técnica Superior de Arquitectura, Departamento de Construcciones Arquitectónicas, Universidad de Sevilla, 41012 Sevilla, Spain
[2] Departmento de Arquitectura, Diseño y Urbanismo, Universidad Autónoma de Tamaulipas, Tampico 89000, Mexico; swldarq_@hotmail.com
[3] Escuela Técnica Superior de Arquitectura, Departamento de Construcciones Arquitectónicas, Universidad de Málaga, 29017 Málaga, Spain; carmenmgonzalez@uma.es
*   Correspondence: pmm@us.es; Tel.: +34-618305559

Abstract: Currently, one in eight people live in neighborhoods with social inequality and around one billion people live in precarious conditions. The significance of where and how to live and in what physical, spatial, social, and urban conditions has become very important for millions of families around the world because of mandatory confinement due to the COVID-19 pandemic. Today, many homes in poor condition do not meet the basic requirements for residential environments in the current framework. Theoretical models for the urban evaluation of this phenomenon are a necessary starting point for urban renewal and sustainability. This study aims to generate a model for evaluating homes in a situation of social inequality (hereinafter Vrs) with indicators on physical, spatial, environmental, and social aspects. The methodology used in this study evaluates housing, taking into consideration habitability factors (physical, spatial, and constructive characteristics), as well as the qualitative characteristics assessing the satisfaction of users with the adaptation and transformation of the housing and its surroundings. The application of 51 indicators distributed in four previous parameters was established for this study. This quantification identifies the deficiencies of the dwellings and sets the guidelines for the establishment of future rehabilitation policies for adapting the dwellings to current and emergency scenarios. The innovation of this study is the construction of a tool for social research surveys designed to include individual indicators from the dwellings' users, to provide a more dependable representation of the problems found in Vrs. The results of this research identified the deficiencies of precarious housing and could be used for applying effective proposals for improvement of habitability and their surroundings in the future. Furthermore, the results showed that when all the indicators were considered, the level of lag reached would be similar to that of a real housing situation, further confirming the suitability of the methodology applied in this investigation.

Keywords: city; indicators; social gap; urban regeneration; pandemic

 check for **updates**

Citation: Mercader-Moyano, P.; Morat-Pérez, O.; Muñoz-González, C. Housing Evaluation Methodology in a Situation of Social Poverty to Guarantee Sustainable Cities: The Satisfaction Dimension for the Case of Mexico. *Sustainability* **2021**, *13*, 11199. https://doi.org/10.3390/su132011199

Academic Editor: Agnieszka Bieda

Received: 27 August 2021
Accepted: 25 September 2021
Published: 12 October 2021

Publisher's Note: MDPI stays neutral with regard to jurisdictional claims in published maps and institutional affiliations.

Copyright: © 2021 by the authors. Licensee MDPI, Basel, Switzerland. This article is an open access article distributed under the terms and conditions of the Creative Commons Attribution (CC BY) license (https://creativecommons.org/licenses/by/4.0/).

1. Introduction

One in eight people worldwide live in socially disadvantaged neighborhoods. This means that, in total, there are about one billion people living in substandard conditions [1], a situation which is expected to worsen in the future.

Between 2000 and 2014, developed countries saw a visible improvement in the situation of these neighborhoods and the urban population living in such situations decreased from 39 to 30%.

The Social Lag Index is a weighted measure that summarizes four indicators of social deprivation (education, health, basic services, and spaces in the home) in a single index

7

which aims to order the observation units according to social deprivation. Nowadays, social lag in housing continues to be a critical factor in the persistence of poverty [2]. In fact, 90% of urban growth occurs in developing countries where every year, urban areas gain 70 million new residents. Over the next two decades, the urban population of the two poorest regions in the world, South Asia and Africa, is expected to double, leading to an increase in the number of precarious settlements. In Africa, where 59% of the urban population lives in disadvantaged neighborhoods, the number of urban dwellers is expected to reach 1.2 billion by 2050 [1].

In Asia and the Pacific, this would affect around 28% of the population, despite successful efforts from Asian governments in recent years to improve the quality of life of the 172 million inhabitants of these neighborhoods [3,4]

In Latin America and the Caribbean, where regulations on Vrs improvement have long been in place, approximately 21% of the population still lives in these buildings, although this figure has dropped by 17% [5]. In Arab countries, these figures range from a minority to 67–94%, depending on the country.

In Europe, soaring housing costs in the richest major cities have led to an increase in the number of urban dwellers unable to afford rent. In fact, over 6% of the urban population in Western Europe lives in extremely precarious conditions [6]. At the same time, in the USA, housing precariousness can be observed among low-income inhabitants in rural settings [7].

One of the Millennium Development Goals for 2020 is to significantly improve the lives of at least 100 million slum-dwellers, ensuring access to basic housing and services for all by 2030.

Therefore, this study aims to create an evaluation model for socially disadvantaged homes, in order to analyze, evaluate, and propose improvements to ensure adequate housing habitability currently and in emergency scenarios. This could lead to the proposal of improvement solutions suiting the real needs of users. Following an in-depth study of the bibliography, the social parameters and indicators for housing satisfaction are identified and combined with other research indicators which could be generalized and replicated in other case studies. In addition, based on the results of this study, a data collection tool for social research on the occupants of the homes studied is designed to address the real needs of the users.

1.1. Socially Disadvantaged Homes

During the state of alarm caused by COVID-19, homes became a refuge from the pandemic, but they also became workspaces, places of leisure, games, etc. However, the lockdown experience has not been the same for everyone and has been greatly conditioned by the type of housing. The experiences of living in single-family or multi-family homes or in situations of social inequality are not comparable [8].

The importance of how and where to live and in what physical, spatial, social, and urban environment conditions has hugely increased for millions of families around the world following the compulsory lockdown enforced during the COVID-19 pandemic. Housing is one of the issues at the center of this battle, as social distancing and good hygiene practices cannot be guaranteed without suitable housing [9].

The social dimension and urban environment have been affected by overcrowded coexistence in Vrs. Increased stress levels were observed as the limited space available in the vast majority of these homes could not meet the needs of their inhabitants. This has also led to an increase in instances of violence against women, transmission of the disease due to having to live in close quarters, and psychological issues in children and adults stemming from the absence of well-lit green open spaces. This pandemic has highlighted the importance of social cohesion and urban development, and therefore the urban agenda should be placed at the center of public policies [10].

Worsening of quality of life and well-being in precarious housing during lockdown has prompted a search for urgent solutions, which have now been added to previous actions set out in international agendas, such as the Sustainable Development Goals for 2030.

For the sustainable development of cities, extensive research has been carried out by experts such as Mona Atia in Morocco [11], D. Rockwood [12] and Mᵃ José Rúa [13] on the issue of action and rehabilitation processes in depressed urban environments experiencing social poverty. To carry out this task, the authorities and local communities in particular, play a key role in urban regeneration and planning human settlements, while the transformative power of urban policies and territorial planning tools for sustainable development is undeniable [14–16].

In Africa, programs have been set up in partnership with local communities to improve the population's living conditions, prioritizing rainwater and sanitation facilities in public spaces [1] In Australia, the New Office of Urbanization developed the country's first 2010–2030 urbanization policies to reduce urban poverty in unplanned settlements [17] In South America, governments are implementing programs for sustainable urban and rural development, such as the ICHP housing project and a water project [18].

In addition, according to the Diagnostic Study of the Right to Proper Housing in 2018 [19], in Mexico there are fourteen million homes in need of new construction or substantial improvement. This represents 45% of the overall housing stock, although, in rural areas, this figure can reach up to 97%.

At the urban level, socially disadvantaged areas are considered to be those occupied by groups of people in situations of marginalization and social exclusion, and the issues affecting them are rarely addressed in policies, especially in developing countries [20]. These settlements form as a result of internal migrations of population in search of better job opportunities brought about by the injection of foreign capital and the transition from agricultural to industrial and service economies [21]. Whereas habitability conditions should be tackled in order to improve living conditions in housing, ignoring social objectives and focusing on the development of cities could result in a negative cycle of imbalances within the structure of the city [12], to the detriment of the global objectives of achieving resilient cities.

Therefore, an urban study of social lag is important for policy formulation, especially as few studies have attempted in-depth examinations of this issue using real data in situ, as highlighted by Sebastián Galiania [22]. Different governments have selected a range of approaches for the improvement of the quality of life of the occupants of these neighborhoods. One example of this was the proposed relocation of these neighborhoods to other places in the 1970s, usually to cheaper land where homes could be rebuilt [23]. However, in the 1980s, programs emerged which were based on a series of policies for the improvement of urban infrastructure and services. The current situation is the direct result of these initiatives for improvement.

At present, theoretical models are a necessary starting point for assessing the habitability of homes on an international scale. However, attempts to implement this practice in real cases in a given physical and socioeconomic context could lead to different housing models, considered unsuitable for the case study [13].

An example of this can be found in the Right to Housing Plan, drawn up to relieve pressure on the housing situation in Barcelona. Taking into consideration access to urban resources and equipment services, this plan aimed to reverse inequality by improving habitability conditions in the housing and its surroundings [24]. Certain proposals presented aimed to expand the housing stock in order to reduce real estate speculation, while others focused on Spanish and European policies for urban regeneration and rehabilitation of these neighborhoods, restoring their importance within the city. However, since these neighborhoods were in central areas, the housing fell prey to speculators [25].

Given the worsening quality of housing, poverty, and unemployment conditions in the Netherlands and Great Britain, housing diversification plans have tended to recommend demolishing, selling, or updating these dwellings. In this case, the main objective was not just to improve housing but, more so, to improve the economies of cities to the advantage of urban areas in the regional housing market [26].

The habitability of housing, which inherently conditions the well-being of individuals and their environment, is essential for ensuring sustainable global development [27]. Habitability and quality of life are conditions determined by the psychosocial profile of individual families, and are seen as habits, behaviors, and ways of being consolidated over time. Quality of life in their homes is an influential factor in the way of life of its residents, covering specific characteristics to meet their needs, while also providing some degree of satisfaction.

As described above, although the issue of habitability is global, its intensity is determined by the particular economic and social characteristics of individual regions. This is the case in Mexico, where the situation has worsened, especially in city centers. At present, housing in Mexico is an urban problem stemming from the housing policies put in place to guarantee families access to this constitutional right [28]. Marginalized sectors of the population outside the economic structure of the industrial city and the lack of purchasing power have prevented the population from accessing decent housing. According to calculations made in recent years, one million homes in Mexico need retrofitting or adaptation. In addition, 220,000 new homes must be built every year in order to cope with population growth. Since the 1950s, the population of Mexico City has increased from 3.4 to 21.3 million inhabitants and the city is expected to be one of the top ten in the world population growth ranking for the year 2035, as seen in Figure 1 [29].

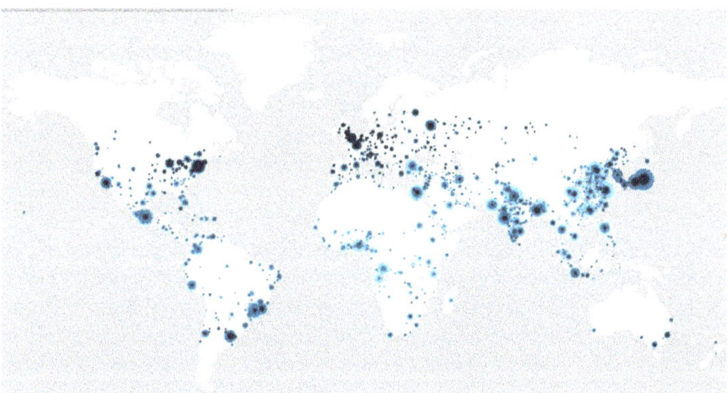

Figure 1. World city populations 1950–2035. Dark Blue (current population), light blue (future population). © Colaboradores de OpenStreetMap.

1.2. Urban Sustainability Indicators

Given the large number of factors to be considered and correlated in the growth of cities, the assessment process for this situation is complex. It requires the use of indicators, which are tools that provide concise information through the description and analysis of complex situations [13]. The task of establishing a list of suitable indicators is challenging, as these are becoming increasingly important tools for drawing up policies on management performance in key problems for contemporary cities [30].

The indicators, i.e., parameters that provide information or a value derived from them, describe the status of a given phenomenon [31], summarize a situation, and are developed for specific objectives. When applied in cities, they make it possible to evaluate certain processes or aspects of a reality geared toward previously defined objectives, identifying improvements, changes, etc. [13,32].

Among the current wide range of indicators for different purposes, we find the BUES comprehensive evaluation model which uses multidisciplinary indicators on physical, urban, environmental, and social conditions to formulate policies on vulnerable housing [33].

Other indicators are used in the evaluation of the vulnerability of an urban environment to climate change [34,35] the energy reconditioning of social housing [36,37] and housing spaces and their construction materials [38]. Therefore, given the multiple scientific, political, and social fields involved, establishing a methodology to standardize the indicators is a complex task.

Most studies have reviewed indicators suited to the case study for existing sustainable development, ascertaining which indicators are most appropriate with the help of experts in the field [13,39].

These indicators, which are influenced by the scale and unique features of an urban study environment [40,41] are of importance in renewal policies. According to the studies by Marta Braulio et al. [42], these indicators vary from one region to another and must be suited to the specific conditions of the context of the study region if they are to be used as tools.

At present, these indicators are of importance in decision making on renewal policies in European countries. This is the case of the Sustainable Growth Operational Programs (2014–2020) for smart, sustainable, and inclusive growth and the 2018–2021 Housing Plan, which reinforce the Spanish Ministry of Development's commitment to rehabilitation and energy efficiency. For example, in social housing, dating from 1939 to 1979 in Zaragoza (Spain), physical performance indicators were developed in collaboration with the local administration at the neighborhood and district levels. This made it possible to identify the buildings with the worst performance in energy efficiency, sound insulation, accessibility, etc., with a view to retrofitting [43].

In Barcelona, these indicators are geared toward the retrofitting of homes for energy savings [44]. This is also the case in Malaga, where Spanish and European policies for urban regeneration and rehabilitation of neighborhoods have been implemented, as seen in the case studies of the areas of Trinidad and Perchel between 2005 and 2012 [25]. The approval of the General Plan of Urban Planning for Malaga jeopardized the traditional neighborhoods, which were targeted by major speculative interests, prompting mass protests from city residents.

Other international regeneration plans for homes and neighborhoods, including that for the new Urban Habitat Agenda III (NAU) in Quito in Ecuador, address the issue of human settlements [45]. In Colombia, for instance, a subsidy was awarded for 100,000 homes [46] under the national development plan (2014–2018) aimed at the population in extreme poverty.

Despite all of these policies, the participation of administrations, and intervention programs for social housing, it has not yet been possible to halt the development of informal housing built by the population in extreme poverty. These studies clearly highlight why public policies should focus on solving the needs studied by an evaluation tool. Target users must be willing to embrace the improvement solutions if these are to succeed. If attention is paid only to the development of cities, but neglecting social objectives, the structure of these cities is subjected to a negative cycle of imbalances to the detriment of the global goals of achieving resilient cities. This research develops a tool which incorporates the indicators to be used, following an in-depth selection process from those internationally available and combined with those obtained from in situ fieldwork on Vrs.

2. Material and Methods

The evaluation methodology in this study focused on housing, taking into consideration habitability factors (physical, spatial, and constructive characteristics), as well as on the qualitative characteristics assessing user satisfaction regarding the adaptation and transformation of housing and its surroundings. The following five phases in the evaluation model meet the lower-level objectives established hierarchically in order to guarantee the main objective, as seen in Figure 2:

➢ The first phase is based on a review of international, national, and local scientific studies on the use of indicators for detecting and analyzing dwellings in situations of social lag.

➢ The second phase describes the study area and sample size in order to define appropriate indicators, based on the analysis in areas of social inequality.

➢ In the third phase, the hierarchy matrix of indicators applicable to the Vrs is generated through fieldwork in the social context of the sample determined in the previous phase, with social, physical-spatial, and surrounding indicators from relevant international and national bibliographical and institutional sources [5,19,47–49]. The indicators of physical-spatial characteristics of the homes and the urban context of the neighborhood and social considerations of the users are all used as a starting point for the development of the data collection tool. Qgis software version 3.6.2 is used to display the results obtained by territory.

➢ The fourth phase uses the database obtained to evaluate and quantify user satisfaction for the social, physical-spatial, and housing environmental indicators. This makes it possible to establish a habitability scale, as well as guidelines for rehabilitation and urban regeneration. For the purposes of comparing results, the level of social lag is represented in five strata (very low, low, medium, high, and very high), determined through statistical sampling using computer programs such as Statistical Package for the social Sciences SPSS (IBM) and statistics for excel XLSTAT.

➢ Finally, in the fifth phase, these statistical data are used to formulate the evaluation model, containing all the indicators to be considered, including the results obtained from the research. The evaluation model is validated in a real case in a situation of social inequality in Mexico. Recommendations and actions are proposed for the reduction of social lag based on the results obtained.

The methodology is described in the subsections below.

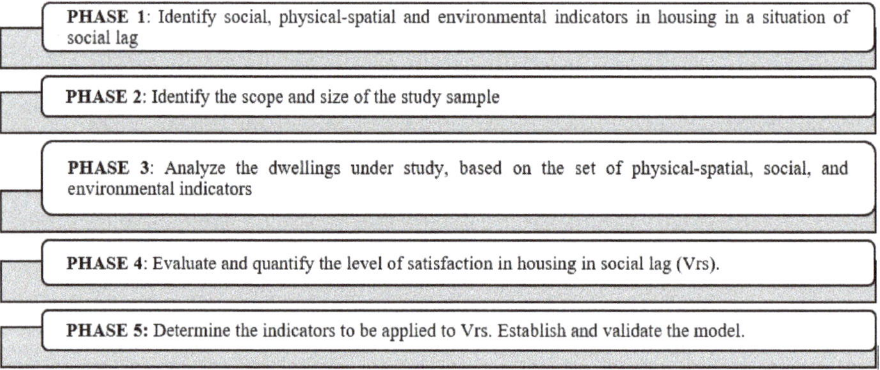

Figure 2. The process for developing a set of urban sustainability indicators. Source, author.

2.1. Phase 1: Identifying Social, Physical-Spatial, and Urban Environment Indicators for Housing Satisfaction

In this phase of the methodology, the satisfaction parameters and indicators for housing are determined based on the social, physical, spatial, and inter-urban aspects. These are obtained from the analysis of bibliographical sources, from the ethnographic analysis carried out on the housing occupants, and from the research contrasting the previous points, for which a number of steps is necessary, see Figure 3.

<table>
<tr><td rowspan="2">Phase 1:</td><td>• Process</td></tr>
<tr><td>

• Step 1 Identification of satisfaction indicators and parameters for housing from the SOCIAL DIMENSION, using bibliographical and institutional sources

• Step 2 Identification of parameters and indicators of housing satisfaction from the PHYSICAL-SPACE AND ENVIRONMENTAL DIMENSIONS, using bibliographical and institutional sources

• Step 3 Study of the social, physical-spatial and environmental dimension based on indicators valued through ETHNOGRAPHICAL RESEARCH of home users through surveys carried out by experts.

• Step 4 Determination of SOCIAL, PHYSICAL-SPACE AND URBAN ENVIRONMENT PARAMETERS AND INDICATORS for the satisfaction level evaluation model, based on the results obtained from social and bibliographical exploration.

</td></tr>
</table>

Figure 3. Detailed steps of Phase 1 of the process in order to obtain a list of indicators for housing satisfaction. Source, author.

Similar studies selected in the analysis of indicators from other bibliographical sources have considered social, physical-spatial, and environmental aspects. For the social dimension, academic models have been used in the evaluation of residential habitats, notably the indicators for social diagnosis by Maite Muñoz [50], the quality of life in the home [47,51,52], the social lag indices of the National Council for Social Development Policy Evaluation [19] and its social measurement dimensions [53].

In addition, parameters and indicators of the social dimension have been identified using international institutional sources including: Sustainable development of communities: Indicators for city services and quality of life [48]; UN-HABITAT indicators [54]; indicators for social diagnosis [50]; the index of quality of life in housing by María Salas [55]; and study dimensions for precarious housing [5]. The social, physical, spatial, and urban environment indicators were all identified based on the studies and bibliographical models analyzed (see Appendix A. The frequency, the number of times in which the study models were repeated, and their link with the Vrs were established. An ethnographic analysis was carried out to identify the indicators best suited to the different case studies. A questionnaire was drawn up with the indicators identified, as well as their frequency and relation to the Vrs in all three aspects studied, i.e., physical-spatial, social, and urban environment (see Appendix B). Finally, any indicators considered to be important by over 50% of the sample were selected and included in the final survey tool to assess satisfaction in Vrs. In the same questionnaire, users were asked questions for each dimension about which aspect would improve their satisfaction and well-being in housing. Any results of this social exploration that exceeded 60% of the answers were included as new indicators for this tool to assess the level of social lag.

The indicators of the physical, spatial, and urban environment dimensions were rated following a trichotomic scale, that is, with "1", "2", and "3" responses representing highly deficient, deficient, and adequate levels, respectively. The assessment of the social dimension followed a dichotomous scale, with "yes" or "no" answers, justified in the analysis of other studies [5,47,48,50]. In this case, the aim was to detect the shortcomings of housing in order to include them as indicators; therefore, the response scale was not decisive, as shown in Table 1.

Where n, n1, and ni, are the number of dwellings in the sample. R is the response of home users in the ethnographic analysis.

This phase of the methodology was completed with the social, physical-spatial, and urban environment parameters of the model evaluating the level of social lag, based on the results obtained from social research, satisfaction level, and adaptation of the existing bibliographical sources (Table 2).

Table 1. Questionnaire scheme of indicators applicable to Vrs.

Space Dimension Indicators	n	n1	Ni		Assessment
Bedroom					
Ventilation-window				1	Very lacking
Solar Lighting				2	Deficient
Space				3	Suitable
Kitchen					
Garage					
Garden					

Which aspects of your home's structure, spaces, and equipment do you think would improve the well-being of family members? Consider as indicator if R $\geq$ 60%

Physical Dimension Indicators	n	n1	Ni		Assessment
Sanitary Sewerage					
Electricity				1	Very lacking
Water				2	Deficient
Gas				3	Suitable
Wall finish					
Floor finish					
Roof finish					

Which aspects of your home's facilities do you think would improve family well-being? Consider as indicator if R $\geq$ 60%

Urban Environment Indicators	n	n1	Ni		Assessment
Urban lighting					
Bench				1	Very lacking
School				2	Deficient
Primary school				3	Suitable
Medical Care					
Market					
Commercial area					
Children´s park, sports tracks					

Which facilities in the neighborhood do you think would improve your family's satisfaction and well-being? Consider as indicator if R $\geq$ 60%

Social Dimension Indicators	n	n1	Ni		Assessment
Participation in social movements					
Participation in community support crews				1	YES
Access to educational networks for children and/or adults				2	NO
Access to public transport					
Security, relationships (perception of the neighborhood)					
Security, relationships (perception of the neighborhood)					
Your children go to school in the neighborhood					
There is a health center nearby					
You have access to gardens, courts, or other spaces					
You feel settled in your neighborhood					
You feel you belong in the neighborhood					
You identify with the neighborhood					

Which aspects of the neighborhood do you think would improve your family's satisfaction and well-being in terms of equipment? Consider as indicator if R $\geq$ 60%

Table 2. Evaluation model: System of parameters and indicators for the evaluation of satisfaction in Vrs. Source, author.

Dimension	Parameters	Indicators	Indicator Code
Social	Access to urban resources	Access to community health services	3
		Access to education (schools)	5
		Access to public transport	8
		Access to recreation and recreation spaces	101
	Social cohesion	Security, perception in the neighborhood	0
		Access to educational networks for children and adults	10
		Neighborhood organization and participation	11
	Households	Others in the home	24
		Separate animal-human environments	26
		More than 2 occupants per room	83
	Legal	Land ownership	4
Physical	Infrastructure	Other means of drinking water supply	0
		Sanitary drainage connection to municipal network	1
		Connection to municipal drinking water network	5
		Mains connection	32
	Furniture and Facilities	Bathroom inside the house	0
		Laundry facilities in yard	0
		Access to gas for cooking	103
		Use of water and electricity saver system	0
		Use of water heater	0
		Drinking water container	0
		Bathroom with shower	0
		Indoor toilet	0
		Kitchen sink	0
		Suitable doors	0
		Suitable windows	0
	Housing envelope	Fourth extra bedroom	0
		Wall finish	0
		Floor finish	0
		Roof finish	0
Spatial	Inside	Sunlight through the window	0
		Dining space	0
		Living space	135
		Window ventilation inside the house	169
		Separate spaces between bedrooms	195
	Outside	Space for extension	45
		Pedestrian access to the home	145
		Garden area in the home	0
Urban Environment	Equipment	Playground in the neighborhood	0
		Community medical care	0
		Paved roads	60
		Elementary school	67
		Commercial area	68
		Church	70
		Bank	114
		Areas for neighborhood sports	151
		Green spaces in the neighborhood	159
	Services	Street lighting	64
		Guardhouse	65
		Waste collection	66
		Nearby transport	72

The evaluation model obtained was composed of 51 indicators structured into four dimensions: social (11 indicators), physical (19 indicators), spatial (8 indicators), and urban environment (13 indicators). The key innovation of this phase of the methodology was the introduction of indicators proposed by technicians who knew the area under study and the home occupants themselves.

The numbering in the "indicator code" column corresponds to the numbering that appears in the general list from the study of the existing literature (see Appendix A) while "0" corresponds to the indicators proposed by the research. Indicators shaded in gray were obtained from social research.

2.2. Phase 2: Identification of the Scope of Study and Scale of the Study Sample

An area with a high Vrs index where major population growth is expected was selected in order to establish the scope of the study and scale of the sample. This step increases the global impact of the application of the Vrs evaluation model proposed. The level of social lag in the field of study, and the physical environment and surroundings, should be considered in order to base the study on logical processes and on-site data collection of the actual case study, outlining a specific area (Figure 4).

Phase 2:

• **Process**
- **Step 1**. Analysis of the urban a interconnection of interest, identifying the level of social lag from STATISTICAL SOURCES.
- **Step 2**. Definition of study area in social lag within the urban fabric based on INSTITUTIONAL AND GEOGRAPHICAL INFORMATION.
- **Step 3**. Definition of the sample size using STATISTICAL METHODS.

Figure 4. Detailed steps of Phase 2: Determination of the scope of study and scale of the study sample. Source, author.

The use of qualitative mixed method, in this case, provides richer and more varied results thanks to more dynamic inquiries, physical-spatial, and social issues. This, in turn, ensures more reliable results from the data research, by integrating qualitative and quantitative data based on collection, analysis, and integration. According to Sampieri [56], one of the reasons for using mixed methods (qualitative and quantitative) is that the weaknesses of either method is counteracted, and their individual strengths are reinforced. Taking the aforementioned into consideration and following J. C. López Alvarenga [57], the following calculation formula was used to determine the sample size from the population size (Formula (1)):

Formula (1) is the sample calculation formula based on population as follows:

$$\mathrm{n} = \frac{Z^2 \times P\,(1-P)}{e^2} \Big/ 1 + \frac{Z^2 \times P\,(1-P)}{e^2 \times N} \tag{1}$$

where n is the sample size, Z is the confidence level value, N is the population size, e is the margin of error, and P is the percentage of the population with the study characteristic. The result of n must have a level of confidence of 90% with an error of $\pm 10\%$, for the qualitative questions of the study based on surveys.

As this study aims to analyze dwellings within a specific area, with a finite population known to expert technicians, the sampling technique used was discretionary rather than probabilistic. The selection was determined by the physical and spatial characteristics of the housing studied and its occupants, allowing the habitability conditions and the perception of the immediate environment to be established. Study subjects were selected by the technicians after the relevant on-site examinations. It did not matter if the final sample was small, as long as the characteristics mentioned above were highly variable. The study sample was, then, determined and verified through application to the case study.

2.3. Phase 3: Analysis of the Housing Studied, Based on the Set of Physical-Spatial, Social, and Environmental Indicators

At this phase (Figure 5), the conditions of the dwellings under study (social, space, physical, and urban environment) were initially analyzed through in situ data collection.

<table>
<tr><td rowspan="4">Phase 3:</td><td>• Process</td></tr>
<tr><td>• Step 1. Survey of physical-spatial and environmental characteristics of the dwellings under study from field data collection instruments, SURVEYS.</td></tr>
<tr><td>• Step 2. Data collection of the physical-spatial and environmental dimension of the objects of study from surveys using the LIKERT scale to measure levels of satisfaction.</td></tr>
<tr><td>• Step 3. Geostatistical analysis by QGIS on the results of social, physical-spatial and urban environment indicators obtained.</td></tr>
</table>

Figure 5. Detailed steps of Phase 3: Analysis of the housing studied, based on the set of physical-spatial, social, and environmental indicators. Source, author.

The survey of physical-spatial and environmental characteristics of the dwellings was carried out using field data collection instruments (see Appendix C), i.e., files designed for this purpose containing the graphic information, construction elements, and materials and distribution of spaces of each dwelling.

At this phase, a survey model was also designed which included the system of parameters and indicators detailed in Phase 1 (Table 2) and allowed the level of satisfaction in the Vrs of the defined sample to be assessed (see Appendix D). The Likert psychometric scale (system of indicators used in questionnaires for research to evaluate the opinions and attitudes of a person, measuring the level of agreement or disagreement of the respondent) was chosen in order to obtain information which statistically showed a mean or average trend. This scale was considered to be suitable for measuring the degree of satisfaction of the users of the socially disadvantaged homes studied, establishing the validity and confidence of the results obtained for the spatial, physical and urban environment dimension. This follows the methodology used in another study [58] which recommended an evaluation system with five response categories, i.e., totally unnecessary, somewhat necessary, moderately necessary, necessary, and very necessary.

For the social dimension, the two response categories established were YES and NO. The results obtained were used to identify the indicators with the highest level of need among the population. Then, the group of indicators most often with the highest need was statistically analyzed for each dimension, which translated into the requirement to satisfy existing needs.

The results of individual sample units were synthesized in georeferenced graphics using the QGIS version 3.6.2 program (Figure 6).

The indicators of the satisfaction evaluation system for the Vrs must be linked to the existing support policies of the study area. Although this relationship with existing policies is not binding, as regards the possible results obtained, it does suggest a correlation between the indicator and the aid model and recommends the most appropriate support policy, depending on the indicators most in need. This is important, as one of the biggest problems in the Vrs is the owners' lack of knowledge of support programs for the improvement of their homes.

2.4. Phase 4: Evaluation and Quantification of the Level of Satisfaction in Housing in Social Lag

At this point, based on the information obtained from the previous phase, the attributes of the indicator system were mathematically weighted to quantify and evaluate the level of satisfaction in the Vrs. Once the users of the homes rated these and the level of satisfaction of the different dimensions was established through the Likert survey, the data obtained were treated statistically through a numerical stratification of values and satisfaction levels. The steps of Phase 4 are shown in Figure 7.

Figure 6. Georeferenced maps of the sample. QGIS version 3.6.2. Source: author.

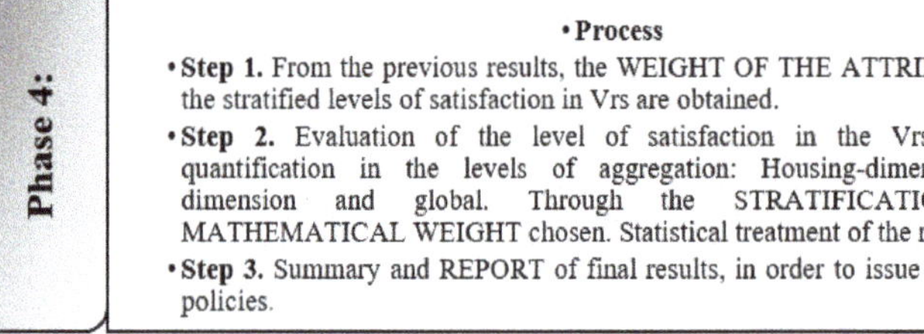

Figure 7. Detailed steps of Phase 4: Evaluation and quantification of the level of satisfaction in housing in social lag. Source: author.

The weight of the attributes and the stratified levels of satisfaction were determined mathematically based on the results of the previous evaluation of the characteristics relating to the physical-spatial, social, and urban environment dimensions treated with the Likert scale method in socially disadvantaged housing.

To do so, the results obtained were homogenized to consider the attributes of the indicators. In this case, a value was assigned to the data generated by transforming and combining these attributes to create a set of results. Then, a weight was assigned to the indicators, from 0 to 1, with each of them contributing a value to individual components or dimensions. These results were assessed on a scale from 1 to 5 by the user, the occupant of the dwelling, who determined the level of needs to be satisfied in ascending order (see Appendix F). The level of need ranged from Number 1, determining the lowest level of need, considered unnecessary to Number 5, assigned to analyzed indicators which were considered very necessary. Each level of satisfaction was weighted based on a mathematical equation determined by experts in the field from the University of Seville. This mathematical technique uses absolute frequencies to determine an expected value that goes from 0 to 1, and therefore the previous satisfaction scale between 1 and 5 represented 100% of the total dimension.

Cronbach's alpha method was used [59] to validate this psychometric scale tool (Likert scale), in order to measure the correlation between the items. It was assumed that they measured the same variable and were highly correlated, and the approximate result of 0.8–1 was considered to be highly acceptable based on the results of other studies by S.

Welch [60], J.A. Gliem [61], Victor Corral-Verdugo et al. [62]. The software used was IBM Statistical Package for the Social Sciences (SPSS).

Formula (2) for Cronbach's alpha procedure to obtain the results from the correlations between the items:

$$\alpha = \left[\frac{k}{k-1}\right]\left[1 - \sum_{i=1}^{k}\frac{S^2 i}{S_t^2}\right]$$ (2)

where S^2 is the variance of the item or indicator, S_t^2 is the variance of the total values observed, and K is the number of questions or items.

Once the tool was validated, a data distribution normality test was carried out using XLSTAT software for the purpose of making comparisons, including data distribution (median, average, maximum, and minimum).

These methodologies aim to obtain a score of the valuations from the data collected in the field using the Likert scale surveys. These indicators must be weighted by assigning a specific value matching the importance of the characteristic being measured in order to quantify and evaluate the characteristics and dimensions of the sample units.

For this, the five levels of satisfaction initially measured on the Likert scale represented a quantitative ratio number, obtained through Formulas (3) and (4). Then, a specific value (between 0 and 1) was established, with 1 indicating a higher level of dissatisfaction. As Table 3 shows, the degree of satisfaction by dwelling (vertical reading) and by indicator (horizontal reading) could be obtained for all dwellings analyzed.

Table 3. Summary of previous results of dimensions on evaluation of housing satisfaction in social lag and its symbols. 1, Unnecessary (very satisfied); 2, somewhat necessary (satisfied); 3, medium (moderately satisfied); 4, necessary (dissatisfied); 5, very necessary (completely dissatisfied).

Indicators by Dimension	Space Characteristics															
	N	n1	n2	Ni												
						Previous Results										
A	3	3		4	2	5	3	1	1	5	5	1	3	5	5	
B	3															IAi
C	5				Item-dimension											IBi
D	5															
E	5															
F	1															
G	1															
I	4				Item-dimension											
Ivn	Ivn1	Ivn2	Ivni													$\sum$ global

This formula must be weighted by function (attributed to the user), as the weights and values were variable, guaranteeing that the data obtained from the in situ measurements depended directly on the results of the dwelling occupants, and not on the researcher, as recommended by J. L. P. Gómez [63].

Thus, help was sought from experts from the field of mathematics and statistics in the Faculty of Mathematics of the University of Seville [64] to create a weighting formula for a new approach to the weighting strategies known to date.

Formulas (3) and (4), weighting for the level of satisfaction in socially disadvantaged housing were as follows:

$$Iv = \frac{\sum v.o - n.c}{n.c\,(L) - n.c}$$ (3)

where Iv is the index for housing, $n.c$ is the number of indicators in the column, L is the upper limit of the assessment, and $v.o$ is the levels observed.

$$Ii = \frac{\sum v.o - n.i}{n.i\,(L) - n.i}$$ (4)

where Iv is the index by item, *n.i* is the number of indicators in the row, *L* is the upper limit of the valuation in the row, and *v.o* is the levels observed (refers to the sum of the levels in the row in this case).

The data for the values of the degree of satisfaction per item per dimension and per housing dimension were defined by applying these formulas. An average value was established for both dimensions, determining the value of the dimension studied ($\sum$ *global*).

Considering the items for the set of dwellings studied, in order to calculate the degree of satisfaction for each of the dimensions, the results were treated and sorted in ascending order (from lowest to highest) on an Excel sheet in order to calculate the amplitude (A) of the statistical dataset (the amplitude calculates the difference between the highest and lowest value scores). This value was added to the minimum value and these data were considered the first level of satisfaction. Then, this procedure was repeated with the remaining values until the existing upper limit of the group, as shown in Table 4, was reached. Stratum 1 indicates maximum satisfaction while Stratum 5 represents the level of minimum satisfaction.

Table 4. Degree of satisfaction sorted in ascending order and stratification by levels according to dimension.

Dwelling	Observations from Lowest to Highest	Stratum	Stratification by Dimension			
			A	(Ivni − Ivn)	Stratum	Satisfaction Level
Av	Ivn	1	Ivn	M = Ivn + (Ivni − Ivn)	1	Very satisfied
Bv	Ivn1	2	M	N = M + (Ivni − Ivn)	2	Satisfied
Cv	Ivn2	3	N	O = N + (Ivni − Ivn)	3	Moderately satisfied
Dv	Ivn3	4	O	P = O + (Ivni − Ivn)	4	Dissatisfied
Xvi	Ivni	5	P	Ivni	5	Very dissatisfied

Where (Av, Xi) is study dwellings, Ivn is the minimum value of observations, and eIvni is the maximum value of observations Ivni.

The left hand table shows the number of dwellings and satisfaction data for housing and stratum. The right hand table shows the calculation of the amplitude and level of satisfaction by stratification and dimension.

Phase 4 of this investigation concluded by establishing the degree of satisfaction for each indicator in each dimension and global dimension.

2.5. Phase 5: Determination of the Indicators Applicable to Vrs. Establishment and Validation of the Model

Finally, the model was applied to a real case in order to validate this tool for the evaluation of housing in social inequality, as a strategy for urban renewal and regeneration (Figure 8).

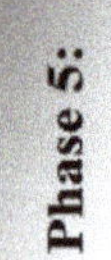

Figure 8. Detailed steps of Phase 5. Determination of the indicators applicable to Vrs. Source, author.

This study focused on Mexico, where currently more than half of the housing stock needs improvement and almost 47% of homes are in situations of social inequality. Because it is also one of the cities with the highest population growth predictions, the model for assessing socially disadvantaged housing is expected to have a greater impact on a global scale.

The analysis of social lag in the Mexican Republic at the national, state, and local levels showed that Mexico City, Hidalgo, and Pachuca de Soto displayed the highest levels of social lag. The level of social lag for each of the thirty-two federal entities within Mexico is shown in Figure 9.

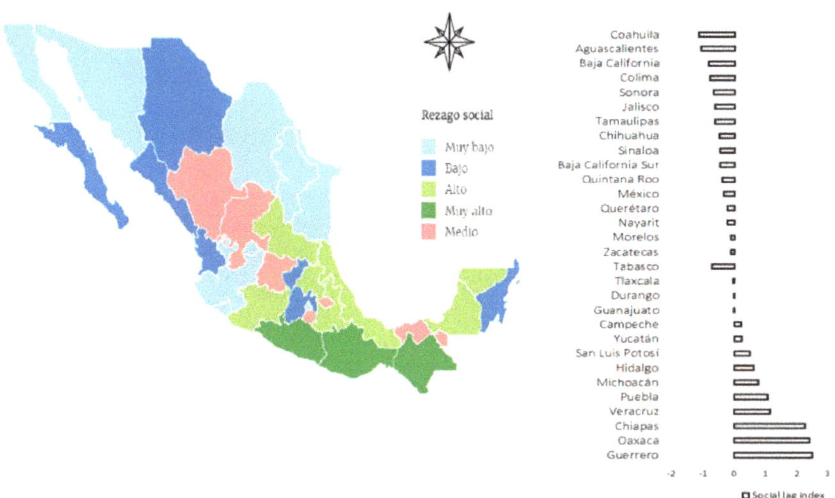

Figure 9. Left: Social lag by federal entity in the Mexican Republic, in 5 levels of stratification, i.e., very low, low, medium, high, and very high; Right: social lag indices of the federal entities of the Mexican Republic. Red (Hidalgo). Source, author.

As can be seen in Figure 9, in terms of the level of social lag, the State of Hidalgo is seventh among the thirty-two states that make up the Mexican Republic. According to the National Institute of Statistics, Geography, and Informatics [49] it has one of the highest levels of social lag, despite a decrease in recent years. The municipal government is in charge of managing the state, made up of 84 municipalities including the capital city, Pachuca de Soto, which has areas of social inequality in the periphery and where an area has been outlined for the application of the study methodology. Of the 84 municipalities in the State of Hidalgo, 12. municipalities display a very low level of social lag, followed by 31 municipalities with a low level, 25 municipalities with a medium level, and 16 municipalities with a high level [49]. With almost 53 million poor people in the country, these levels are comparable to a European city like Barcelona.

The starting point for this research was the socially disadvantaged housing, which according to CONEVAL, was defined in areas based on the construction of a single index with 5 indicators, considering information on education, access to health services, quality of housing, basic services in housing, and household assets.

The study sample selected was the city of Pachuca, which has 277,325 inhabitants and is classed as an average city according to the national urban system. Currently, this is a physically fragmented city, with social disintegration due to migratory flows, and it absorbs the population of the federal capital. Pachuca is one of the territories most affected by the new socio-spatial dynamics, most notably the construction of extensive urban peripheries, hence, the interest in studying this geographical area. The study area within the city is an urban settlement with high levels of social inequality and eligible for subsidies. The Cruz del Cerrito neighborhood is one of the oldest in the urban peripheries of the City of Pachuca. It dates from approximately the 1950s, a time when Mexico was moving away from rural activity in favor of industrial activity in the city. An area of this population was outlined, considering the level of social lag, the physical environment, and surroundings, in order to carry out a study based on logical processes and data collection from the case study. A homogeneous sample with the same characteristics was selected from the field of study.

In order to ensure an accurate representation of this study, based on the previous formula (Formula (1)) the sample size (n) should be 21. With 21 dwellings, the level of confidence of the sample is 90% with an error of ±10%. Finally, 23 dwellings were selected for this case study, representing 80% of the population. The study sample was characterized as dwellings with deed-holding owners, since ownership is essential to ensuring eligibility for subsidies at local and national level.

Formula (1):

$$n = \frac{Z^2 \times P\,(\,1\,-\,P\,)}{e^2} \,/\, 1 + \frac{Z^2 \times P\,(\,1\,-\,P\,)}{e^2 \times N} \tag{5}$$

where Z is the confidence level value, in this case 90% = 1.65; N is the population size, in this case = 29; E is the margin of error, in this case 10%; and P is the percentage of the population with the applicable study characteristic, in this case 0.5 as constant.

The dwellings selected are domestic units occupied by one or several families, sometimes sharing a common access. These are small houses, built using precarious or recycled materials. Most are built of cement blocks covered with corrugated iron or fiber cement. The floors are also finished in cement, without any type of flooring and the metal windows are single glazed. Figure 10 shows the most representative housing examples within the study sample.

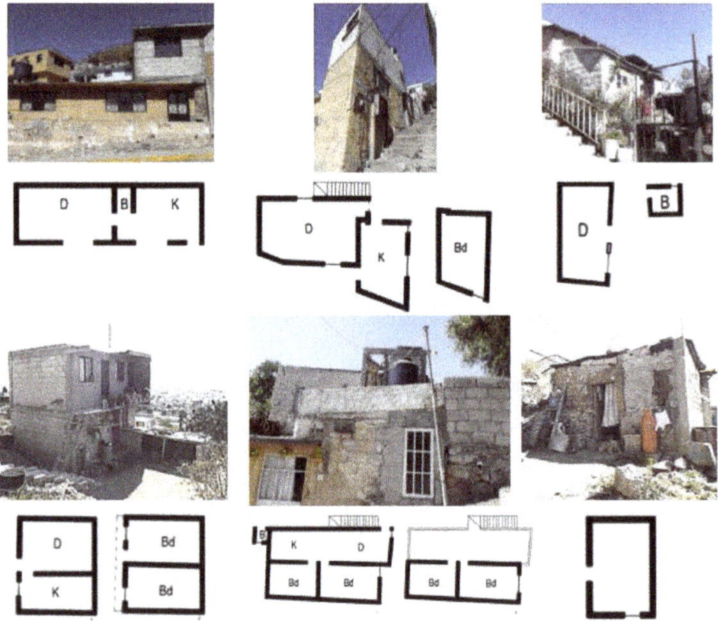

Figure 10. Examples of case studies. K (Kitchen), B (Bathroom), D (dining room), Bd (Bedroom).

3. Results

The results obtained from the analysis and evaluation of the study sample in the different phases of the methodology above are described in this section.

The results on social, physical-spatial, and environmental indicators were based on the ethnographic analysis (see Appendix E).

Figure 11 shows the results of the social research on the spatial dimension, providing trichotomic data on the needs of the users of the 23 dwellings. These results highlight users' needs for more space in their home, as many would like to have additional rooms for use as bedrooms. In addition, most of them would like more open space in their homes, which is very limited due to the spatial dimensions of the different areas in the dwellings.

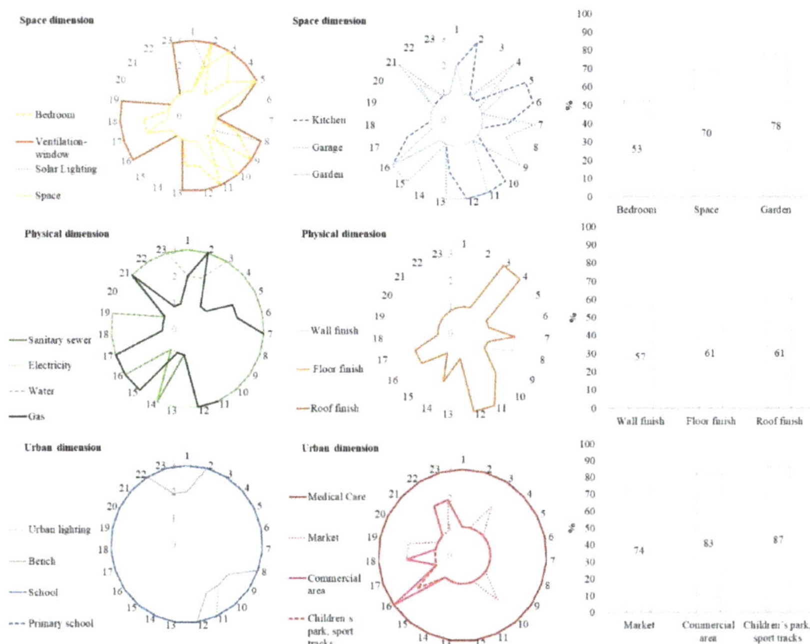

Figure 11. Results of the response of the users of the 23 dwellings studied with spatial, physical, and urban dimensions based on seven preliminary indicators, considering whether they perceive these as adequate (3), deficient (2), or highly deficient (1).

For the physical dimension, determined by seven indicators (sanitary sewerage, electricity, water, gas, and wall, floor and roof finishes), users would improve the finishes on walls, floors, and roofs. It was observed that housing users perceived greater deficiencies in finishes in cases where recycled materials had been used as protection; 61% of housing users felt the need for wall and floor finishes and 57% for wall cladding.

The results of the urban dimension were determined by eight indicators (urban lighting, bench, school, primary school, medical care, market, commercial area, and children´s park). In this case, users would improve recreational areas (87%), commercial areas (83%), and markets (74%).

Finally, Table 5 shows the results of social dimension indicators based on social research and considered the most deficient of the indicators stated.

Other social, physical-spatial, and urban environment indicators from bibliographical sources were added to these indicators. In order to simplify these, those with a greater weight in the valuation of housing were identified. The purpose of this was to choose the indicators relevant to the objective of the investigation. Finally, a model for assessing satisfaction for socially disadvantaged housing was designed (see Table 1).

The results of the main housing deficiencies were obtained from the physical survey of physical-spatial characteristics of the dwellings studied, with the tool described above.

As regards urban services, only 78% of the dwellings had access to piped drinking water, while the rest had a water tank close to the house. In addition, 52% of the dwellings had access to a gas supply and the rest used wood or coal. Furthermore, only 74% of the dwellings had access to urban sanitation, while the remaining 26% had no access at all.

Regarding the physical conditions of the partition elements, 87% of the dwellings were built with cement blocks, but only 23% were finished off, while 8% incorporated recycled materials in their walls.

Table 5. Indicators to incorporate for the evaluation of satisfaction in Vrs.

Dimension	Indicators of Social Research
Space	Bathroom with toilet Bathroom with bath Kitchen with sink Suitable doors and windows Garden area
Physical	Water heater Water tank Wall finish Floor finish Roof finish
Urban Environment	Commercial areas Recreation and leisure areas
Social Environment	Social participation, cohesion Transport

For the finishes of the dwelling floors, 26% had earth floors, 70% concrete, and 4% stone; 52% of the roofs were concrete, 39% sheet metal, and 11% recycled materials.

Next, we describe the results of evaluation and quantification of the level of satisfaction in socially disadvantaged housing.

The levels of housing-dimension aggregation were calculated, as were those on item dimension using the formulas from Phase 4 and the results obtained from the evaluation tool regarding the level of satisfaction in socially disadvantaged housing in the case study (see Appendix F).

The results of the evaluation of satisfaction of the different dimensions in the Vrs showed a dissatisfied (I) level of satisfaction for the spatial dimension, while the physical dimension and medium-sized urban environment were moderately satisfied (MS), and the social dimension was dissatisfied (I). The overall result was moderately satisfied (MS), as can be seen in Table 6a.

Table 6. Result of the dissatisfaction value for the item-by-dimension segregation level.

	Degree of Dissatisfaction	Weight	Final Value	Satisfaction Level			
a	Lower Limit	Upper Limit	Value	Stratum	Importance	Final Value Weighted	Satisfaction Level
Space dimension	0.25	0.97	0.69	4			I
					0.3	0.207	
Physical dimension	0.13	1.00	0.63	3			MS
					0.29	0.1827	
Urban environment dimension	0.25	0.63	0.43	3			MS
					0.2	0.086	
Social dimension	0	1.00	0.45	4			I
					0.21	0.0945	
			0.55			0.5702	MS

	Degree of Dissatisfaction	Weight	Final Value	Satisfaction Level			
b	Lower Limit	Upper Limit	Value	Stratum	Importance	Final Value Weighted	Satisfaction Level
Space dimension	0.35	1.00	0.64	3			MS
					0.34	0.2176	
Physical dimension	0.1	0.80	0.35	2			I
					0.19	0.0665	
Urban environment dimension	0.34	0.64	0.42	2			I
					0.23	0.0966	
Social dimension	0	0.51	0.51	3			MS
					0.24	0.1224	
			0.48			0.5	MS

The table shows the summary of results for the satisfaction evaluation of individual dimensions in the Vrs, as well as the weighted final value and the level of satisfaction in the two right hand columns.

The evaluation process was repeated using only indicators from bibliographical and institutional sources in order to compare the impact and contribution of this study to the methods traditionally used. The indicators obtained from the social research proposed in this study were excluded, and each dimension was evaluated again (Table 6b).

The comparison of results (Table 6) shows that the novelty of this study, as compared with earlier studies, is that it considers the particular situation of individual families. From the needs of the user in the first case, it is possible to observe a high level of social lag in social and spatial dimensions, while the level of satisfaction is moderately satisfied when considering only the state indicators. Despite the similar overall results, variations are observed among the data of the different dimensions. Thus, it is clear that social development programs at the federal, state, and municipal levels have carried out general analyses of the situation of socially disadvantaged housing, without delving further into the true needs of families, which can be identified following the methodology used in this research.

The importance of this study is that by including these new YES indicators, recommendations can be obtained to improve the satisfaction of the Vrs by study dimension. In turn, urban rehabilitation and regeneration policies designed are accepted by the user and alleviate the social lag (see Appendix G).

On the basis of the results, the programs defined are linked to the sample country and could meet the needs of the indicator guide, based on the results of the satisfaction assessment of Vrs.

Analysis of the programs of the sample country showed that these programs focus on the expansion and improvement of socially disadvantaged housing.

As these programs do not provide solutions for each indicator in deficit, it is necessary to follow procedures which sample users cannot carry out due to their lack of knowledge.

Furthermore, programs to improve the urban environment should clearly be carried out directly by the state or municipal government.

It was also observed that a person with the appropriate knowledge in this field was needed to explain to the users the necessary steps for achieving the objectives of housing improvement. The participation of other organizations and upper-secondary education was clearly needed with specialized social service actions to contribute to improving the conditions of habitability, health, and social integration and cohesion.

Thanks to the results obtained in all the phases of the methodology applied to the sample, it was possible to visualize the real physical and spatial conditions of the houses.

4. Conclusions and Discussion

One of the negative consequences for society worldwide, as a result of the current COVID-19 crisis, is that it has highlighted how its impact is greater in the more socially vulnerable and poorer neighborhoods of our cities. Overcrowding and increased stress during lockdowns, in spaces which are far from meeting the minimum habitability and hygiene conditions, have become a further concern to be added to the social dimension and the urban environment of these neighborhoods and the absence of public spaces or green areas for leisure. According to media reports, this has led to an increase in gender-based violence and has even affected mental health. This pandemic has emphasized the importance of social cohesion. Urban development and urban agendas should be at the center of public policies, taking into consideration the real needs of the user. Thus far, these have not been examined in similar studies, and therefore this research constitutes an innovation in the field of the creation and management of indicators for sustainable development of our cities. The quality of life and well-being in precarious housing has been greatly compromised by the lockdown situation during this global pandemic.

While international agendas have already warned of the need for urgent solutions within the 2030 sustainable development goals, it is now a pressing priority. This study provides a methodology which has proved its validity in a neighborhood in a precarious situation in Mexico, and which prior to the pandemic already had a high index (67%) of social vulnerability in housing. In fact, 67% of the almost 9.5 million Mexicans living in extreme poverty would represent 1.43 times more than the population of Madrid, Spain (6642.00 inhabitants).

Among endless potential solutions, this research has developed an alternative from observations of the needs and perceived deficits from the users of this type of housing. To the best of our knowledge, this has never before been taken into consideration with a response based on spatial needs and physical, social, and urban environments. This is combined with indicators identified through a painstaking analysis of national and international institutional and bibliographical sources linked to precarious housing. The ultimate goal is to provide the relevant guidelines for the reform of existing public policies with effective solutions to suit the real needs of the target users.

In this study, user satisfaction in socially disadvantaged housing was analyzed using indicators quantifying physical, spatial, environmental, and social aspects. Determining the shortcomings of this typology established a guide for urban rehabilitation and regeneration in the current situation and in emergency scenarios. The application of 51 indicators distributed in four previous parameters was established for this study.

The results of this study have identified the deficiencies of precarious housing in order to apply effective proposals to improve habitability and surroundings in the future.

One of the innovations of this study is the construction of a tool for social research surveys designed to include individual indicators from dwelling users, to provide a more dependable representation of the problems found in Vrs. One of the biggest contributions is social exploration based on the inclusion of the community, assessing the deficiencies in the characteristics of their own homes. Likewise, the evaluation of the level of satisfaction in socially disadvantaged housing, based on the evaluation by aggregation level, has allowed us to observe results both at the dimension level (spatial, physical, urban and social environments) and at the global level, as well as providing the satisfaction status of each dwelling by dimension.

This tool, built with the participation of a socially disadvantaged community and national-international global indicators, has the potential to analyze homes and user satisfaction anywhere. Likewise, the results of this study were compared with the results of the area studied by the CONEVAL evaluation commission of social development policies. In both cases, the results obtained indicated a degree of average satisfaction at the global level. At the same time, these results were compared with those which would be obtained if only the indicators specific to the local subsidies were applied. These results showed that, when all the indicators (bibliography, contributed by the technician and obtained from social research) were considered, the level of lag reached was similar to that of the real housing situation, further confirming the suitability of the methodology applied in this investigation.

However, a notable limitation of this study was the time needed to carry out the social research and the effort to explain the application of the surveys directly to the users, as continuous visits to the place of study were required.

The case study results highlighted the degree of dissatisfaction of socially disadvantaged housing for the different dimensions; the results showed a moderately dissatisfied value (high social lag) mainly in the spatial and physical dimension, followed by urban and social environments. These results are in keeping with the data obtained from the initial surveys where users reported green spaces as one of the most deficient aspects of their home environment, together with the relationship between spaces and their function within the dwelling.

This study concludes that the methodology, techniques, and procedures for evaluation, weighting, and valuation of indicators serve as an application guide for policies to rehabilitate precarious housing, both rural and urban.

This study serves as a basis for proposals for rehabilitation and improvement of social disadvantaged housing. It has considered a population that is living in and linked to the social fabric of a neighborhood, fully aware of their needs over time and with the same family development.

Socially lagging housing is a definition coined in housing policies in Mexico but also refers to precarious housing, which exists worldwide. The indicator guide can be applied anywhere in the world, not only in Mexico, based on the specific needs, customs, and culture of the place under study. There is a pressing need for policymakers to stop thinking collectively, that is, thinking that everyone requires the same solution either for political or financial convenience, as in the case of cities such as Malaga or Barcelona.

The study models reviewed in the state-of-the-art, authors, and procedures which have been widely referenced, are mostly weighted using multicriteria. This is based on the authors' own experiences, and the procedures correspond to techniques by agreement. It should be noted that the result may be affected by the arbitrary assignment of the importance values to the study dimensions.

The evaluation of satisfaction in housing in social lag has been linked to the level of well-being and quality of life. These criteria are based on Abraham Maslow's theory of motivation, presenting housing within the five levels to achieve self-realization. The state-of-the-art shows that some authors relate these three terms as if they were all the same, but satisfaction in this research refers specifically to habitability, to the housing envelope, and the relationship with its immediate environment, considering physical-spatial and social dimensions. While other studies analyze psychological aspects, which are valid of course, architecture itself deals with the psychological impact stemming from elements such as texture or color.

Author Contributions: Conceptualization, P.M.-M.; Data curation, P.M.-M.; Formal analysis, P.M.-M.; Funding acquisition, P.M.-M.; Investigation, P.M.-M. and O.M.-P.; Methodology, P.M.-M.; Resources, O.M.-P.; Software, O.M.-P.; Supervision, C.M.-G.; Writing–original draft, P.M.-M.; Writing-review & editing, C.M.-G. All authors performed the experiments. All authors analysed the data, and wrote, reviewed, and approved the final manuscript. All authors have read and agreed to the published version of the manuscript.

Funding: This work is part of the research project (reference no. 3719/0632) "Implementation of eco-efficient and urban health measures in the refurbishment of buildings and neighborhoods for the prescription of new products in the Construction Sector (Stage II)", the University of Seville and was funded by MAPEI SPAIN, S.A; the paper was funded by UAT, support for publication costs, patents, and postdoctoral support of the Autonomy University of Tamaulipas, México; the study was funded by "Funding for the international mobility of a PhD student for the development of co-supervised thesis" (reference 2.2.3 (2018)).

Institutional Review Board Statement: The study was conducted according to the guidelines of the Declaration of Helsinki. All participants, were previously informed.

Informed Consent Statement: Informed consent was obtained from all subjects involved in the study.

Data Availability Statement: The data are not publicly available due to ethical reasons.

Acknowledgments: The authors would like to thank Jose Adán Espuna from the Autonomous University of Tamaulipas, Mexico, for his support in the research carried out.

Conflicts of Interest: The authors declare no conflict of interest.

Appendix A. Indicators Linked to Evaluation Methodologies of Previous Studies. Indicators of Social, Physical, Spatial and Urban Environment Dimension

> ## MODELS OF EVALUATION OF PRIOR INVESTIGATIONS SOCIAL DIMENSION

- Mod. 1 Academic model for the evaluation of residential habitat. Falabella
- Mod. 2 Indicators for social diagnosis. MM Muñoz
- Mod. 3 Quality of life in rural housing. Sánchez y Jiménez
- Mod. 4 Measurement of social habitat in Mexico. Hernández y Velásquez
- Mod. 5 Residential Satisfaction Index of S.H.F. Y CIDOC.
- Mod. 6 Quality of life index linked to housing. (ICVV)
- Mod. 7 Social Lag Index. CONEVAL
- Mod. 8 Quality of life in the city ISO 37120
- Mod. 9 Post occupancy evaluación (POE)
- Mod. 10 UN-HÁBITAT (GUO) Indicator
- Mod. 11 Study dimensions for precarious housing UN-HÁBITAT GLOBAL III

Orange shading shows indicators more frequently or that have a relationship with some other indicator

Table A1. Indicators of the social dimension.

INDICATORS OF THE SOCIAL DIMENSION		Mod.1	Mod.2	Mod.3	Mod.4	Mod.5	Mod.6	Mod.7	Mod.8	Mod.9	Mod.10	Mod.11
1	Access to housing											
2	Access to employment											
3	Access to health service											
4	Land tenure											
5	Access to education											
6	Pride of belonged to the neighbourhood											
7	Outdoor activities											
8	Access to public transport											
9	Obtaining own achievements											
10	Educational networks											
11	Neighbourhood Organizations											
12	Social movements											
13	Religious											
14	Professional organizations											
15	Non-governmental organizations											
16	Policies											
17	Trade unions											
18	Other											
19	Language, culture, ethnicity											

20	Internment													
21	Special features													
22	Other													
23	Privacy													
24	Rest areas													
25	Family relationships													
26	Separate animal-human environments													
27	Hygiene													
28	Security													
29	Relationships with the community													
30	Sense of community													
31	Social cohesion													
32	Efficiency of means of production													
33	sustainable domestic life													
34	respect for the sociocultural and ecological context													
35	Social equity in the community and housing													
36	Respect and conservation of customs													
37	Planning of housing needs													
38	Social Security													
39	Support for agricultural production													
40	Human well-being													
41	Personal growth													
42	Sense of affiliation													
43	Sense of parting													
44	Comfort													
45	Aesthetic delight													
46	Order													
47	Tranquillity													
48	Silence													
49	Temperature													
50	Light													
51	Colour													
52	Contrast													
53	Identity													
54	Status													
55	Belonging													
56	Uprooting													
57	Spatial Provision													
58	Communicability													

59	Practicality	
60	Effectiveness	
61	Comfort	
62	Amplitude	
63	Dynamism	
64	Adaptability	
65	Displacement	
66	Security	
69	Interaction	
70	Modulation	
71	Neighbourhood Organization	
72	Set perception	
73	Social and community equipment	
74	Perception of the city	
75	Coexistence with the community	
76	Illiterate population	
77	Illiterate population that does not attend school	
78	Population aged 15 to 29 with basic education	
79	Population of 29 years or more with primary	
80	Population without health services	
81	Homes that do not have a refrigerator	
82	Homes that do not have a washing machine	
83	Average occupants per room	
84	Unemployment	
85	Young employees	
86	Number of new patents/100 thousand hab.	
87	Poverty	
88	Population with primary education/100 thousand Ha	
89	Fire/100 thousand inhabitants	
90	Deaths from disasters/100,000 inhabitants	
91	Number of emergency calls	
92	Level of participation in municipal elections	
93	% Of women elected in municipal positions	
94	Index of convictions for corruption/100 thousand hab.	
95	% Of women employed by Gob. Municipal	
96	Life expectancy rate	
97	No. Of hospital beds/100 Thousand Hab.	
98	Mortality in children under 5 years/100 born	
99	Suicide rate/100,000 inhabitants	

100	Mental health professionals/100Mil Hab.													
101	M2 of open recreation space/person													
102	M2 of closed recreation space/person													
103	No. Of property crimes/100Mil Hab.													
104	Violent crimes/100 Thousand Hab.													
105	Emergency response time													
106	% of poor people/100 Thousand Hab.													
107	No. of homes without legalization													
108	You have green areas/100 Thousand Hab.													
109	No. Of jobs per home													
110	Delivery and reception at the end of construction.													
111	Delivery process and final adjustments.													
112	Cost of energy, water and waste, cleaning													
113	Rent and insurance cost													
114	Initial cost of construction													
115	Operating and maintenance costs													
116	Amenities and amenities													
117	Adaptability to changes due to rehabilitation													
118	Tenure security level													
119	% of regularized homes													
120	% Of women and men who run a home													
121	% of population able to read and write more than 15													
122	% of enrolment from primary to professional													
123	Water price/100 lts in Dlls in expensive season													
124	Water consumption per day/hrs/inhabitant													
125	Title of land, housing or both													
126	Rental contract													
127	Customary tenure													
128	% Of men and women evicted the last. 10 years													
129	% Of bosses and households who do not believe they are evicted													

MODELS OF EVALUATION OF PRIOR INVESTIGATIONS PHYSICAL-SPACE DIMENSION AND URBAN ENVIRONMENT

- Mod. 1 Methodology for the design of quality of life index (Burgoin Rooms)
- Mod. 2 Parameters and indicators for the assessment of habitability (Alarcon and Justinian)
- Mod. 3 Parameters and indicators of the spatial and environmental physical dimension (Hdz and Velasquez)
- Mod. 4 Indicators of the physical-spatial dimension and environment for social diagnosis (MM Muñoz)
- Mod. 5 Residential satisfaction index of S.H.F. And CIDOC.
- Mod. 6 Housing quality index. ICAVI (INFONAVIT)
- Mod. 7 Accident satisfaction index. ISA (INFONAVIT)
- Mod. 8 Index of quality of life linked to housing. (ICVV)
- Mod. 9 Quality of life in the city ISO 37120.
- Mod. 10 Post occupancy evaluation (POE).
- Mod. 11 UN-HABITAT (GUO) Indicators
- Mod. 12 Study dimensions for precarious housing UN-HABITAT GLOBAL III
- Mod. 13 CONEVAL Social Delay Index

The indicators are valued by four criteria: indicators linked to housing in social lag (Vrs), not representative, redundant and inaccurate. These are previously analyzed by their frequency and their relationship between them, to determine under the four criteria above the most appropriate to be part of the satisfaction assessment tool in Vrs.

Table A2. Assessment for the selection of indicators.

	ASSESSMENT FOR THE SELECTION OF INDICATORS.	
	AVAILABILITY-RELIABILITY-QUALITY	**COMPARISON FEATURE**
1	**Linked to housing in social backwardness**	Consistent, accessible, easily interpreted, applicable, feasible
2	**Not representative**	Not relevant for the research objectives.
3	**Redundant**	Repetitive data with others presented in the study models
4	**Imprecise**	There is no specific breakdown for your measurement
	Indicator more frequently or that has a relationship with some other indicator	

Table A3. Physical-space and urban environment indicators.

PHYSICAL-SPACE AND URBAN ENVIRONMENT INDICATORS.	Mod.1	Mod.2	Mod.3	Mod.4	Mod.5	Mod.6	Mod.7	Mod.8	Mod.9	Mod.10	Mod.11	Mod.12	Mod.13
Excreta disposal													
1 Poet connected to sewer													
2 Poceta connected to septic tank													
3 Poceta offline													
4 Latrine													
5 It has no pot or well													
Rooms with shower													
6 One to five													
Type of housing													
7 Fifth or fifth house													
8 Home													
9 Ranch													
10 Refuge													
11 Other class													
Materials in walls													
12 Block or brick													
13 unframed block or brick													
14 Sawn timber													
15 Adobe, tapia bahareque frisado													
16 Adobe, wall, barefrisk													
17 Palms, boards or similar													
18 Cement or concrete													
19 Flooring materials													
20 Marble, mosaic, granite													
Flooring materials													
21 Marble, mosaic, granite													
22 Cement or concrete													
23 land													
24 Others													
Roofing Materials													
25 Platabanda													
26 Roof tiles													
27 Asphalt sheet													
28 Metal sheet													
29 Palm, board or similar													

30	Other class	
	Number of rooms	
31	One to five	
32	Has	
33	Does not have	
	Water supply	
34	Aqueduct or pipe	
35	Public battery or pond	
36	Well with pipe or pump	
37	Well or protected spring	
38	Others	
39	Cooking fuel	
40	Gas	
41	Electricity	
42	Kerosene	
43	Firewood or carbon	
44	Infrastructure (From bad to very good. 5 levels)	
45	Allegation (From bad to very good. 5 levels)	
46	Overcrowding (From bad to very good. 5 levels)	
47	Community equipment (from bad to very good)	
48	Expansion space (from bad to very good)	
49	Overcrowding (5 levels)	
	Space	
50	Number of bedrooms	
51	housing surface	
52	Number of bathrooms	
53	parking drawers	
	Shape	
54	Ground surface	
55	Number of floors	
	Overcrowding	
56	Inhabitants/Bedrooms	
	Service	
167	Building Cleaning	
168	Building Security	
	Operational management	
169	Temperature	
170	ventilation	
171	Acoustics	

		Physical systems
172	Illumination	
173	Air conditioner	
174	Durability	
	Durable structures	
175	Durable structure outside risk zone	
	Access to safe water	
176	% Housing with drinking water	
177	% Solid waste landfills	
178	% Closed solid waste	
179	% Solid waste landfill	
	Home Connections	
180	% Of homes connected to drinking water	
181	% Of homes connected to the sewer	
182	% Of homes connected to electricity	
183	% Of homes connected to the phone	
	Risk areas	
184	% of population subject to risk areas	
185	Transportation time in round trip minutes	
	Access to drinking water	
186	% Of people with connection to drinking water	
187	% of homes with a shared tap max 2	
188	Access to water from other sources	
	Access to basic sanitation	
189	% Of population connected to the drainage service	
190	% Of population connected to a septic tank	
191	% Of connected to a latrine with connection	
192	% Of connected to a non-public latrine	
	Housing durability	
193	Construction quality in flooring materials	
194	Construction quality in wall materials	
195	Construction quality in roofing materials	
	Enough area to live	
196	% Of homes with at least 3 per room	
197	Quality and spaces in the house	
198	% of houses with dirt floor	
199	Average number of occupants per room	
200	Basic housing services	

201	Unexcused homes																
202	Housing without connection to drinking water																
203	Homes without connection to sanitary sewer																
204	Housing without connection to electricity																

Appendix B. Data Collection Instrument for Social Exploration on Indicators. This Questionnaire Presents Existing Indicators Related to Socially Lagging Housing

Table A4. Data collection instrument for social exploration on indicators.

A series of characteristics that your home has or should have is presented as a list, mark with an X the degree of deficit that you find in the home that you and your family live in, or if the appearance is appropriate. In the second part, check the box with an X after reflecting your answer			
SOCIAL COMPONENT TO EVALUATE BY HEAD OF FAMILY			
NUMBER:	**Informant Name:**		
BETWEEN STREETS:			
COORDINATES:	**Date:**		
TOPOGRAPHY:			
INDICATORS	**NEEDS FOR YOUR SATISFACTION**		
ASPECTS SATISFACTORY TO EVALUATE WITH RESPECT	**SUITABLE**	**DEFICITARY VERY**	**DEFICITARY**
SPACE DIMENSION			
ROOMS/INHABITANTS RELATIONSHIP			
WINDOW VENTILATION			
SOLAR LIGHTING			
RELATIONSHIP OF SPACES AND THEIR FUNCTION			
KITCHEN SPACE			
PARKING AREA			
GARDEN AREA			
PHYSICAL DIMENSION			
ACCESS TO SANITARY DRAINAGE			
ACCESS TO ELECTRICAL ENERGY			
ACCESS TO DRINKING WATER BY CONNECTION			
ACCESS TO A GAS INSTALLATION			
FINISHES IN WALLS			
FINISHES IN FLOORS			
COVERED AS PROTECTION FOR INCLEMENCIES			
URBAN ENVIRONMENT			
ACCESS TO PUBLIC LIGHTING			
BANQUETS			

PRE-SCHOOL			
PRIMARY SCHOOL			
CLINIC			
COMMERCIAL AREA			
SPREADING (GAME AREAS, ETC COURTS)			
ACCESS TO GREEN AREAS			
ACCESS TO PUBLIC TRANSPORTATION			

SOCIAL DIMENSION	EVALUATION OF SOCIAL CONDITIONS	
	YES	NO
HAS NEIGHBORHOOD PARTICIPATION		
PARTICIPATE IN SOCIAL MOVEMENTS		
PARTICIPATE IN CLEANING AND MAINTENANCE BRICKS		
PARTICIPATE IN EDUCATIONAL NETWORKS OF ANY KIND		
EASILY ACCESS TO TRANSPORT		
HAVE A GOOD PERCEPTION OF THE COLONY (SECURITY)		
YOUR CHILDREN GO TO SCHOOL IN THE COLONY		
ACCOUNT WITH SOME NEAR HEALTH CENTER		
YOU HAVE ACCESS TO GARDENS, COURTS, OR OTHER		
FEEL ROOTED TO YOUR NEIGHBORHOOD (FOR THE TIME IT LEADS)		
FEEL IDENTIFIED WITH YOUR NEIGHBORHOOD (PROUD)		

Appendix C. Instrument for Data Collection of Physical-Spatial and Urban Environment Characteristics for Sample Units

Table A5. Instrument for data collection of physical-spatial and urban environment characteristics.

HOUSING NUMBER	Regular Lot Irregular:	Photography Direction:				
	Measures:	Head of the family				
SQUARE	Coordinates					
		Housing spaces and occupational structure		Physical, spatial and urban environment characteristics of housing			
		🚹🚺	Existed	FLOOR	WALLS	ROOF	DOOR/ WINDOWS
		Garage					
		Yard					

	Room					
	Dining room					
	Kitchen					
	Bath. inside					
	Shared bathroom					
	Occupants/no. Bedrooms					
	Cto. Round					
	Serv patio					
	Space for extension					
	Infrastructure and urban environment					
	Drain connection					
	Drinking water connection					
	Cooking fuel					
	Distance to transport					
	Observation					

Physical, spatial and urban environment characteristics:

1 Connection to municipal sanitary sewer network	13 block walls seated with mortar	25 Concrete Deck
2 Sanitary sewer connection to septic tank	14 stacked block or partition walls	26 Laminate cover
3 Use of latrine with drain connection	15 Walls with any type of coating	27 Wood and tile roof
4 Use of latrine with septic tank	16 adobe walls	28 Mixed deck
5 Connection to municipal drinking water network	17 Stone walls seated with mortar	29 Palm cover
6 Access to drinking water by community well	18 stacked stone walls	30 Cover with recycling materials
7 Access to drinking water by auto-tank distribution	19 Wall of recycled material (laminate, wood	31 It has a service patio
8 Access to rainwater	20 Mixed Composition Walls	32 Urban transport distance
9 Access to drinking water by cistern or public battery	21 Ground floors	33 Paved Roads

10 Access to gas l.p for cooking	22 cement floors	34 Roads in dirt roads
11 Access to firewood for cooking	23 ceramic floors	
12 red partition walls seated with mortar	24 floors with tile or stone	

Appendix D. Survey Instrument Used in the Model of Evaluation of Indicators for the Evaluation of Housing Satisfaction in Social Lag

Table A6. Survey instrument used in the model of evaluation of indicators.

Please read carefully each aspect that you have or should have your home, marking with a line the most important box according to the need of you and your family						
INDICATORS TO EVALUATE						
STREET: TEPEYAC						
NUMBER: WITHOUT NUMBER HOUSING COUNT: 21						
BETWEEN STREETS: CALLEJON DE TEPEYAC AND SAN NICOLAS						
COORDINATES:						
TOPOGRAPHY: ACCIDENTED						
INDICATORS		**SATISFACTION FROM NEEDS**				
INFORMATOR NAME: **Sr. Antonio Garcia Reyes**		**Do you think it is necessary to increase the satisfaction of you and your family considering in your home?**				
Indicator code	**SATISFACTING ASPECTS TO EVALUATE REGARDING THE SPACE, PHYSICAL, URBAN AND SOCIAL ENVIRONMENT**	Completely unnecessary	Little necessary	Medium necessary	Necessary	Very necessary
	SPACE DIMENSION					
0	Sun lighting inside the house					
169	Window ventilation inside the house					
195	Separate spaces between bedrooms and other housing					
135	Living space					
0	Dining space					
145	Pedestrian access to your home					
45	Space for extension					
0	Garden area in the house					
	PHYSICAL DIMENSION					
1	Sanitary drainage connection to municipal network					
5	Connection to municipal drinking water network					

0	Other means of drinking water supply					
32	Mains connection					
103	Access to gas for cooking					
134	Use of water and electricity saving systems system					
0	Use of water heater					
0	Drinking water container (Tinaco)					
0	Bathroom with shower					
0	Toilet with toilet					
0	Bathroom inside the house					
0	Kitchen sink					
0	Suitable doors.					
0	Suitable windows					
0	Laundry in service yard					
0	Wall coverings					
0	Floor coverings					
0	Roof Coatings					
0	Fourth extra bedroom					
	URBAN ENVIRONMENT DIMENSION					
159	Green areas in the neighbourhood					
0	Playground in the neighbourhood					
151	Area for sports in the neighbourhood					
114	Sidewalks					
68	Commercial area					
60	Paved roads					
70	church					
67	Basic Education School					
0	Community medical care					
66	Waste collection					
65	Guardhouse					
64	Street lighting					
72	Nearby transport					
	SOCIAL DIMENSION	**ESCALA**	**DICOTOMICA**			
		YES	NO			
0	Perceive Security in the colony					
10	Has access to educational networks children and adults					
11	neighbourhood organization and participation					
3	You have access to community health service					
5	Has access to basic education schools					
101	Has access to recreation and recreation spaces					

8	You have access to public transport					
26	It has separate animal-human environments					
24	You have rest in your home					
83	There are more than 2 occupants per room in your home					
4	Land tenure (Deed or in process)					

Appendix E. Results on Social, Physical-Spatial and Environmental Indicators Based on Ethnographic Exploration Based on a Survey Aimed at the Inhabitants of the Study Housing

Table A7. Results for housing tricotomic scale survey.

INDICADORES DE DIMENSION ESPACIAL	RESULTS FOR HOUSING TRICOTOMIC SCALE SURVEY																							Homes under study
	1	2	3	4	5	6	7	8	9	10	11	12	13	14	15	16	17	18	19	20	21	22	23	
DIMENSION ESPACIAL																								
BEDROOM	1	3	1	2	3	1	1	2	3	1	3	2	2	1	1	1	2	2	1	1	1	1	1	1 Very lacking
VENTILATION/ WINDOWS	3	3	3	3	3	2	1	3	3	3	3	3	3	1	1	3	3	3	3	1	1	1	3	2 Deficient
SOLAR LIGHTING	3	2	3	2	3	2	1	3	1	3	3	3	3	1	1	3	3	3	3	1	1	1	3	3 Suitable
SPACE RELATIONSHIP-FUNCTION	1	2	3	2	3	1	1	1	1	3	2	1	1	1	1	1	2	1	1	1	1	1	1	
KITCHEN	2	3	1	1	3	3	2	1	1	3	3	3	2	1	1	3	2	1	1	1	1	1	1	
GARAGE	2	3	1	3	1	1	3	2	3	1	3	3	3	1	3	3	1	1	1	1	3	1	1	
JARDEN	2	1	1	2	2	1	1	1	1	1	1	1	1	1	1	3	1	1	1	1	2	1	1	

What aspects within your home regarding its structure, spaces and equipment would you consider that would improve the well-being of your family?			Indicators according to social exploration results
1	Bathroom with shower and toilet bowl	13	Bathroom with shower and toilet	**Bathroom with shower**
2	Watering can and toilet for the bathroom	14	Watering can and windows	**Toilet with toilet**
3	Sink and a quarter more	15	Sink, toilet, windows and shower	**Kitchen area with sink**
4		16	A quarter more	**Suitable doors and windows**
5	Watering can and toilet for the bathroom	17	A quarter more	
6	A quarter more	18	Sink, toilet and shower	
7	A quarter more	19	One more room, sink, toilet, doors and windows	
8	Put glass on the windows, toilet and shower	20	sink, toilet and shower	

9	Toilet with cup and watering can	21	Suitable windows and doors
10	Wider windows, toilet and shower	22	Sink, toilet and shower
11	A quarter more	23	Toilet, shower, laundry
12	Bathroom with shower and toilet		

RESULTS FOR HOUSING TRICOTOMIC SCALE SURVEY

PHYSICAL DIMENSION INDICATORS	1	2	3	4	5	6	7	8	9	10	11	12	13	14	15	16	17	18	19	20	21	22	23	Homes under study
SANITARY DRAINAGE	3	3	3	3	3	3	3	3	3	3	3	3	1	3	1	3	3	3	3	1	3	3	3	**1** Very lacking
INST. ELECTRICAL	3	3	3	3	3	3	3	3	3	3	3	3	3	3	1	3	3	3	3	1	3	3	3	**2** Deficient
INST. OF WATER	2	2	3	3	3	3	3	3	3	3	3	3	1	1	1	3	3	3	3	1	3	3	3	**3** Suitable
INST. OF GAS	2	3	1	1	2	2	3	3	3	3	3	3	1	1	3	3	3	1	1	1	3	1	1	
FINISHES IN WALLS	1	1	3	3	1	1	2	2	1	2	3	3	1	2	1	2	2	1	1	1	1	1	1	
FINISHES IN FLOORS	1	1	3	3	1	1	2	1	1	2	3	3	1	2	1	2	2	1	1	1	1	1	1	
FINISHES IN COVERS	1	1	3	3	1	1	2	1	1	2	3	3	1	2	1	2	2	1	1	1	1	1	1	

What aspects within your home regarding its structure, spaces and equipment would you consider that would improve the well-being of your family?		Indicators according to social exploration results
1	Flattened walls and paint, water heater	13 Water heater	Water heater
2	Tile floor	14 Water heater and water tank	
3	Water heater	15 Concrete tile and slab floor	Drinking water container
4	Tinaco and water heater	16 Water tank and water heater17 Water heater	Water heater
5	Flattened walls, water heater and water tank	17 Water heater	Drinking water container
6	Water heater and flattened walls	18 Water heater and water tank	
7	Water heater and water tank	19 Flattened walls and floor tile, heater and water tank	
8	Water heater and water tank	20 concrete slab, water heater and water tank	
9	Wall finishes	21 Concrete slab, water heater and water tank	
10	Tinaco and water heater	22 Tinaco, water heater and flattened	
11	Water heater	23 Water tank and water heater	
12	Flattened walls and paint, water heater		

URBAN DIMENSION INDICATORS	RESULTS FOR HOUSING TRICOTOMIC SCALE SURVEY																								Homes under study
	1	2	3	4	5	6	7	8	9	10	11	12	13	14	15	16	17	18	19	20	21	22	23	
STREET LIGHTING	3	3	3	3	3	3	3	3	2	2	3	3	3	3	3	3	3	3	3	3	3	3	3	**1** Very lacking
SIDEWALK	2	3	3	3	3	3	3	3	2	2	2	3	3	3	3	3	3	3	3	3	3	3	2	**2** Deficient
PRE-SCHOOL	3	3	3	3	3	3	3	3	3	3	3	3	3	3	3	3	3	3	3	3	3	3	3	**3** Suitable
PRIMARY SCHOOL	3	3	3	3	3	3	3	3	3	3	3	3	3	3	3	3	3	3	3	3	3	3	3	
COMMUNITY MEDICAL CARE	3	3	3	3	3	3	3	3	3	3	3	3	3	3	3	3	3	3	3	3	3	3	3	
MARKET	1	1	2	1	1	1	1	1	1	2	1	1	1	1	1	3	1	2	2	1	1	1	2	
COMMERCIAL AREA	1	1	1	1	1	1	1	1	1	1	1	1	1	1	1	3	1	2	1	1	1	2	2	
RECREATION (GAME AREAS)	1	1	1	1	1	1	1	1	1	1	1	1	1	1	1	2	1	1	1	1	1	2	2	

What aspects of the neighbourhood do you think would improve the satisfaction and well-being of your family in terms of equipment?		Indicators according to social exploration results
1 A market and green areas	13 Green Areas	Green areas
2 Playground, green areas	14 Market and green areas	Sports area
3 Green areas and sports fields	15 sports fields and green areas	Playground for children
4 Green areas and sports fields	16 children's play areas and sports courts	green areas
5 Green areas and playgrounds	17 green areas and playgrounds	
6 Paint on garnishes, green areas	18 sports fields and green areas	
7 Sports courts and playgrounds	19 Children's play area and green areas	
8 green areas	20 Courts and green areas	
9 Children's games and sports courts	21 Best conditions on sidewalks	
10 Common parking and green areas	22 sports fields and green areas	
11 Sports courts and green areas	23 Children's play area and green areas	
12 Security booth and green areas		

SOCIAL DIMENSION INDICATORS	RESULTS FOR HOUSING DICOTOMIC SCALE SURVEY																								Homes under study
	1	2	3	4	5	6	7	8	9	10	11	12	13	14	15	16	7	18	9	20	21	22	23	
PARTICIPATION IN SOCIAL MOVEMENTS																								**1** YES

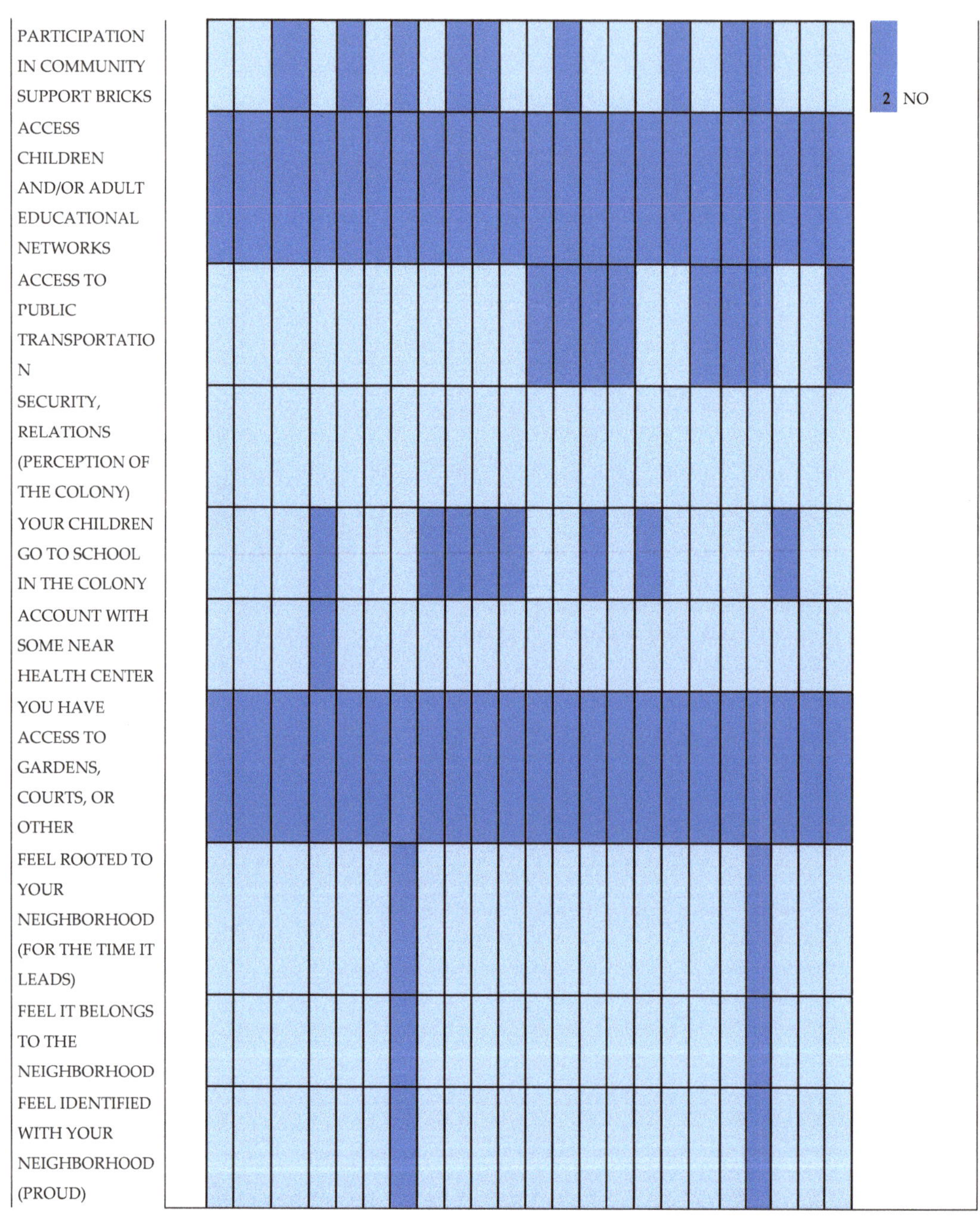
PARTICIPATION IN COMMUNITY SUPPORT BRICKS
ACCESS CHILDREN AND/OR ADULT EDUCATIONAL NETWORKS
ACCESS TO PUBLIC TRANSPORTATION
SECURITY, RELATIONS (PERCEPTION OF THE COLONY)
YOUR CHILDREN GO TO SCHOOL IN THE COLONY
ACCOUNT WITH SOME NEAR HEALTH CENTER
YOU HAVE ACCESS TO GARDENS, COURTS, OR OTHER
FEEL ROOTED TO YOUR NEIGHBORHOOD (FOR THE TIME IT LEADS)
FEEL IT BELONGS TO THE NEIGHBORHOOD
FEEL IDENTIFIED WITH YOUR NEIGHBORHOOD (PROUD)
2 NO

What aspects of the neighbourhood do you think would improve your family's satisfaction and well-being in terms of equipment?		Indicators according to social exploration results
1	That people participate in community tasks	Community participation
2	That transport be closer	
3	Best organization to do community work	Distance accessible transport
4	That people participate in community tasks	
5	That people participate in community tasks	
6	That people participate in community tasks	
7	That people participate in community tasks	
8	That people participate in community tasks	
9	That the transport was closer	
10	That the transport was closer	
11	That people participate in community tasks	
12	That the transport was closer	

Appendix F. Results Obtained in the Application of the Survey Described in the Previous Points Consisting in the Issuance of Responses of the Heads of Household of the Dwelling under Study Based on the Indicators of the System of Evaluation of the Level of Satisfaction in Housing in Social Lag in Five Categories of Possible Answers (1: Unnecessary, 2: Little Necessary, 3: Moderately 4: Necessary and 5: Very Necessary). In the Case of Social Dimension, Two Categories of Possible Answers (Yes, No)

Table A8. Results obtained in the application of the survey described in the previous points consisting in the issuance of responses of the heads of household of the dwelling under study based on the indicators of the system of evaluation of the level of satisfaction in Vrs.

Code	DIMENSIONS AND INDICATORS	1	2	3	4	5	6	7	8	9	10	11	12	13	14	15	16	17	18	19	20	21	22	23
	SPACE CHARACTERISTICS																							
	DIMENSION ESPACIAL — PREVIOUS RESULTS																							
0	Window ventilation inside the house	3	3	3	4	2	4	5	3	2	2	2	1	1	5	5	2	2	1	3	5	5	5	3
169	Sun lighting inside the house	3	4	2	4	2	4	5	2	5	2	2	2	1	5	5	2	2	1	3	5	5	5	3
195	Separate spaces between bedrooms and other housing	5	4	2	4	1	5	5	5	5	2	4	5	5	5	5	5	4	5	5	5	5	5	5
135	Living space	5	4	2	5	5	5	5	3	4	5	4	5	5	5	5	2	5	4	4	2	2	5	5
0	Dining space	5	4	4	5	5	5	5	4	4	5	4	5	5	5	4	4	5	4	4	2	2	4	5
145	Pedestrian access to your home	1	4	4	4	3	5	4	4	5	5	4	1	4	5	5	4	5	5	5	5	5	5	4
45	Space for extension	1	1	1	1	1	1	1	3	1	1	3	4	1	1	1	1	1	3	4	4	4	5	2
	Garden area in the house	4	5	5	4	5	5	5	5	5	5	5	5	5	5	5	3	5	5	5	5	4	5	5
	PHYSICAL DIMENSION																							
1	Sanitary drainage connection to municipal network	1	1	1	1	1	1	1	1	1	1	1	1	5	1	5	1	1	1	1	5	1	1	1
5	Connection to municipal drinking water network	4	4	1	1	1	1	1	1	3	3	1	1	5	5	5	1	1	3	3	5	1	1	1
0	Other means of drinking water supply	5	5	1	1	1	1	1	1	1	1	1	1	5	5	5	1	1	1	1	5	1	1	1

	Indicator																							
32	Mains connection	1	1	1	1	1	1	1	1	1	1	1	1	3	1	5	1	1	1	1	5	3	1	1
103	Access to gas for cooking	4	1	5	5	4	4	3	5	1	5	1	2	5	5	3	3	3	5	5	5	1	5	5
134	Use of water and electricity saving systems	1	1	4	4	5	5	5	2	2	2	2	2	2	2	2	5	5	5	1	5	1	1	1
	Use of water heater	5	5	5	5	5	5	5	5	5	5	5	5	5	5	5	5	5	5	5	5	5	5	5
	Drinking water container (Tinaco)	5	5	4	5	5	4	5	5	1	5	1	1	1	5	5	5	1	4	5	5	5	5	5
	Bathroom with shower	5	5	3	5	1	5	5	5	5	2	5	5	5	5	5	5	4	5	5	5	5	5	5
	Toilet with toilet	5	5	3	5	1	5	5	5	5	2	5	1	1	5	5	5	4	5	5	5	5	5	5
0	Bathroom inside the house	1	1	5	2	5	5	5	5	5	5	1	3	5	5	5	5	5	3	1	1	1	1	5
	Kitchen sink	4	1	5	5	3	2	5	5	5	2	3	1	4	5	5	3	4	5	5	5	5	5	5
	Suitable doors.	4	3	1	4	3	5	5	5	5		4	5	5	5	5	3	5	3	1	1	2	2	5
	Suitable windows	1	2	4	1	5	5	5	5	4	4	1	4	5	5	5	3	5	2	5	3	3	5	5
0	Laundry in service yard	1	1	1	1	4	5	5	4	3	3	1	4	5	5	5	4	4	1	1	1	1	5	5
	Wall coverings	5	5	2	3	5	5	4	4	5	4	2	2	5	4	5	4	4	5	5	5	5	5	5
	Floor coverings	5	5	2	3	5	5	4	5	5	4	2	2	5	4	5	4	4	5	5	5	5	5	5
	Roof Coatings	5	5	2	3	5	5	4	5	5	4	2	2	5	4	5	4	4	5	5	5	5	5	5
0	Fourth extra bedroom	5	4	5	4	3	5	5	5	3	5	3	5	5	5	5	5	5	4	5	5	5	5	5
	URBAN ENVIRONMENT DIMENSION																							
159	Green areas in the neighbourhood	5	5	5	5	5	5	5	5	5	5	5	5	5	5	5	2	5	5	5	5	5	4	4
0	Playground in the neighbourhood	5	5	5	5	5	5	5	5	5	5	5	5	5	5	5	4	5	5	5	5	5	5	5
151	Area for sports in the neighbourhood	5	5	5	5	5	5	5	5	5	5	5	5	5	5	5	5	5	5	5	5	5	5	5
114	Sidewalks	1	1	1	1	1	1	1	1	4	2	1	1	2	1	1	1	1	1	2	1	1	1	1
68	Commercial area	5	5	5	5	5	5	5	5	5	5	5	5	5	5	5	2	5	4	5	5	5	4	4
60	Paved roads	1	1	1	1	1	1	1	1	1	1	1	4	5	5	5	1	1	1	1	1	1	1	5
70	church	1	1	1	1	1	1	1	1	1	1	1	1	1	1	1	1	1	1	1	1	1	1	1
67	Basic Education School	1	1	1	1	1	1	1	1	2	1	1	1	2	1	2	1	1	1	1	1	1	1	1

0	Community medical care	1	1		1	1	1	1	1	2	1	1	1	2	1	2	1	1	1	1	1	1	1
66	Waste collection	3	3	3	3	3	3	3	3	3	3	3	4	4	4	4	4	3	3	3	3	3	4
65	Guardhouse	2	2	2	2	2	2	2	2	5	2	2	4	4	4	4	2	2	2	2	2	2	2
64	Street lighting	1	1	1	1	2	2	2	3	3	4	1	1	1	2	2	1	1	1	2	1	3	1
72	Nearby transport	1	1	1	1	1	1	1	3	3	4	4	4	4	4	5	1	1	1	1	1	1	3
	SOCIAL DIMENSION																						
0	Security, perception in the colony	Y	Y	Y	Y	Y	Y	Y	Y	No	Y	Y	Y	Y	Y	Y	Y	Y	Y	Y	Y	Y	Y
10	Access to educational networks children and adults	No	No	No	No	No	No	No	No	No	No	No	No	No	No	No	No	No	No	No	No	No	No
11	Neighbourhood Organization and Participation	Y	Y	No	Y	No	Y	No	Y	No	No	Y	Y	No	Y	Y	Y	No	Y	No	No	Y	Y
3	Access to community health service	Y	Y	Y	No	Y	Y	Y	Y	Y	Y	Y	Y	Y	Y	Y	Y	Y	Y	Y	Y	Y	Y
5	Access to basic education schools	Y	Y	Y	No	Y	Y	Y	No	No	No	No	Y	Y	No	Y	No	Y	Y	Y	Y	No	Y
101	Access to recreation and recreation spaces	No	No	No	No	No	No	No	No	No	No	No	No	No	No	No	No	No	No	No	No	No	No
8	Access to public transport	Y	Y	Y	Y	Y	Y	Y	Y	Y	Y	Y	No	No	No	No	Y	Y	No	No	No	Y	Y
26	Separate animal-human environments	Y	Y	No	No	No	No	No	No	Y	No	Y	No	No	No	No	No	No	Y	Y	Y	Y	No
24	Rest in the house	No	Y	No	No	No	No	No	Y	No	Y	No	No	No	No	Y	No	Y	Y	Y	Y	No	No
83	2 or less occupants per room	Y	No	No	No	No	No	No	No	Y	Y	Y	No	No	No	Y	Y	Y	Y	No	No	Y	No
4	Land tenure	Y	Y	Y	Y	Y	Y	Y	Y	Y	Y	Y	Y	Y	Y	Y	Y	Y	Y	Y	Y	Y	Y

Appendix G. Proposal on Recommendations to Improve Housing Satisfaction in Social Lag by Study Dimensions. The Actions to be Taken as a Proposal on Recommendations to the Solutions of the Results Obtained on Very High Dissatisfaction and Needs of Socially Lagging Housing Are Shown, the Dimensions and Indicators Marked in an Orange Box that Require Priority Action Are Shown

Table A9. Proposal on recommendations to improve housing satisfaction in social lag by study dimensions.

DIMENSION	PARAMETER	INDICATORS WITH HIGH NEED AND VERY LOW SATISFACTION	LINE OF ACTION	MEASURES SOLUTION	ORGANISM
SOCIAL	Cohesion Social	Security, perception in the colony Access to educational networks children and adults Neighbourhood Organization and Participation	Encourage the cultural and educational activity of the community among young children and adults	Educational networks based on actions of government agencies, with active participation of the community through an agency	State, municipal, DIF Civil organization, Secretary
	Access to the resources urban	Access to community health service Access to basic education schools Access to recreation and recreation spaces Access to public transport	Rehabilitation of green areas under the land use regulation in areas intended for it	Under the regulation of parks and gardens of the State of Hidalgo, regulation of municipal land use and with community participation with prior information on procedures to rehabilitate green areas and gardens	Ministry of Public Works municipal, civil organization community secretary environment and resources natural
	Housing	Human Animals Separate Environments Rest in the house More than 2 occupants per room	Enabling housing in social lag	Characterization of housing, project and budget, enabling measures with sustainable materials, easy execution, durable and low maintenance costs	Universities, application of material research sustainable. SEDATU_FONAHPO. Housing state commission.
	Legal	Land trend			

PHYSIC AL	Infrastructu re	Sanitary drainage connection to municipal network Connection to municipal drinking water network Other means of drinking water supply Mains connection			
	Mobility and Facilities	Access to gas for cooking Use of water and electricity saving systems Use of water heater Drinking water container (Tinaco) Bathroom with shower Toilet with toilet Bathroom inside the house Kitchen sink Suitable door Suitable windows Laundry in service yard	Enabling adequate indoor bathroom spaces in socially lagging homes, as well as adequate equipment for basic toilet and kitchen needs. Rehabilitation in openings for doors and windows appropriate to their function.	Within the characterization of the house, analysis of the facilities and basic equipment of the homes under study, identify the deficiencies and project solutions, make a budget with the condition of having them more efficient equipment and materials in saving water and energy	SEDATU (Development Secreta agricultural, territorial and urba medium of its support dispersin for rehabilitation in areas in FOHAPO social lag (Fund national of populous rooms) conavi (national commission of housing) State commissions and municipal housing.
	Mobility and Facilities	Wall covering Coating in floors Roof Coatings Fourth extra bedroom	Characterization, projection and budget of the dimensions of the deficiencies in high need	Refurbishment of coverings in walls, floors and roofs with sustainable materials and low maintenance costs	SEDA TU, FONHAPO, CONAV state and municipal commission living place.
SPACIAL	INDOOR	Sun lighting inside the house Window ventilation inside the housing	Enabling spaces in homes with this priority need based on projects and budget	Enabling spaces with adequate horizontal circulation in order to have a better function for the tailoring of its inhabitants	SEDATU, FONHAPO, CONAV state and municipal commission living place

		Separate spaces between bedrooms and other housing Living space Dining space			
	OUTDOOR	Pedestrian access to your home Spaces for expansion Garden area in the house	Refurbishment of access to homes and gardens	Refurbishment actions with community action from the civil organization and municipality	Government Public Works Secre… Municipal for access and garden… civil organization
URBAN ENVIRONMET	Equipment	Green areas in the neighbourhood Playground is in the neighbourhood Area for sports in the neighbourhood Area for sports in the neighbourhood Sidewalks Commercial area Paved City Churches Elementary school Community medical care	Enabling existing spaces for these purposes under the land use regulation, project of spaces for sport.	Refurbishment of green areas from civil organization and actions with the secretary of public works and gardens of the municipality.	Municipal Public Works Secreta… community civil organization, s… of the environment and natural…
	Services	Waste collection Guardhouse Street lighting Nearby transport			

References

1. UN-Habitat. *Slum Almanac 2015–2016. Tracking Improvement in the Lives of Slum Dwellers*; UN-Habitat: Nairobi, Kenya, 2017.
2. United Nations. 12 March 2019. *Millennium Development Goals.* Available online: https://www.un.org/millenniumgoals/environ. shtml (accessed on 15 February 2021).
3. World Bank. *Urban Poverty, World Bank Urban Papers*; World Bank: Washington, DC, USA, 2008.
4. UN-Habitat. Slums and Cities Prosperity Index; CPI: 2014. Available online: https://unhabitat.org/programme/city-prosperity-initiative (accessed on 15 February 2021).
5. UN-Habitat. *Conferencia Hábitat III. La Nueva Agenda Urbana*; Naciones Unidas: Quito, Ecuador, 2016.
6. Ball, M. Housing provision in 21st century Europe. *Habitat Int.* **2016**, *54*, 182–188. [CrossRef]
7. Talukdar, D. Cost of being a slum dweller in Nairobi: Living under dismal conditions but still paying a housing rent premium. *World Dev.* **2018**, *109*, 42–56. [CrossRef]
8. Shamsuddin, S. Resilience resistance: The challenges and implications of urban resilience implementation. *Cities* **2020**, *103*, 102763. [CrossRef] [PubMed]
9. ONU HABITAT. El plan estratégico 2020–2023. 9 May 2020. Available online: https://unhabitat.org/sites/default/files/2019/1 2/strategic_plan_esp_web.pdf (accessed on 1 March 2021).
10. Scott, M.; Parkinson, A.; Waldron, R.; Redmond, D. Planning for historic urban environments under austerity conditions: Insights from post-crash Ireland. *Cities* **2020**, *103*, 102788. [CrossRef] [PubMed]
11. Atia, M. Refusing a "City without Slums": Moroccan slum dwellers' nonmovements and the art of presence. *Cities* **2019**, in press. [CrossRef]
12. Rastegar, M.; Hatami, H.; Mirjafari, R. Role of social capital in improving the quality of life and social justice in Mashhad, Iran. *Sustain. Cities Soc.* **2017**, *34*, 109–113. [CrossRef]
13. Ruá M., J.; Huedo, P.-F. A simplified model to assess vulnerable areas for urban regeneration. *Sustain. Cities Soc.* **2019**, *46*, 101–144. [CrossRef]
14. Barrios, A.; González, E.; Mariñas, J.; Molina, M. (Re) Programa (Re) Habitation + (Re) Generation +(Re) Programming. In *The Recycling and the Sustainable Management of the Andalusian Housing Stock. Management of Habitable Surrondings from the Criteria of Active Aging, Gender and Urban Habitability*; University of Seville: Sevilla, Spain, 2015.
15. Serrano-Jimenez, A.; Barrios-Padura, A.; Molina-Huelva, M. Towards a feasible strategy in Mediterranean building renovation through a multidisciplinary approach. *Sustain. Cities Soc.* **2017**, *32*, 532–546. [CrossRef]
16. Luxán García De Diego, M.; Sánchez-Guevara Sánchez, C.; Román López, E.; Barbero Barrera, M.; Gómez Muñoz, G. *Re-Habilitación Exprés para Hogares Vunerables. Soluciones de Bajo Coste*; Fundación gasNatural fenosa: Barcelona, Spain, 2017.
17. UNICEF. *A Snapshot of Water and Sanitation in the Pacific*; UNICEF: New York, NY, USA, 2013. Available online: https://www.unicef.org/eap/reports/snapshot-water-and-sanitation-pacific (accessed on 1 March 2021).
18. United Nations. HABITAD III: Jamaica National Report 2014. In Proceedings of the Third United Nations Conference on Housing and Sustainable Urban Development, Quito, Ecuador, 17–20 October 2016.
19. CONEVAL. May de 2019. Consejo de Nacional de Evaluación de las Políticas de Desarrollo Social. Available online: https://www.coneval.org.mx (accessed on 26 February 2021).
20. Kanwal Zahra, A.A. Marginality and social exclusion in Punjab, Pakistan: A threat to urban sustainability. *Sustain. Cities Soc.* **2018**, *37*, 203–212. [CrossRef]
21. Rockwood, D. Tran Duc Quang.Urban immigrant worker housing research and design for Da Nang, Viet Nam. *Susteinable Cities Soc.* **2016**, *26*, 108–118. [CrossRef]
22. Galiania, S.; Gertlerb, P.; Undurrag, R.; Cooper, R.; Martínez, S.; Ross, A. Shelter from the storm: Upgrading housing infrastructure in Latin American slums. *J. Urban Econ.* **2017**, *98*, 187–213. [CrossRef]
23. Jaitman, L. *Evaluation of Slum Upgranding Programs: A Literature Review*; International Development Bank: Washington, DC, USA, 2015.
24. Estrada Casarín, C.; Obregón Davis, S. *El Rezago Habitacional: Alternativas Desde lo Local*; ITESO: Jalisco, Mexico, 2017.
25. Hernández Aja, A.; Matesanz Parellada, Á.; García Madruga, C.; Rodríguez Suárez, I. Análisis de las Políticas Estatales y Europeas de Regeneración Urbana y Rehabilitación de Barrios. *Ciudad. Territ. Estud. Territ.* **2012**, *46*, 182–183.
26. Kleinhans, R. Social implications of housing diversification in urban renewal: A review of recent literature. *J. Hous. Built Environ.* **2004**, *4*, 367–390. [CrossRef]
27. Saldaña-Márquez, H.; Gómez Soberón, J.; Arredondo-Rea, P.; Gámez-Gacía, R.; Corral-Higuera, D. Sustainable social housing: The comparison of the Mexican funding program for housing solutions and building sustainability rating systems. *Buinding Environ.* **2018**, *133*, 103–313. [CrossRef]
28. Abramo, P. The com-fused city: Market and production of the urban structure in the great Latin American cities. *Eure* **2012**, *38*, 35–69.
29. Smith, D. World City Populations. 19 March 2019. Available online: http://luminocity3d.org/WorldCity/#2/34.5/17.2 (accessed on 6 February 2021).

30. Higueras, E.; Macias, M.; Rivas, P. *Sustainable Buidling Conference. SB10mad. Metodología para la Evaluación de la Sostenibilidad en Nuevas Planificaciones Urbanas*; Selección de criterios e indicadores: Madrid, Spain, 2010; Available online: https://vapaavuori.net/2010/09/22/sustainable-building-2010-conference-espoo/ (accessed on 6 February 2021).
31. OECD. 1993. Available online: http://enrin.grida.no/htmls/armenia/ (accessed on 1 March 2021).
32. Verma, P.; Raghubanshi, A. Urban sustainability indicators: Challenges and opportunities. *Ecol. Indic.* **2018**, *93*, 282–291. [CrossRef]
33. Mercader-Moyano, P.; Morat, O.; Serrano, J. Urban and social vulnerability assessment in the built environment: An interdisciplinary index-methodology towards feasible planning and policy-making under a crisis context. *Sustain. Cities Soc.* **2021**, *73*, 103082. [CrossRef]
34. Tapia, C.A. Profiling urban vulnerabilities to climate change: An indicator-based vul- nerability assessment for European cities. *Ecol. Indic.* **2017**, *78*, 142–155. [CrossRef]
35. Tyler, S.N. Indicators of urban climate resilience: A contextual approach. *Environ. Sci. Policy. Environ. Sci. Policy* **2016**, *66*, 420–426. [CrossRef]
36. López-Mesa, B. Nuevos enfoques en la rehabilitación energética de la vivienda hacia la convergencia europea. In *La vivienda social en Zaragoza.1939–1979*; Prensas de la Universidad Zaragoza: Zaragoza, Spain, 2018.
37. Luxan, M. *Re-Habilitación Exprés para Hogares Vulnerables. Soluciones de Bajo Coste*; Fundación Gas Natural fenosa: Madrid, Spain, 2018.
38. Le, L.H.; Da, A.D.; Dang, H.Q. Building up a System of Indicators to Measure Social Housing Quality in Vietnam. *Procedia Eng.* **2016**, *142*, 116–123. [CrossRef]
39. Mohammed Amenn, R.M. Urban sustainability assessment framework development: The ranking and wighting of sustainability indicators using analytic hierarchy process. *Sustain. Cities Soc.* **2019**, *44*, 359–366.
40. Lynch, A.J.; Mosbajh, S. MImproving local measures of sustainability: A study of built environment indicators in the United States. *Cities* **2017**, *60*, 301–313. [CrossRef]
41. Graw, V. Mapping marginality hotspots: Geographical targeting for poverty reduction. *SSRN J.* **2012**. [CrossRef]
42. Braulio-Gonzalo, M.; Bovea, M.; Rua, M. Sustainability on the urban scale: Proposal of a structure of indicators for the Spanich context. *Environ. Impact Assess. Rev.* **2015**, *53*, 16–30. [CrossRef]
43. Monzón, M.; López-Mesa, B. Buildings performance indicators to prioritise multi-family housing renovations. *Sustain. Cities Soc.* **2018**, *38*, 109–122. [CrossRef]
44. Generalitat de Cataluña. *Estrategias Para el Desarrollo Sostenible de Cataluña-Mobilidad*; Departamento de Medio Ambiente: Barcelona, Spain, 2009.
45. Naciones Unidas. Nueva Agenda Urbana de Hábitat III. Quito, Ecuador. 2016. Available online: http://habitat3.org/wp-content/uploads/NUA-Spanish.pdf (accessed on 10 February 2021).
46. Cabrera Chaparro, A. *Suelo Urbano en Bogotá: Vivienda para Población de Bajo Ingreso*; Universidad Politécnica de Valencia: Valencia, Spain, 2015; Available online: https://riunet.upv.es/handle/10251/60452 (accessed on 10 February 2021).
47. Hernández, G.S.V. Vivienda y Calidad de vida. Medición del Hábitat Social en el México Occidental; Revista Bitácora Urbano Territorial. 2014. Available online: https://revistas.unal.edu.co/index.php/bitacora/article/view/31463/pdf_31. (accessed on 10 February 2021).
48. ISO 37120:2015. *Sustainable Development of Communities—Indicators for City Services and Quality of Life*; ISO: Geneva, Switzerland, 2015.
49. INEGI. *Datos del INEGI (Instituto Nacional de Estadística, Geografía e Informática) Empleados por CONEVAL para el Corte Intercensal*; CONEVAL: México City, Mexico, 2018.
50. Mrtin Muñoz, M. *Manual de Indicadores para el Diagnóstico Social*; Colegios Oficiales de Diplomados en TRabajo Social y Asistentes Sociales de la Comunidad Autónoma Vasca: Madrid, Spain, 1996.
51. Mellace, F. *Tecnología en la Vivienda Rural: Tucamán, Argentina. II Seminario y Taller Iberoamericano Sobre Vivienda Rural y Calidad de Vivienda en los Asentamientos Rurales*; Universidad Autónoma de San Luis Potosí: San Luis Potosí, Mexico, 2000; pp. 308–316.
52. Sánchez Quintana, C.; Jiménez Rosa, E. La vivienda rural. Su complejidad y estudio desde diversas disciplinas. *Rev. Luna Azul* **2010**, *30*, 174–196.
53. Consejo Nacional de Evaluación de la Política de Desarrollo Social. *Estudio Diagnóstico del Derecho a la Vivienda Digna y Decorosa*; Consejo Nacional de Evaluación de la Política de Desarrollo Social: Mexico City, México, 2018.
54. UN-Habitat. *State of the World´s Cities 2012/2013*; Routledge: New York, NY, USA, 2013.
55. Salas-Bourgoin, M. Propuesta de índice de calidad de vida en la vivienda. *Cuad. Cendes* **2012**, *29*, 57–78.
56. Hernández Sampieri, R.F. *Metodología de la Investigación*; McGraw-Hill: México City, Mexico, 2006; Volume 3.
57. López Alvarenga, J.B. How to estimate the sample size of a study. *Rev. Mex.* **2010**, *54*, 375–379.
58. Maldonado Luna, S. Manual práctico para el diseño de la Escala Likert. *Trillas* **2007**, *2*. [CrossRef]
59. Cronbach, L. Coefficient alpha and the internal structure of tests. *Pshychometrika* **1951**, *16*, 297–334. [CrossRef]
60. Welch, S.; Comer, J. *Quantitative Methods for Public Administration: Techniques and Applications*; Houghton Mifflin Harcourt: Charles, IL, USA, 1988.

61. Gliem, J.; Gliem, R. Calculating, interpreting, and reporting Cronbach´s alpha reliability coefficient for Likert-type scales. In Proceedings of the Midwest Research-to-Practice conference in Adult, Continuing, and Community Education, Columbus, OH, USA, 2003; Available online: https://scholarworks.iupui.edu/bitstream/handle/1805/344/gliem+&+gliem.pdf?sequence=1 (accessed on 10 February 2021).
62. Corral-Verdugo, V.; Barrón, M.; Tapia-Fonllem, A.C. Housing habitability, stress and family violence. *Psyecology* **2011**, *2*, 3–14. [CrossRef]
63. Gómez, J. Estrategias de ponderación de la respuesta en encuestas de satisfacción de usuarios de servicios. *Metodol. Encuestas* **2002**, *4*, 175–193.
64. Muñoz Pichardo, J.M. Propuesta de Ponderación para el Nivel de Satisfacción en la Vivienda en Rezago Social. Grupo de Estadística de la Facultad de Matemáticas de la Universidad de Sevilla. 2018. Available online: https://www.coneval.org.mx/SalaPrensa/Comunicadosprensa/Documents/Indice-de-rezago-social-2010.pdf (accessed on 10 February 2021).

 *sustainability*

Article

Optimization of a Combination of Thermal Insulation and Cool Roof for the Refurbishment of Social Housing in Southern Spain

Carlos-Antonio Domínguez-Torres [1], Helena Domínguez-Torres [2] and Antonio Domínguez-Delgado [3,*]

[1] Escuela Técnica Superior de Arquitectura, Universidad de Sevilla, Avda. Reina Mercedes 2, 41012 Sevilla, Spain; cdtorres@us.es
[2] Facultad de Ciencias Económicas y Empresariales, Universidad de Sevilla, Avda. Ramón y Cajal 1, 41018 Sevilla, Spain; hdtorres@us.es
[3] Department of Applied Mathematics 1, Escuela Técnica Superior de Arquitectura, Universidad de Sevilla, Avda. Reina Mercedes 2, 41012 Sevilla, Spain
* Correspondence: domdel@us.es

Citation: Domínguez-Torres, C.A.; Domínguez-Torres, H.; Domínguez-Delgado,A. Optimization of a Combination of Thermal Insulation and Cool Roof for the Refurbishment of Social Housing in Southern Spain. *Sustainability* **2021**, *13*, 10738. https://doi.org/10.3390/su131910738

Academic Editor: Chi-Ming Lai

Received: 24 August 2021
Accepted: 22 September 2021
Published: 27 September 2021

Publisher's Note: MDPI stays neutral with regard to jurisdictional claims in published maps and institutional affiliations.

Copyright: © 2021 by the authors. Licensee MDPI, Basel, Switzerland. This article is an open access article distributed under the terms and conditions of the Creative Commons Attribution (CC BY) license (https://creativecommons.org/licenses/by/4.0/).

Abstract: Social housing built in the middle of the last century in Spain suffers from poor thermal insulation conditions that cause situations of discomfort and energy poverty. For this reason, the energetic refurbishment of the envelope of this social building stock is necessary to overcome these situations and reduce energy consumption aimed at achieving interior comfort for its occupants. The goal of this work is to optimize a constructive solution that combines cool roof techniques with the use of thermal insulation applied to the refurbishment of the roof of buildings belonging to a quarter of social housing in Seville, Spain. The optimization analysis is based on the computation of the energy performance of the roofs when the energy retrofitting measure is applied, considering a variety of combinations of solar reflective coatings and insulation layer thickness, performing a dynamic analysis that accounts for the aging effect of the cool coats on the monthly roof energy performance and on the economic balance for the whole life cycle (LC) span. Economic and energy optimization analysis show that a suitable combination of cool roof emissivity and insulation layer thickness produces significant savings in the operational energy and in the economic profitability of the proposed retrofitting measure: the optimum combination obtained provides for the entire life cycle timespan an energy savings of 5.71 GJ/m^2 and a cost savings equivalent to the 63.1% of the total costs when compared to the non-refurbished roof. The application of a time-dependent pattern for the changes on time produced by the aging effect on the cool roof emissivity, and its effects on the optimization of the combination of cool roof and insulation layer, can be considered novel in literature, both from an energy and an economic point of view.

Keywords: optimization; cool roof; thermal insulation; aging effect; social housing; life-cycle cost analysis

1. Introduction

The building sector is responsible for around 40% of final energy consumption in Europe and is a major source of CO_2 emissions. In order to reduce these emissions, in the last few decades, efforts have been made, consisting largely of the promulgation of regulations for the implementation of energy savings and efficiency measures, such as the Spanish norm DB-HE-2019 [1], which establishes new standards for the reduction of energy consumption in buildings, in accordance with the directives set by the EU for the H2030 [2], which establishes as goals for 2030 an improvement in energy efficiency equal to or greater than 32.5%, cuts in greenhouse gas emissions equal to or greater than 40%, and a share for renewable energy equal to or greater than 32%, all from 1990 levels [2].

55

In climates characterized by mild winters and very long hot seasons, such as the case of southern Spain and much of the Mediterranean area, a large amount of energy consumption is necessary to obtain the internal comfort of buildings during the cooling season. This poses a serious problem in the case of the social park housing built in southern Spain in the mid-twentieth century, in which envelopes often lack thermal insulation.

The energy efficiency of roofs has been object of a wide research focused on the enhancement and the optimization of its energy performance. This is largely justified by the fact that the roof is the envelope component that receives the most solar radiation in summer and in winter is responsible of an important percentage of the heat losses. Two widely used types of techniques aim to improve the thermal performance of roofs: those that modify their thermal properties, how to add or change thermal insulation, and those who carry out roof surface treatments, how to install radiant barriers or cool roofs. Taking into account that, in the geographical context considered here, in summer, most of the time, the sky is clear, and, in winter, the temperatures are not too cold, the cool roof technique seems advisable to lower the temperature of the roof, while the combination with a layer of insulation can help to achieve indoor comfort conditions and reduce energy consumption for this purpose.

Yu et al. [3] analyzed the optimum thickness of the thermal insulation installed on the roof of a residential building, and they studied different colors for the outer roof surface; to carry out the analysis, a life-cycle cost analysis (LCCA) was performed by considering solar-air degree-hours and four cities representing different China geographical regions; authors concluded that the effect of the colors on the optimal insulation thickness is different for each city. Gentle et al. [4] studied the energy behavior of a roof, analyzing different combinations of solar albedo, thermal emittance, and thermal resistance of a roof; a cost benefit assessment of the different combinations was performed that concluded the importance of high values for the roof albedo. Ramamurthy et al. [5] used the minimization of yearly costs for heating and cooling in order to determine the optimal combination of roof solar albedo and insulation thickness; they concluded that an albedo greater than 0.7, combined with a polyiso foam insulation of 18 cm, significantly reduces thermal flux through the roof when applied to new constructions; likewise, they found that, if installations costs are included in the analysis, the optimal combination is reached for a 4-inch insulation thickness with a payback period of 13 years. Arumugam et al. [6] studied the combination of solar reflective coating, radiant barriers, and thermal insulation for a roof by using EnergyPlus; a total of five climatic zones in India and 88 different roof combinations were analyzed; through an economic analysis, they found that the use of solar reflective coating and radiant barriers on the roof reduced the need for insulation layer for all the climatic regions analyzed.

Other studies have focused on optimizing the properties of the roof. Piselli et al. [7], by using dynamic energy computation, developed a method to perform an optimization analysis; the objective of this analysis was to optimize the solar reflectivity of the roof surface to minimize yearly energy loads for conditioning the building; the analysis was carried out by considering five cities of Italy located in different climate zones as the case study; from the performed optimization analysis, it was concluded that the the climate context is the most influential factor on the optimum roof solar reflectance.

Shi and Zhang [8] studied the effect on the reduction of annual energy loads of different solar reflectance and long-wave emissivity combinations for the outer surface of the building envelope; the study was carried out in 35 different climatic zones of the world and concluded that the greatest potential for energy savings is obtained for tropical climates and for high solar reflectance and high long-wave emissivity values.

Furthermore, a large number of studies deal with usefulness of cool roofs to improve indoor comfort conditions and to reduce energy consumption [9–22], while other studies analyze the benefits derived from the the attenuation of the heat island effect in cities, as well as achieving greater comfort in them [13,14,17].

It is worth noting that the performances of the building in summer and winter condition depend on the global behavior of the building, contending with global the performance of the roof and the infills. This means that the evaluation of thermal lack must also be extended to the performances of the infills. On this last topic, the literature is wide and complex, particularly for the fact that the nonstructural elements, such as the infills, must maintain the integrity under earthquake effect. Some works related with this topic are Kibert [23], Vailati and Monti [24], Manfredi and Masi [25], and Vailati et al. [26].

Likewise, some other recent works related to the research presented here and that may be of interest are the developed by Zingre et al. [27], Hernández-Pérez [28], and De Masi et al. [29].

Regarding the economic perspective, Levinson and Akbari [30] demonstrated the ability of cool roofs to achieve significant monetary savings when used in commercial building retrofits in the USA. Boixo et al. [19] stated that a large-scale implementation of cool roofs in the whole region of Andalusia, southern Spain, would have the ability to save yearly around 59 million euros in electricity costs, considering for this calculation only dwellings with flat roofs and the use of electricity as energy source, which would imply a reduction of about 136,000 metric tons on CO_2 emissions.

Nevertheless, as stated in Reference [31], in the literature, there are few studies that analyze the cost-effectiveness of cool roofs and, therefore, the combination of cool roof and insulating layer, based on an LCCA. Jo et al. [32] performed a research to investigate the impact on energy consumption and savings of replacing, in a commercial building, the original dark material of the roof with a solar reflective coating; after conducting a cost benefit analysis over a 20-year period, including installation and maintenance costs, the authors found monthly reductions of 1.3–1.9% and 2.6–3.8% in electricity consumption after a roof retrofit of 50% and 100%, respectively. Sproul et al. [33] performed an economic comparative analysis on the effect of some colors, specifically white, green, and black flat roofs in the United States; after a 50-year life-cycle analysis (LCA) that included installation, replacement, and maintenance costs, they concluded that the best economics results are obtained for the white color. Zhang et al. [34], after performing an LCA, demonstrated that a combination of roof ventilation and cool paint produces an annual energy savings for cooling of 109.94 USD/m^2, with a payback period of 2 and 6 months for unventilated and ventilated roofs, respectively, under the weather conditions in Singapore. Hernández-Pérez [28] carried out a comparison of the cost-effectiveness of reflective white and colored roofs versus conventional gray roofs in six cities in Mexico. A 10-year LCCA concluded that, for the case of no thermal insulation, gray roofs are less cost effective than white and colored reflective roofs for all the considered locations. Yuan et al. [35] performed an optimization analysis of the combination of thermal insulation thickness and surface reflectivity through a 10-year LCCA; as result of the analysis, optimum combinations were proposed for each of the six considered regions in Japan. Saafi and Daouas [31] performed a 20-year LCCA in order to find the optimum combination of solar reflective coatings and insulation thickness for roofs of residential buildings in Tunisia; authors demonstrated that the use of these combinations are cost-effective and lead to economic savings of up to 44.53 Tunisien dirhams per m^2 with a payback period equal to 3.4 years; moderate values of the insulation thickness, combined with values of the solar reflectivity as high as possible, are recommended for use in the Tunisian climate; in this study, the effect of loss of reflectivity of the cool roof is taken into account when carrying out the life-cycle cost analysis; nevertheless, this effect is assumed to be constant throughout the whole life-cycle timespan.

The optimization analysis carried out in the present work tries to establish the cost-effectiveness of an Energy Saving Refurbishment Measure (ESRM) consisting of the combination of an insulation layer and a roof cool coat when used to refurbish roofs belonging to the social housing built in Spain in the middle of the past century and to optimize this combination both from an energy and economic point of view under the climate conditions of Southern Spain.

Taking into account the obsolete energy conditions of the building park considered, the present analysis is performed with the objective of providing guidelines to decision-makers for the implementation of refurbishment actions on the social housing aimed at reducing situations of internal thermal distress, energy poverty, and associated unhealthy situations. Thus, the improvements on energy efficiency that result from the optimization of the combination of thermal insulation and cool roofs for the refurbishment of social housing can be framed in Buildings Regeneration Strategy to Climatic Change Mitigation, Energy, and Social Poverty.

Likewise, the present optimization analysis is proposed in order to appraise in a deeper way the combined effect of insulation layer thickness variability and different values of solar reflectivity for the roof coat and the effect on this combined system, together with the aging effect of the cool paint on the energy performance of the refurbished roof, as well as their implications on the economic and energy LCCA of these combinations when they are implemented in the retrofitting of the social housing park under consideration.

To carry out the optimization analysis, the integration of the aging pattern for the cool coat with the proposed variability in the combinations of insulation layer thickness and roof reflectivity requires a specific simulation code allowing the accurate calculation on thermal loads for the refurbished roofs. Usually, commercial packages for building energy simulation (BES) do not have this ability. This software limitation may be the cause of the lack of studies that address the time-dependent loss of solar reflectivity due to the aging effect of the cool coat. To the best of our knowledge, there exists a gap in the literature with regard to studies that perform optimization analysis by considering annual variable costs due to the aging effect, when performing the LCCA of the proposed combination. Such a gap is especially important in the geographic area analyzed in this research where cooling demand is usually very high in the hot season.

This way, the thermal performance of a combination of insulation layer and cool roof is analyzed through a comprehensive numerical study when such a combination is used to refurbish roofs from residential buildings belonging to the social housing built in Seville, Spain, in the middle decades of the 21st century before the enactment of the first Spanish legislation intended at regulating energy demand in buildings, the NBE-CT-79 [36], in 1979.

The main objective of this work is to perform an energy and economic optimization of the proposed refurbishment measure within a broad framework considering the effect on the LCCA of the real weather of Seville, a wide range of combinations for the insulation layer thickness, and the cool coat reflectivity, as well as the aging effect of the cool coating, the energy source, and equipment type used to condition the buildings and current standard economic parameters in Spain and Europe.

Finally, it is important to note that the findings of the present research can be extended to other places with climatic conditions similar to those existing in the south of Spain, as are as are almost all southern European regions.

2. Methodology

Compared to previous research aimed to estimate the optimum combination of cool roof reflectivity and insulation thickness, the methodology followed in this work takes into account the time-dependent loss of reflectivity of cool paint produced by the aging effect and the cyclical restoration of the solar reflectivity by power washing of the cool roof. In this way, in order to estimate the impact on thermal loads of the evolutionary change in roof reflectivity in combination with different insulation thickness, a dynamic finite difference model is developed that calculates the monthly thermal loads for the entire LCA timespan, taking into account the monthly updated reflectivity estimated following the patterns shown in Reference [37].

The LCCA is performed by taking into account the costs of surface reflective coating material, its installation, and the power washing when done, as well as the costs of the insulation material and its installation. To compute the costs for the whole LCA span, the energy cost is computed monthly for every studied scenario, considering the changes in the energy consumption produced by the cool coating reflectivity changes along the LCA span, and then the costs are estimated under the assumption of the use of three different sources of energy to obtain thermal comfort: heat pumps for heating and cooling, air conditioners to cool and gas boilers to heat, and, finally, air conditioners to cool and electrical cheating.

The thermal load calculations are used to find out the most advantageous combination of roof surface reflectivity and insulation thickness in order to reduce the energy load needed to obtain indoor comfort in the studied buildings and the considered climate framework for the whole LCA time period. This way, the economic analysis based in the energy results is used to conclude the optimum combination that provides the minimum total cost for the LCA timespan.

2.1. Energy Roof Model

In the energy analysis, the hypothesis of one-dimensional heat conduction through the roof is considered. It is assumed that the roof is made up of M parallel layers meeting the following hypothesis: the material of every layer is homogeneous and isotropic, thermal properties of materials remain constant under temperature changes, and there is no internal heat generation. Then, for the multilayer heat conduction through the roof, the linear heat conduction equation,

$$\frac{\partial T_j(x,t)}{\partial t} = \alpha_j \frac{\partial^2 T_j(x,t)}{\partial x^2}, \quad j = 1, \ldots, M, \tag{1}$$

can be applied in each layer j, with T_j being the temperature in layer j, and α_j the thermal diffusivity of the layer material j [38].

Equation (1) requires the definition of boundary conditions. The boundary conditions for the external and internal surfaces of roof are obtained from the energy balance equation corresponding to each surface.

The outer surface of the roof is affected by the solar radiation I^{SW}, the long-wave exchange with the sky q^{LW}, the heat exchange by convection between the outer surface and the ambient air $q_{c,ext}$, and, finally, by the heat conduction through the outer layer. Thus, the energy balance for this surface can be written as:

$$k \frac{\partial T}{\partial \vec{n}} + q_{c,ext} + \alpha^{SW} \cdot I^{SW} + q^{LW} = 0, \tag{2}$$

where $\vec{n}$ is the outward normal vector to the roof outside surface, k is the thermal conductivity of the outer layer material, and α^{SW} is the solar absorptivity of the external roof surface.

The external convection heat transfer $q_{c,ext}$ is computed as

$$q_{c,ext} = h_{c,ext}(T_a - T),$$

where T_a is the outdoor air dry-bulb temperature, T the temperature of the outer surface of the roof, and the correlation

$$h_{c,ext} = 8.18 + 2.28\,V_R \ [\mathrm{W/m^2\,K}]$$

proposed by Hagishima and Tanimoto [39], based on experiments carried out on a roof, has been adopted. Here, V_R is the wind velocity above the roof in m/s.

The long-wave radiation exchange between the roof external surface and the sky is calculated as:

$$q^{LW} = Q^{LW}_{sky} - Q^{LW}_{w},$$

with Q^{LW}_{w} being the thermal radiation intensity emitted by the external surface of the roof, and Q^{LW}_{sky} being the sky down-welling long-wave radiation.

For computing Q_w^{LW}, the law of Stefan–Boltzmann establishes that

$$Q_w^{LW} = \varepsilon \sigma T^4, \tag{3}$$

where ε is the surface emissivity, T the surface temperature in Kelvin degrees, and $\sigma = 5.67 \times 10^{-8}$ [W/m^2 K^4] is the Stefan–Boltzmann constant.

Q_{sky}^{LW} is estimated through the expression:

$$Q_{sky}^{LW} = \varepsilon_{sky} \sigma T_a^4, \tag{4}$$

where ε_{sky} is the sky emissivity that, following Walton [40] and Clark et al. [41], has been computed as:

$$\varepsilon_{sky} = (0.787 + 0.764 \ln(\frac{T_{dp}}{273}))(1 + \frac{224}{10^4}n - \frac{35}{10^4}n^2 + \frac{28}{10^5}n^3),$$

where T_{dp} is the dew point temperature in Kelvin, and n is the fraction of sky covered by clouds in tenths.

Other sources of thermal radiation, such as plant masses or nearby buildings higher than the building where the studied roof is placed, have not been considered in the present study, for the sake of brevity.

For the roof inside surface, the energy balance is given by

$$\kappa \frac{\partial T}{\partial \vec{n}} + q_{cr,int} = 0, \tag{5}$$

where T is the roof interior surface temperature, $\vec{n}$ is the outward normal vector to the roof interior surface, κ is the thermal conductivity of the roof internal layer, and $q_{cr,int}$ is the combined convective-radiant heat transfer balance between the roof interior surface and the indoor temperature computed as:

$$q_{cr,int} = h_{cr,int}(T - T_{room}), \quad (\text{W/m}^2). \tag{6}$$

Here, $h_{cr,int}$ is the combined convective-radiant heat transfer coefficient defined as the sum of the convective $h_{conv,int}$ and the radiant $h_{rad,int}$ heat transfer coefficients. Following the correlations established in Reference [42] for horizontal surfaces, the convective heat transfer coefficients are taken as $h_{conv,int} = 0.92\,(\text{W/m}^2)$ for downward heat flow and as $h_{conv,int} = 4.04\,(\text{W/m}^2)$ for upward heat flow. The radiant heat transfer coefficient is taken as $h_{rad,int} = 5.72\,\varepsilon\,(\text{W/m}^2)$, where ε is the emissivity of the roof inside surface. These values are recommended by ASHRAE [43], and they are often used in the calculation of heat transfer in buildings [44,45].

Finally, the M roof layers are assumed to be in good contact; therefore, the interlayer thermal resistance is neglected.

2.2. Changes in Reflectivity Due to Aging Effect and Maintenance

To consider realistic conditions, the aging effect of the cool coating must be considered. Some authors have pointed out that the factors that result in a loss of reflectivity in cool paints are mainly the weatherization of the cool roofs, the characteristics of the paint, and the material on which the paint is applied [22,46]. Other studies quantified the lost reflectivity and the rate at which this loss occurs. This way, in Reference [47], a loss of the solar reflectivity for white roofs of about 10–30% was found; this reduction was found mostly during the first year, whilst, in Reference [48], it was established that the solar reflectance of a cool paint decreased about 11% after being subjected to 400 h of artificial accelerated weathering. In Bretz and Akbary [49], it is stated that the decrease in the solar reflectivity of the the cool coat reflectivity occurred mostly in the first year, and, very likely, in the first months, and, in Reference [50], from an in-field study, it was shown that, for all reflective roofing types analyzed, in the first two years, about 95% of the loss of solar reflectivity occurred, and about 98% of the total aging effect took place in the interval of three years after the installation.

Finally, from in-field measurements [49,50], it has been found that a power washing of the cool roof could restore up to 90–100% of initial values for the solar reflectivity of the cool coat. In Reference [49], the cold coat was washed one year after its installation, and, in Reference [50], no data is provided on the period of time in which the wash was performed.

According to the aging behavior mostly supported by previous literature, to compute the heat flux through the roof, the following pattern for the reflectivity loss of the cool coating has been considered: during the first year, the solar reflectivity suffers a loss of the 20% of its original value, with most of this reduction occurring in the first few months; in the second year, a reflectivity loss equal to 8.5% is taken, and, during the third year, a loss equal to 0.9% is considered. After these first three years, the solar reflectivity decrease is assumed to stabilize [49], until a loss of 30% is reached for the whole LC timespan. With these assumptions, the aging pattern stated in Reference [50] is fulfilled. On the other hand, according to the aforementioned literature, the solar reflectivity is considered to be restored to 90% of its initial value after a power washing and then following the same rate of aging loss that occurs after the first application of the cool layer.

Finally, the pattern established in Reference [37] for the time evolution of the solar reflectivity, that collects the foregoing, is used in the computations. This pattern is illustrated in Figures 1 and 2. In Figure 1, the evolution of the solar reflectivity of a cool roof with an initial value equal to 0.9 is shown for a triennial washing maintenance.

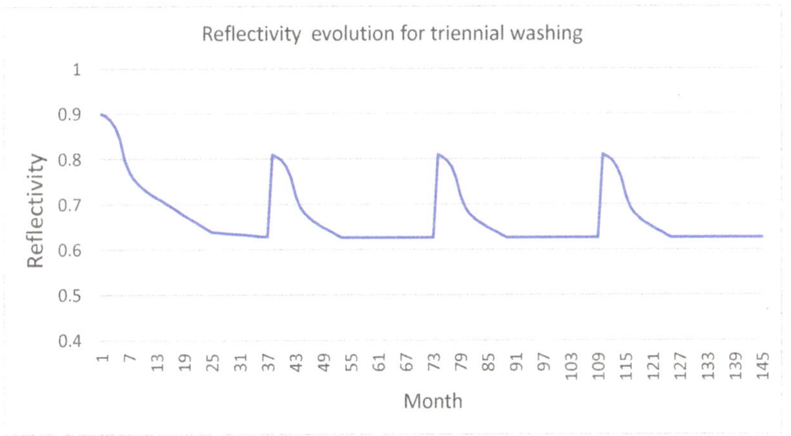

Figure 1. Evolution of solar reflectivity for an initial value equal to 0.9 and a triennial washing maintenance.

In Figure 2a, the monthly and accumulated loss of the solar reflectivity is shown for a period of 42 months, and, in Figure 2b,c, the assumed evolution of the solar reflectivity without maintenance and for a decennial washing, respectively, are shown for a period of 20 years.

An immediate consequence of the evolution over time of the exterior reflectivity is the need to calculate the heat transfer through the roof on a monthly basis for the whole LCA timespan. For this, the reflectivity value in Equation (2) must be updated for each moment of time, following the aging pattern described above. That is, the heat transfer through the roof cannot be considered the same for all the years of the LCA timespan, which is a frequent assumption in the literature.

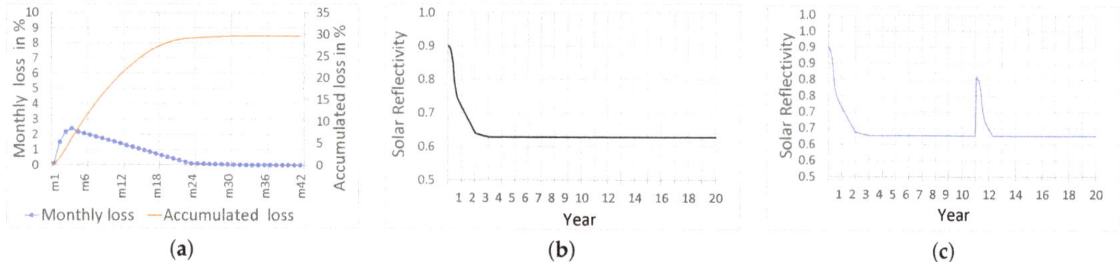

Figure 2. (**a**) Percent loss of solar reflectivity, monthly and accumulated. (**b**) Solar reflectivity evolution for the aged case without maintenance. (**c**) Solar reflectivity evolution for the aged case with decennial washing.

2.3. Thermal Load Calculations

The key point of the developed analysis is the accurate computation of the heat transfer through the roof to the building indoor and the associated thermal loads to obtain indoor comfort. To do that, the load at time t is calculated using the expression entered above:

$$q(t) = h_{cr,int}(T(t) - T_{room}) \quad (W/m^2), \tag{7}$$

which takes into consideration the convective and radiant thermal flux from the internal roof surface to the building indoor. $q(t)$ yields positive values when the heat flux enters to the building, and negative values when heat flows out of the building.

Then, if weather hourly values are considered, as it is usual in building energy software, the hourly load is derived from (7) for every hourly interval. This way, daily, monthly, and yearly loads are straightforwardly obtained by integrating the hourly loads on the timespan of interest.

The thermal loads due to the heat transfer through the roof, including the cooling and heating loads, are computed varying the absorptivity of the cool paint from 0.1 to 0.5, with this being the usual value range to consider a roof as a cool roof [51]; for computations, steps equal to 0.1 are considered for the absorptivity. The insulation thickness ranges from 0 to 10 cm, with the latter being the maximum value usually used for the roof retrofitting in southern Spain. For the computations, steps of 1 cm for the thickness of the insulation layer are considered.

We assume a fixed value of the indoor temperature in each season of the year. The conditioning system is considered to be in continuous mode with a fixed temperature set point at 24 °C in the cooling season, at 20 °C in the heating season, and at 22.5 °C for the intermediate seasons, according with the comfort temperatures established in the Spanish regulations for thermal installations in buildings [52].

In order to estimate the energy loads due exclusively to the heat transfer through the roof, we follow the approach from Reference [4,31,53], and, then, the zone beneath the roof is assumed to have the envelope adiabatic, except for the roof.

The input data needed to compute the thermal loads are:

- Meteorological data: air temperature, solar radiation, cloud cover, wind speed, and relative humidity. These data can be hourly or any other frequency smaller than one hour.
- Roof data: solar and thermal reflectivity of the outer roof surface and geometrical and thermophysical properties of the roof layers.
- Indoor data: indoor air temperature.

On the other hand, with regard to the aim of assessing the energy and economic performance of the refurbishment measure through the lifetime period of 20 years according to the service time of the roof coating, taken into consideration is the effect stemming from the action of environmental agents as rain, dust, air particles, moisture, or sun, as well as their own aging, on the initial reflectivity properties of the cool paint, as described in Section 2.2.

To estimate the thermal loads, the following cool coating maintenance scenarios were analyzed: no washing, and decennial, quinquennial, and triennial washing. Taking into account the social character of the buildings in which the roof reform is applied, it seems appropriate to consider an annual roof wash unlikely; realistically, only no maintenance, or washing every five or ten years, scenarios should be considered.

The thermal loads were estimated using the open source *FreeFem++* software [54]. Hourly loads, including heating and cooling loads, were computed for the whole LC time interval.

2.4. Case Study

The case study considered is a building in the city of Seville. The city is located in the Guadalquivir valley in southern Spain; see Figure 3.

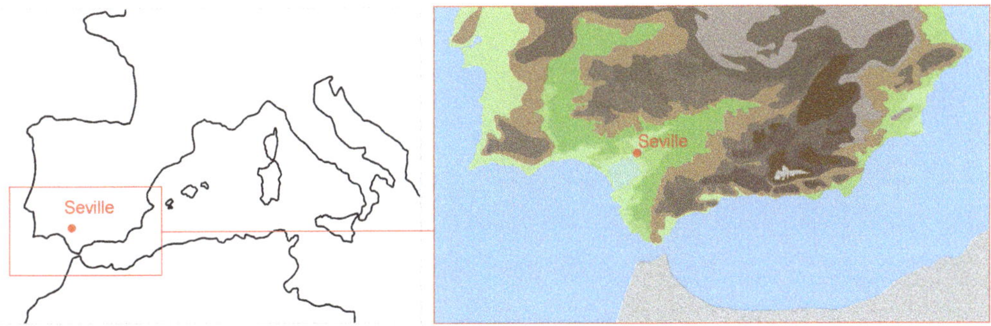

Figure 3. Geographical framework and location of the city of Seville (Spain).

Its location in a valley with a low altitude above the sea, its meridional latitude, and its dry climate, together with the high levels of solar radiation, cause a long hot season, spanning months beyond summer which, together with its soft winters, make this city an appropriate place for the application of cool roofing techniques.

The climatic chart of Seville is shown in Figure 4 from data from the Spanish State Meteorological Agency [55]. The average annual temperature is 19.2 °C, with maximum average temperatures up to 40 °C in the months of July and August and a minimum average of 5.7 °C in January. The normal incident solar radiation reaches a maximum daily mean equal to 8.3 kWh/m^2 in July and a minimum daily mean equal to 2.3 kWh/m^2 in December. Winters are mild while summers are warm, dry, and sunny. On the other hand, autumns and springs are characterized both by moderate temperatures and by being the seasons in which most of the rainfall occurs. According to the described characteristics, the studied area climate is classified as Mediterranean, following the Köppen-Geige climate classification.

It is noteworthy to observe the temperature differential between the temperature of the sky and the ambiance temperatures. As can be noted in Figure 4, this difference is up to 15 °C throughout the year with respect the mean ambiance temperature, which is an indicator of the potential of the longwave heat exchange with the sky to reduce the roof surface temperature.

The studied building belongs to a quarter of social housing called La Juncal Figure 5 built in the sixties of the last century that exhibits the constructive characteristics common in the social housing built in Seville before 1979, when the first Spanish legislation designed to regulate energy demand in buildings was enacted.

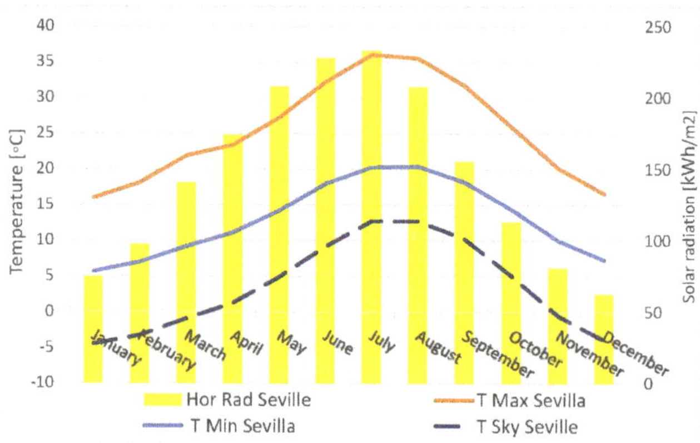

Figure 4. Climatic chart of Seville.

Figure 5. La Juncal (Seville): aerial views.

In Figure 6a, the layout of a typical roof belonging to the social housing studied is shown, and, in Table 1, the dimensions and thermophysical values of the components of the roofs are reported. This constructive configuration for roofs is the usual in the social housing built in the sixties of the past century at southern Spain [56]. For the outer surface of the roof, the absorptivity solar radiation value is taken equal to 0.8, with the latter being a typical value of the material making up this layer. This roof will henceforth be called the reference roof.

Table 1. Thermophysical characteristics of the reference roof.

Layer	Description	Thickness (m)	Density (kg/m^3)	Specific Heat (J/kg K)	Conductivity (W/m K)
1 (Out.)	Ceramic tiles	0.005	2000	800	1.00
2	Mortar	0.01	2000	1000	1.40
3	Protective Layer	0.015	1150	1000	0.23
4	Mortar	0.01	2000	1000	1.40
5	Carbon cinders	0.1	640	657	1.40
6	Concrete vault	0.3	1330	1000	1.32
7 (Ins.)	Gypsum plaster	0.01	1000	1000	0.32

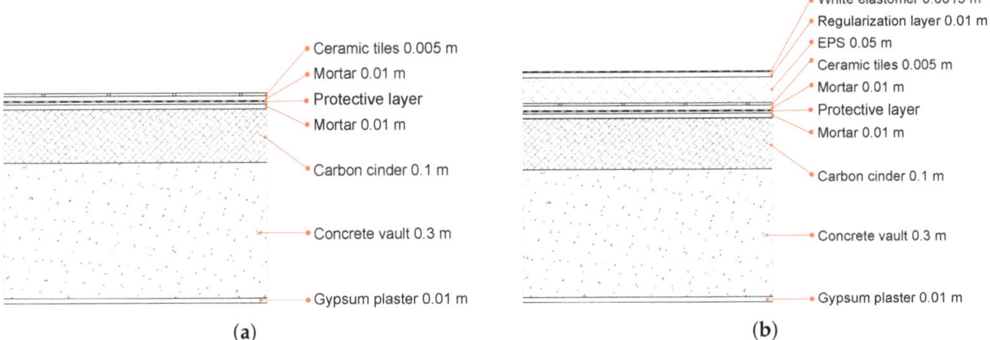

Figure 6. (**a**) Layout of the reference roof. (**b**) Layout of the retrofitted roof.

2.5. Energy Saving Refurbishment Measures

In order to improve the energy performance of the reference roof, a combined refurbishment measure was applied to the outside layer of the roof. This measure consisted of the installation of an EPS insulation and the application of an external layer of white elastomer cool paint to the outer surface. In Figure 6b, the layout of the retrofitted roof is shown, and, in Table 2, the dimensions and thermophysical values of the components of the retrofit are reported. The combination of the insulation thickness and the value of the cool paint absorptivity is then analyzed to optimize the energy and economical performance of the roof, taking into consideration the time variation of the cool coating reflectivity discussed above.

Table 2. Thermophysical characteristics of the refurbished roof.

Layer	Description	Thickness (m)	Density (kg/m^3)	Specific Heat (J/kg K)	Conductivity (W/m K)
1(Out.)	White elastomer	0.0015	1150	1000	0.23
2	Regularization layer	0.01	2000	1000	1.40
3	EPS	0 to 0.1	30	1210	0.04
4–10 (Ins.)	Same as layers 1 to 7 of the reference roof (Table 1)				

2.6. Model Validation

In order to verify the reliability of the numerical model used in the present study, a twofold process of validation is performed: first, the model outputs are compared to the experimental values obtained from the monitoring of a full-scale outdoor test cell; and, second, the calculated values from the numerical model are compared to the values provided by a well known and validated building energy simulation (BES) tool, EnergyPlus, when applied to the case-study used here and described in Section 2.4.

The indices used to check the reliability of the numerical model were the root mean square error (RMSE), the mean of the residuals ε_{mean}, the standard deviation of the residuals $SD(\varepsilon_{mean})$, and the r^2 index, which are often used in the process of validation of building energy simulation models [57].

For both validation procedures, a winter week, from 7 to 14 February 2017, and a summer week, from 17 to 24 August 2017, were considered. The climatic data for both weeks were local data recorded by a meteorological station placed on the roof of a test cell in the city of Seville, as explained in Section 2.6.1. The indoor conditions assumed for the thermal load calculations, as in Section 2.3, are considered: an ambient interior temperature kept constant to 20 °C in the winter week, and to 24 °C in the summer week. In Figure 7, the ambiance temperature and solar radiation for both weeks are shown.

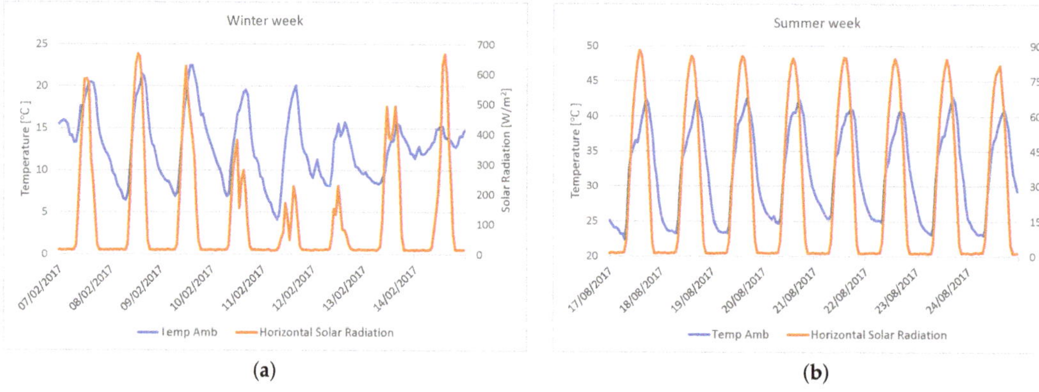

Figure 7. Weather conditions for: (**a**) a winter week (7 to 14 February 2017); (**b**) a summer week (17 to 24 August 2017).

2.6.1. Experimental Validation

For the experimental validation, an outdoor full-scale test cell placed in the center of the city of Seville was used. The interior dimensions of the test cell are 2.40 m wide, 3.20 m deep, and 2.70 m high. The thermophysical values of the materials and the roof stratigraphy are presented in Table 3. The values shown in the table are the nominal ones provided by the Spanish Technical Building Code [58] and the material specifications sheets from the manufacturers. The outer surface of the roof is painted red, and a solar absorptivity of 0.5 and a thermal emissivity of 0.9 are considered for it.

Table 3. Thermophysical characteristics of the roof materials.

Layer	Description	Thickness (m)	Density (kg/m^3)	Specific Heat (J/kg K)	Conductivity (W/m K)
1(Out.)	Galvanized corrugated sheet panels	0.002	7800	460	50.2
2	Ventilated air chamber	0.200	1.184	1007	0.0255
3	Sandwich panel	0.200	40	1884.15	0.017
2	Wool rock panel	0.160	100	840	0.046
3(In.)	Sandwich panel	0.100	40	1884.15	0.017

The test cell is fully instrumented to measure the temperatures of the internal surfaces of the envelope, and data are recorded every 10 min. Thermocouples with an accuracy of ± 0.75 °C and an operating range of -250 to 350 °C are used to monitor the temperature of the inner surfaces. Finally, a local meteorological station located on the roof of the cell is used to record weather data. More details on the test cell and its monitoring equipment can be found in Reference [59].

In Table 4, the values of the statistical indices computed for the temperature of the inside roof surface, are shown.

Table 4. Statistical indicators for the experimental validation.

	RMSE (°C)	ε_{mean} (°C)	$SD(\varepsilon)$ (°C)	r^2
Winter	0.293	0.179	0.213	0.955
Summer	0.301	0.251	0.309	0.961

The results obtained for the indicators shown in Table 4 allow for asserting that the presented model is able to compute the temperatures of the internal roof surface with good precision; the values obtained for the different indicators are similar to those obtained in other works for the validation of energy numerical models for a building's envelope [57].

2.6.2. Comparison with EnergyPlus Software

The EnergyPlus [60] software is a well known BES tool that has been widely used for the calculation of the energy performance of buildings, proven to be very efficient in the computation of the thermal behavior of buildings. An example of the application of EnergyPlus to the analysis of the thermal behavior of roofs can be found in Reference [31].

To carry out the comparison between the numerical model presented here and the EnergyPlus tool, the reference roof of the case study in Section 2.4 and described in Table 1 is used, and the values for the temperature of interior roof surface provided by EnergyPlus are considered as benchmark.

In Table 5, the values of the statistical indices computed for the temperature of the indoor roof surface, are shown.

Table 5. Statistical indicators for the comparison with EnergyPlus.

	RMSE (°C)	ε_{mean} (°C)	$SD(\varepsilon)$ (°C)	r^2
Winter	0.263	0.084	0.107	0.985
Summer	0.287	0.121	0.226	0.981

The values of the indices shown in Table 5 allow to stablish the good fit between the values from the numerical method and from the EnergyPlus tool. As in the case of the experimental validation, these values are similar to those obtained in other validation processes of numerical thermal models for building envelopes [57]. On the other hand, it can be observed that the indicators for the comparison between the models have a slightly better performance than for the experimental validation, a fact also reported in the literature.

3. Energy Optimization Analysis

In this section, an analysis of the energy performance of a variety of roof solar reflectivity and insulation thickness combinations is carried out. The objective of this section is to determine the combination that provides the minimum energy consumption for the whole LCA timespan and to calculate the energy savings of each combination when compared to the reference case.

The analysis is carried out considering a time dependent change of the cool coating solar reflectivity due to the aging effect and different maintenance patterns, as described in Section 2.2.

The solar absorptivity of the cool coat varies from 0.1 to 0.5, with steps equal to 0.1. The insulation thickness ranges from 0 to 0.1 m, with steps equal to 0.1. For the outer surface of the retrofitted and the reference roofs, the thermal emissivity taken is equal to 0.9. The considered maintenance for the cool roof was a power washing every three, five, or ten years. The case where no power washing is performed was also included in the analysis.

The transmission loads through the roofs are calculated for the entire LC timespan taking into account the annual variability of the thermal loads during that period of time due to the changes in time of the cool layer reflectivity produced by the effects of aging and maintenance patterns.

3.1. Thermal Loads Results

In Figures 8 and 9, the total thermal loads (cooling plus heating loads) for the whole LC timespan are shown for different values of solar absorptivity, thickness insulation, and maintenance protocols, together with the total load for the reference roof. As can be seen in these figures, the effect of low absorptivity values on the total thermal loads is very noticeable as it is observed when comparing with the reference case, even in the case of the retrofitted roof without insulation. For all the combinations and all maintenance protocols considered, the total loads are lower than for the reference case, although there is a significant difference between the loads for the uninsulated roof and the insulated ones.

On the other hand, it is observed that the lower the solar absorptivity, the lower is the total load. It is also observed that increasing the thickness of the insulation reduces the loads. So, for every value of the solar absorptivity, the total load decreases as the insulation thickness increases. However, regarding the maintenance protocol, only little differences are found among the total loads, with a slight tendency to decrease as the frequency of the cool coat washes increases. Thus, it can be drawn up that the role played by periodic washing in the total thermal loads is relatively modest.

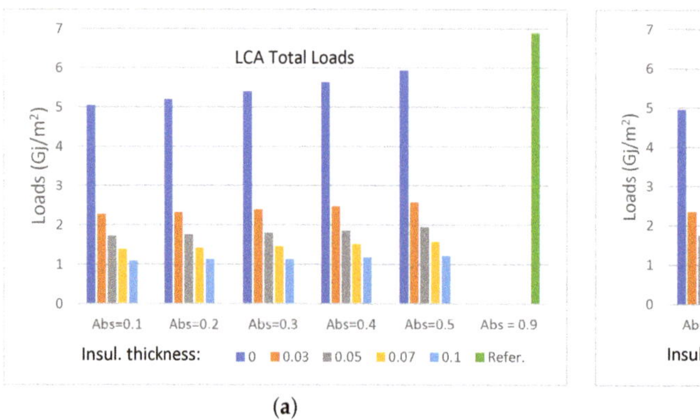

(a)

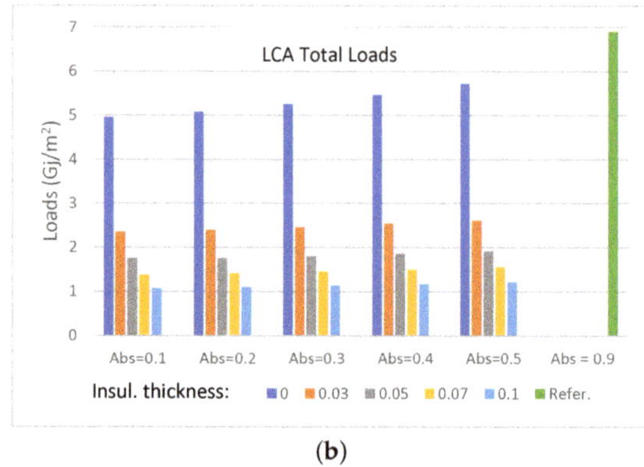

(b)

Figure 8. Total loads (GJ/m^2) for the whole LCA timespan: (**a**) no washing; (**b**) decennial washing.

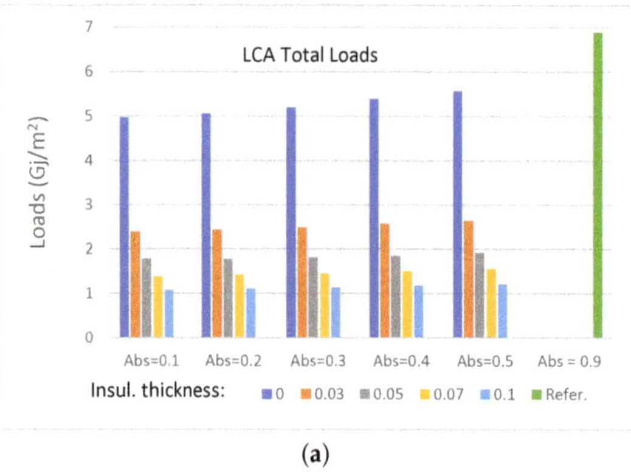

(a)

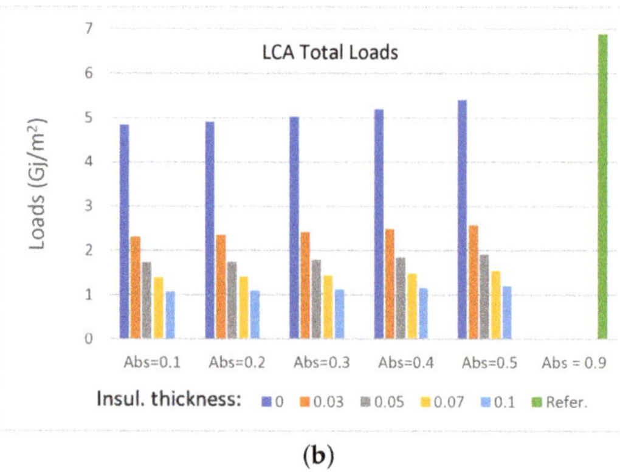

(b)

Figure 9. Total loads (GJ/m^2) for the whole LCA timespan: (**a**) quinquennial washing; (**b**) triennial washing.

Analyzing the heating and the cooling loads separately, it is clear that, for all the cases of absorptivity, thick insulation, and maintenance protocols, the smaller the cool roof absorptivity, the smaller the cooling load, as can be seen in Figures 10 and 11. In addition, it can be noted that the smallest cooling loads are obtained for the case of the triennial wash, which is in accordance to the fact that the original roof solar absorptivity values are restored with a higher frequency than in the other maintenance protocols. Therefore, the cooling loads gradually increase as the frequency of cool roof maintenance decreases and reaches the highest values for the case of no power washing.

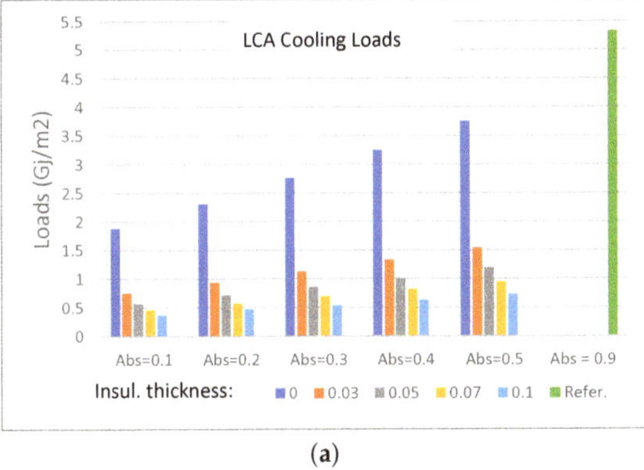 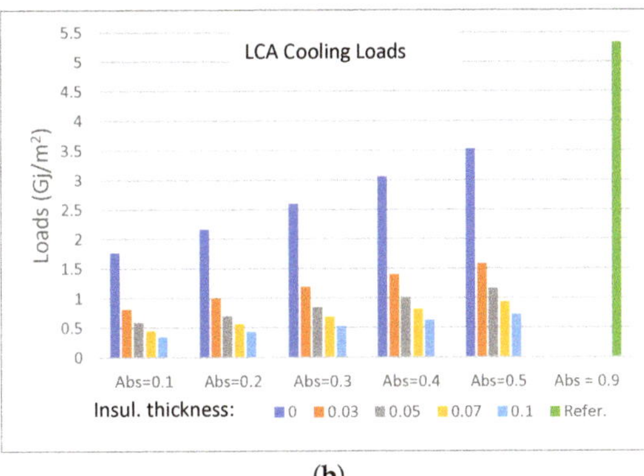

(a) (b)

Figure 10. Cooling loads (GJ/m^2) for the whole LCA timespan: (**a**) no washing; (**b**) decennial washing.

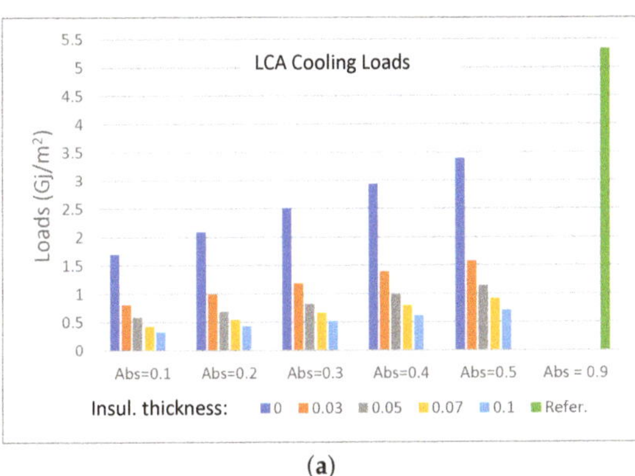 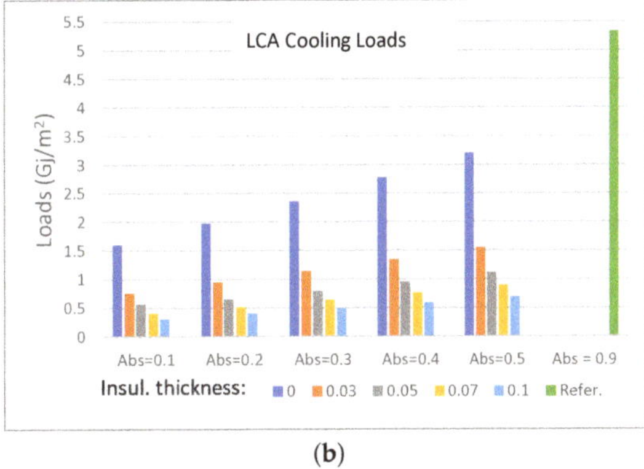

(a) (b)

Figure 11. Cooling loads (GJ/m^2) for the whole LCA timespan: (**a**) quinquennial washing; (**b**) triennial washing.

On the other hand, it can be seen that increasing the thickness of the insulation has a strong effect in reducing the cooling loads, so that the thicker the insulation layer, the lower the cooling load. It can be concluded that, for all the maintenance protocols, the lowest load is obtained for the thickest insulation and the lowest solar absorption. Regarding the maintenance protocol, the most favorable case is the triennial power washing.

Regarding the heating loads, the behavior is the opposite to that described for the cooling loads, as can be observed in Figures 12 and 13. Now, due to solar gains decrease, the retrofitted cool roof without insulation has higher heating loads than the reference case. Clearly, the winter penalty effect on solar gains due to the cool roof is the cause of this. For the roofs with insulation, the only case where the heating loads are greater or close to that of the reference roof is when solar absorptivity equals 0.1. For values greater than 0.1 of the roof absorptivity, the heating loads are lower when compared to the reference case. This result holds for all the insulation thickness and maintenance protocols. Therefore, it can be stated that, although the heating loads increase as the solar absorptivity of the roof decreases, the insulating layer is able to compensate the penalty effect due to the low values of absorptivities, except for the value of the absorptivity equal to 0.1, as discussed before. Again, the role of the insulation layer in reducing the heat flux is obvious, i.e., the thicker the insulation, the lower the heating load. Finally, it is again observed that the effect of the cool coat washing is quite small on the heating loads over the LC timespan.

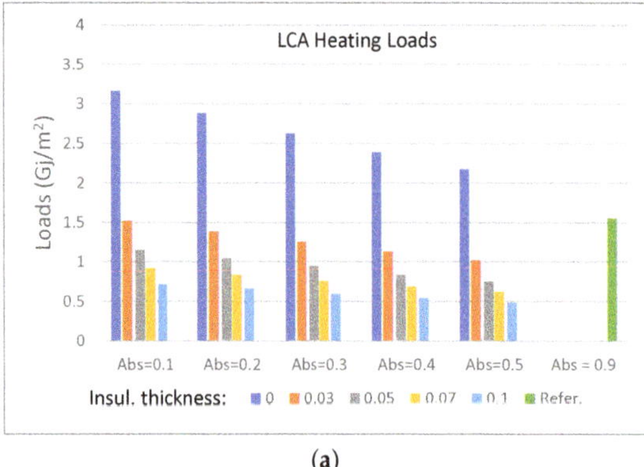
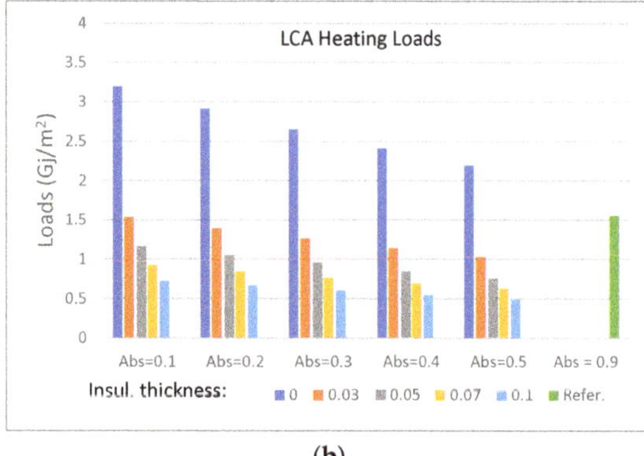

(a)　　　　　　　　　　　　　　**(b)**

Figure 12. Heating loads (GJ/m^2) for the whole LCA timespan: (**a**) no washing; (**b**) decennial washing.

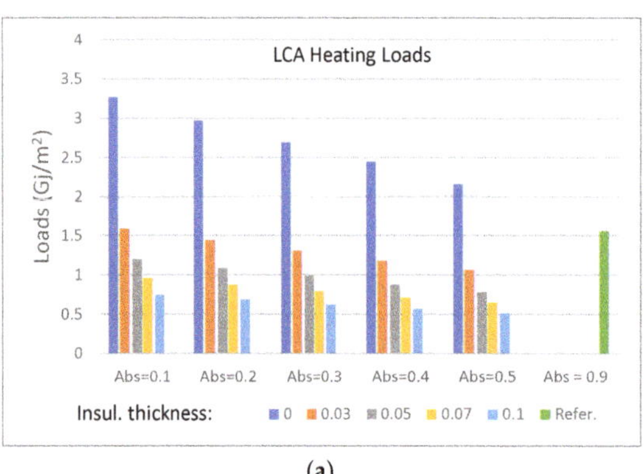
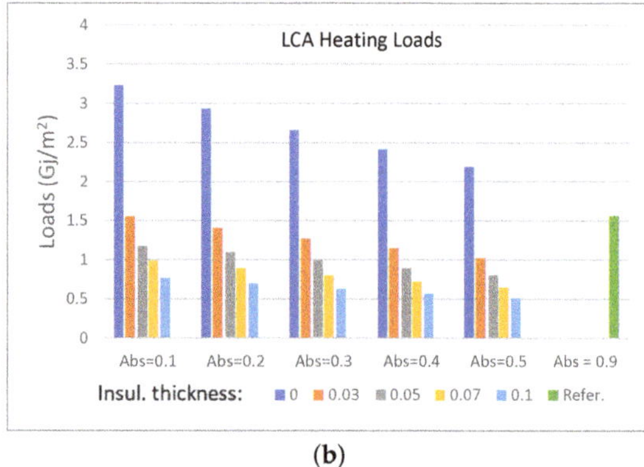

(a)　　　　　　　　　　　　　　**(b)**

Figure 13. Heating loads (GJ/m^2) for the whole LCA timespan: (**a**) quinquennial washing; (**b**) triennial washing.

This opposite behavior of the heating and cooling loads described for the retrofitted roofs is the cause of the small differences observed among the total thermal loads for the whole LC timespan when considering different maintenance scenarios, while the changes in solar reflectivity produced by the aging of the cool coat act in the opposite direction. That is, on the one hand, when power washing is carried out, it produces a decrease in the cooling loads and an increase in the heating loads, while the aging of the cool coat produces the opposite effect in such a way that all these processes combined they result in a certain equilibrium in the resulting total loads. To take into account the aging of the cool coat, the aging pattern discussed in Section 2.2 is used for the thermodynamic computations by using the monthly reflectivity given by the pattern.

In Figures 8 and 9, the most remarkable fact, from a physical point of view, is the difference between roofs equipped with thermal insulation and roofs without it; this is due to the higher heat flux values through roofs without thermal insulation, which implies, in the cold season, a greater loss of heat from the interior and, in the hot season, a greater gain, so that these effects produce a significant thermal load increasing for the roofs without insulation when compared to the equipped with insulation ones. On the other hand, it can be observed that, for the roof without insulation, the reduction of heat gain due to the solar reflective coat compensates the penalty effect on winter; this means that the reduction of thermal conduction through the roof in summer, due to the lower absorption of solar radiation, compensates, in terms of energy consumption to obtain indoor comfort, the increase in energy consumption necessary to achieve indoor comfort conditions in winter, due to the aforementioned lower absorption of solar radiation. This effect is clear in

Figures 10–13, where it can be observed the strong reduction in cooling loads for the roof with cool coat without insulation when compared to the reference case, while, regarding the heating loads, although higher for the reference roof, it has a lower difference on thermal loads when compared to the cool roof; this difference is collected by the total load, which, as said before, is lower for the roof with cool coat.

On the other hand, noteworthy is the energy performance of the roofs with insulation. These roofs are able of reducing thermal conduction in both directions, from inside to outside and vice versa, which produces a significant reduction in thermal loads compared to roofs without thermal insulation, as can been observed in Figures 8 and 9. It is observed, in Figures 10–13, that the higher the thickness of the insulation layer, the less the thermal loads, for both cooling and heating, which is in accordance with the reduction of heat transfer through the roof caused by the insulation layer. On the other hand, as expected, it can be observed the progressive reduction of the cooling loads as the solar absorptivity increases, obviously due to the decreasing solar radiation absorbed by the roof, whilst, for the same reason, the opposite is observed for heat loads. However, what is concluded, as it is shown in Figures 8 and 9, is that, for the climatic conditions and the geographical framework in which the study has been developed, the total load decreases as the reflectivity increases due to the fact that the reduction in thermal flux to the interior achieved in summer, and which impacts on cooling loads, is greater than the increase in thermal flux from the inside to the outside, which occurs in winter, and that is reflected in an increase in heating loads.

3.2. Energy Optimization Results

This section studies the optimum combination of solar reflectivity and insulation thickness in terms of thermal loads for the entire LC time interval. Energy savings of the different combinations when compared to the reference case are also reported.

From the results shown in Figures 8 and 9, it can be concluded that the optimum combination to reduce energy consumption is absorptivity equals 0.1, and insulation thickness equals 0.1. This result holds for all the maintenance scenarios, although the results for the different maintenance scenarios are very close to each other. To clarify this, in Tables 6 and 7, the values of the total thermal loads for the LC timespan are shown.

The similarity of the loads among the different maintenance protocols for the cool roof can be observed in all those cases for which the absorptivity values and the thickness of the insulation layer are the same. Note that, for the optimum combination established, the total thermal load difference between the optimum maintenance scenario, i.e., triennial washing, and the worst scenario, i.e., no washing, is only of 0.01676 GJ/m^2, equivalent to a 1.55% difference.

In Tables 8 and 9, the savings for the total thermal loads of every retrofitted roof when compared to the reference roof are shown. In accordance with the energy results presented above, all the combinations are found to provide energy savings compared to the reference case.

Table 6. Total loads (GJ/m^2) for the LC timespan.

	No Washing					Decennial Washing				
Insulation Thickness	Abs. = 0.1	Abs. = 0.2	Abs. = 0.3	Abs. = 0.4	Abs. = 0.5	Abs. = 0.1	Abs. = 0.2	Abs. = 0.3	Abs. = 0.4	Abs. = 0.5
0.0	5.0350	5.1889	5.3882	5.6335	5.9216	4.9544	5.0798	5.2491	5.4632	5.7190
0.03	2.2712	2.3156	2.3802	2.4625	2.5619	2.3458	2.3863	2.4477	2.5328	2.6110
0.05	1.7131	1.7459	1.7939	1.8426	1.9414	1.7470	1.7480	1.7929	1.8551	1.9163
0.07	1.3835	1.4121	1.4527	1.5012	1.5671	1.3749	1.4066	1.4459	1.4984	1.5588
0.1	1.0802	1.1307	1.1329	1.1745	1.2188	1.0797	1.1010	1.1281	1.1729	1.2136

Table 7. Total loads (GJ/m^2) for the LC timespan.

Insulation Thickness	No Washing					Decennial Washing				
	Abs. = 0.1	Abs. = 0.2	Abs. = 0.3	Abs. = 0.4	Abs. = 0.5	Abs. = 0.1	Abs. = 0.2	Abs. = 0.3	Abs. = 0.4	Abs. = 0.5
0.0	4.9678	5.0592	5.1993	5.3884	5.5577	4.9206	4.9981	5.1178	5.2813	5.4856
0.03	2.3884	2.4287	2.4850	2.5645	2.6415	2.3483	2.3967	2.4519	2.5255	2.6184
0.05	1.7802	1.7655	1.8028	1.8538	1.9162	1.7679	1.7456	1.7813	1.8244	1.8959
0.07	1.3835	1.4135	1.4507	1.5016	1.5618	1.3662	1.3943	1.4319	1.4805	1.5467
0.1	1.0763	1.1103	1.1349	1.1760	1.2168	1.0634	1.0970	1.1220	1.1628	1.2056

Table 8. Total load savings (GJ/m^2) for the LC timespan.

Insulation Thickness	No Washing					Decennial Washing				
	Abs. = 0.1	Abs. = 0.2	Abs. = 0.3	Abs. = 0.4	Abs. = 0.5	Abs. = 0.1	Abs. = 0.2	Abs. = 0.3	Abs. = 0.4	Abs. = 0.5
0.0	2.1181	2.0553	1.9157	1.7359	1.5171	2.0992	2.0029	1.8623	1.6835	1.4685
0.03	4.6168	4.5724	4.5078	4.4255	4.3261	4.5960	4.5474	4.4806	4.3907	4.3049
0.05	5.1749	5.1421	5.0941	5.0454	4.9466	5.1816	5.1463	5.0955	5.0261	4.9663
0.07	5.5045	5.4759	5.4353	5.3868	5.3209	5.5084	5.4781	5.4402	5.3910	5.3296
0.1	5.8078	5.7573	5.7551	5.7135	5.6692	5.8083	5.787	5.7599	5.7229	5.6753

Table 9. Total load savings (GJ/m^2) for the LC timespan.

Insulation Thickness	No Washing					Decennial Washing				
	Abs. = 0.1	Abs. = 0.2	Abs. = 0.3	Abs. = 0.4	Abs. = 0.5	Abs. = 0.1	Abs. = 0.2	Abs. = 0.3	Abs. = 0.4	Abs. = 0.5
0.0	2.1020	2.0138	1.8900	1.7277	1.5865	2.0561	1.9863	1.8642	1.6970	1.4892
0.03	4.5870	4.5369	4.4731	4.3868	4.3050	4.5851	4.5355	4.4786	4.4035	4.3206
0.05	5.1739	5.1433	5.0979	5.0343	4.9684	5.1548	5.1404	5.1013	5.0462	4.9750
0.07	5.5064	5.4790	5.4444	5.3963	5.3328	5.4995	5.4789	5.4482	5.4048	5.3446
0.1	5.8117	5.7777	5.7531	5.7270	5.6806	5.8246	5.7910	5.7663	5.7300	5.6862

From the above, it is straightforward that the greatest energy savings are obtained by the retrofitting combination that uses a cool roof with an absorptivity equal to 0.1 and an insulation layer thickness equal to 0.1 m, with the triennial washing of the cool coat being the case that gives the highest energy saving; on the other hand, the energy savings decrease as the absorptivity increases, and the insulation thickness decreases, according to the previous discussion on thermal loads.

4. Cost Optimization Analysis

The objective of this analysis is to assess the cost-effectiveness of the energy retrofitting measures proposed when compared to the reference case, as well as to find out the combination of cool coating reflectivity and insulation layer thickness that yields the minimum LC cost under the climate conditions considered.

The optimum combination of cool coating reflectivity and insulation thickness is a function of the costs of the cool paint and the insulation material, the application of the cool paint, the insulation layer installation, the cool coat maintenance, the thermal loads due to the energy transfer through the roof, the coefficient of performance (COP) of the heating and cooling systems, the retrofitting materials lifetime, the energy prices, the energy inflation rate, and the monetary discount rate.

Despite the aforementioned tendency of heating and cooling loads to decrease when the thickness of the insulation layer increases, a rise in the reflectivity of the cool coat tends to reduce the cooling load, as well as the solar gain, in winter. Furthermore, the aging and the washing of the cool coat give rise on their own to contrary effects on the loads. So, this time-varying evolution of the thermal loads, combined with the effect of the costs of the retrofitting material and its maintenance, makes it necessary to carry out a detailed analysis of the combined effect on the LCA cost of the different retrofitting combinations in order to obtain the optimum results. The main difference between the present analysis and the usual cost analysis is that, here, the costs are computed monthly for the whole LC time interval, with the goal of assessing the effect on the evolution of costs, along time, produced by changes in the thermal loads associated to the cool coat reflectivity variation, throughout the LC timespan.

4.1. Methodology

First, the assessment of the cost-effectiveness of every ESRM is done. To that aim, the methodology from Reference [61] is used. This methodology performs an LCCA, where all anticipated costs are estimated and discounted to their present worth.

The procedure consists in computing the costs for each future time period using the future value of the variables involved and then discounting each future cost to its present worth. Thereby, the sum of all the present values of the costs yields the life-cycle cost.

Once the LC cost has been determined for each of the considered ESRMs, including different scenarios for maintenance, the cost-effectiveness of each case is assessed by its LC cost: the lower this cost, the more cost effective the ESRM, and the one having the lowest LC cost is selected as the optimum.

The lifetime for the life-cycle analysis is taken as equal to $n = 20$ years, which is the mean service life of the used cool coat and the insulation material [62]. Then, for every considered ESRM, the life-cycle cost, the associated savings, and the payback period are computed.

In the present LCCA, the method P1-P2 introduced in Reference [61] is used with some little changes to account for the time changing character of the energy operating costs associated to the variation of the cool coat solar reflectivity due to the aging and washing of the coat. We use the expression

$$PW_k = 1/(1+d)^k \qquad (8)$$

to compute the present worth (PW) of one monetary unit belonging to the future time interval k (usually years), where d is the market discount rate (fraction per time period).

In the present analysis, k, for $k = 1, \ldots, n$, is taken as the year in which the costs are being computed. Then, if we note as $Q_c(k)$ and $Q_h(k)$ the cooling and heating loads at year k for each ESRM, the energy payment for the whole year k is given by:

$$C_e(k) = \left(\frac{Q_c(k) \times P_{e,c}}{COP_c \times (3.6 \times 10^6)} + \frac{Q_h(k) \times P_{e,h}}{COP_h \times (3.6 \times 10^6)} \right) (1+i)^{k-1}, \qquad (9)$$

where $P_{e,c}$ and $P_{e,h}$ are the current prices of energy used to cool and heat for cooling and heating, respectively, COP_c and COP_h are the coefficients of performance for cooling and heating, and, finally, i is the considered inflation rate for energy costs.

Then, in accordance with expression (8), the present worth of the future energy payment $C_e(k)$ at the period k is given by

$$PW_k(C_e(k)) = C_e(k)/(1+d)^k .$$

It is noteworthy that, although $C_e(k)$ represents the energy costs for year k, it is computed by taking into account the changing values of energy consumption for every specific year k according to the monthly changes in the solar reflectivity of the cool coat.

To compute $C_e(k)$, the energy source considered for cooling is electricity, while, for heating, the use of electricity or gas have been considered, given that they are the usual sources of energy used in the geographical framework under analysis.

To include the maintenance in the cost computation, we name as i_M the inflation rate for this kind of costs. Then, again applying (8), the present worth of the future maintenance payment $C_M(k)$ at the period k is given by

$$PW_k(C_M(k)) = C_M(k)/(1+d)^k = \delta(k)\, C_M(1+i_M)^{k-1}/(1+d)^k,$$

where C_M is the current price of maintenance per unit area of roof surface, and $\delta(k)$ is defined as:

$$\delta(k) = \begin{cases} 1, & \text{if the maintenance is done at period } k; \\ 0, & \text{if the maintenance is not done at period } k. \end{cases}$$

Then, the present worth of the energy total cost for the life-cycle timespan is computed as $C_e = \sum_{k=1}^{n} PW_k(C_e(k))$, whilst the present worth value of maintenance life-cycle total cost is computed as $C_M = \sum_{k=1}^{n} PW_k(C_M(k))$. Now, if we note as $C_{I,cc}$ and $C_{I,ins}$ the initial costs of installation of the cool coat and the insulation layer, the total initial cost for installation is given by $C_I = C_{I,cc} + C_{I,ins}$. Finally, the present worth of the life-cycle total cost per unit area for each considered ESRM and maintenance scenario is given by:

$$C_t = C_e + C_M + C_I.$$

Once C_t is computed for every case, the optimum case is considered to be the one that yields the lower value for C_t.

On the other hand, to compute the possible savings of the proposed ESRMs, the *PW* of the LC total cost must be estimated for the reference case, too. Then, the *PW* of the energy costs for the reference case at period k, $PW_k(C_e^{(ref)}(k))$, is computed by:

$$PW_k(C_e^{(ref)}(k)) = C_e^{(ref)}(k)/(1+d)^k,$$

where $C_e^{(ref)}(k)$ is the energy future cost at period k for the reference roof calculated by:

$$C_e^{(ref)}(k) = \left(\frac{Q_c^{(ref)}(k) \times P_{e,c}}{COP_c \times (3.6 \times 10^6)} + \frac{Q_h^{(ref)}(k) \times P_{e,h}}{COP_h \times (3.6 \times 10^6)} \right)(1+i)^{k-1}, \tag{10}$$

with $Q_c^{(ref)}(k)$ and $Q_h^{(ref)}(k)$ being the cooling and heating loads, respectively, for the reference roofs at the time period k.

Now, we note by $C_I^{(ref)}$ and $C_M^{(ref)}$ the installations and maintenance costs, respectively, of the minimum retrofitting action to be done on the reference roof and which is deemed to be the bituminous paint retrofitting.

Then, the difference between the costs for the cool roofs and the reference case for the whole LC period is estimated through the net savings (NS):

$$NS = \sum_{k=1}^{n} \left[PW_k(C_e^{(ref)}(k)) - PW_k(C_e(k)) \right] + (C_M^{(ref)} - C_M) + (C_I^{(ref)} - C_I). \tag{11}$$

A positive value for *NS* means that the use of the ESRM under consideration yields economic savings when compared to the reference case and that the retrofitting system is cost-efficient. Due to the outlined thermal loads change over time, the net savings (11) cannot be reduced to a simple analytical expression, and its computation must be performed through the yearly computation, according to the yearly changes in thermal loads; this represents a distinguishing feature of our analysis with respect to the usual analysis of cost-effectiveness for retrofitting measures found in the literature.

To compute the period of investment return or payback period (PB), we follow Reference [63] and take PB as the time horizon t, for which the value of NS becomes zero. To compute this value of t, since the terms of the series in Equation (11) are not constant on k, the net savings accumulated for every time horizon k_0 is computed first:

$$NS(k_0) = \sum_{k=1}^{k_0} \left[PW_k(C_e^{(ref)}(k)) - PW_k(C_e(k)) \right] - C_M - C_I.\tag{12}$$

Then, if $NS(k_0) < 0$ and $NS(k_0 + 1) > 0$, the value of the payback period t is estimated as:

$$t = \frac{-NS(k_0)}{NS(k_0 + 1) - NS(k_0)} + k_0.\tag{13}$$

The value of t given by (13) is consistent with the value provided by the usual formulas for the PB found in the literature used under the consideration of constant energy consumption for all the years of the LC period [31,61]. In Appendix C, a comparison between the formula given by (13) and the formula for the computation of the PB from Reference [31,61,63] is performed.

4.2. Variables for the LCA Economic Analysis

The input variables needed to carry out the economic calculations involved in the LCCA are shown in Table 10.

Table 10. Input variables used in the LCA.

Variable	Value	
EPS cost	116	€/m³
Insulation installation cost	13.15	€/m²
Cool paint cost	9.45	€/m²
Washing cost	1.63	€/m²
Bituminous paint cost	8.01	€/m²
Bituminous paint maintenance	0.4	€/m² yearly
Electricity cost	0.2403	kWh
Gas cost	0.0736	kWh
Energy inflation rate	3	%
Discount rate	1.5	%
Lifetime	20	years

The costs of the cool and bituminous paint application, their maintenance, and the insulation material and its installation are taken from Reference [62]. The electricity and natural gas prices are the average final prices for household consumers in Spain during the year 2020 as reported by Eurostat Office [64]. The values of the energy inflation and the discount rates have been taken considering the average values of these parameters under Spanish and European economic conditions, with such values being previously used for the LCA cost analysis in former research [37,65]. Regarding the conditioning equipment, it has been considered that cooling is always done by air conditioning machinery using electricity as the only energy source, while, for heating, we have considered the following cases:

- Heating by air (heat pumps).
- Heating by natural gas.
- Heating by electricity radiators.

The machinery is considered common in the local market and it has a mid-range efficiency. The efficiency indices of the conditioning devices are listed in Table 11.

Table 11. Conditioning equipment efficiency.

Device	Efficiency
Cooling pump	$COP = 3.4$
Heating pump	$COP = 3$
Gas natural heating	$\eta_s = 0.8$
Electrical Radiators heating	$\eta_s = 1$

4.3. Cost Optimization Results

In this section, the economic results from the LCCA for the considered ESRMs are shown. The objective of this analysis is twofold: first, the cost-effectiveness of the ESRMs will be assessed; second, the optimum case for the combination of insulation layer, roof solar absorptivity, and maintenance protocol will be determined.

The NS and PB were calculated by using the economic variables introduced in Section 4.1 and considering the aging and maintenance scenarios presented in Section 2.2. For each ESRM, a positive NS means real savings and, conversely, a negative NS implies that the ESRM has a higher LCA cost than the reference case and, therefore, no real savings. Finally, once C_t is computed for every considered ESRM and for every cool roof maintenance protocol, the lowest value of C_t determines the optimum combination.

On the other hand, the cost of the roof retrofitting must be paid by all the owners of the building, while the benefit of the retrofitting would only be noticeable in the housing under the roof; however, in the event of damage, according Spanish law, the cost of restoring the roof is paid by all the owners of the building; so, since the owners and tenants of the dwellings taken as the case study are characterized by having low incomes, a retrofitting action on the roof is unlikely, except for the existence of some deterioration of the roof that makes such retrofitting necessary; thus, to perform the analysis, it is assumed that the reference roof needs some kind of renovation or maintenance. This way, the application of a bituminous paint is chosen since this is considered to be the most usual retrofitting measure in the socioeconomic context within which the study is performed.

In Table 12, the optimum values with regard to the use of pump for heating and cooling are shown for every protocol maintenance and for the whole LC timespan. These values have been extracted from Table A4 in Appendix B, where all the LC total costs are shown for the different values of both the insulation thickness layer and the roof reflectivity, as well as for all the considered roof maintenance protocols. For this case of considering electricity as the only energy source to conditioning, the estimated LC total cost for the reference case is equal to 174.21 €/m². Then, this value is used to obtain the economic savings and the payback periods shown in the table.

Table 12. Optimum results for cooling and heating by pump.

Cool Roof Washing	Solar Absorptivity	Optimum Thickness (m)	Minimum Total Cost (€/m²)	Maximum total Savings (€/m²)	Payback Period (Years)
No	0.1	0.08	64.27	109.94	4.54
Decennial	0.1	0.08	66.51	107.70	4.78
Quinquennial	0.1	0.08	68.33	105.88	5.01
Triennial	0.1	0.08	72.82	101.39	5.48

As can be observed in Table 12, when a pump is used for heating and cooling, the cool roof absorptivity that provides the minimum total cost is equal to 0.1, and the insulation thickness is equal to 0.08 for all the maintenance scenarios. The optimum result is obtained for the non-maintenance scenario with a total cost equal to 64.27 €/m², a total savings equal to 109.94 €/m², and a payback period of 4.54 years.

In Table 13, the optimum values regarding the use of gas for heating and of a pump for cooling are shown for every protocol maintenance and for the whole LC timespan. These values have been sourced from Table A5 in Appendix B, where all the LC total costs are shown for the different values of both the insulation thickness layer and the roof reflectivity, as well as for all the roof maintenance protocols. Within the framework of this energy source for conditioning, the LC total cost for the reference case is equal to 180.07 €/m². For the cases of no-maintenance and decennial maintenance, the combination that provides the maximum savings is that which has a solar absorptivity equal to 0.1 and an insulation thickness equal to 0.08, while, for the quinquennial maintenance, the combination yielding the maximum savings is that which has a solar absorptivity equal to 0.1 and an insulation thickness equal to 0.09 and, finally, for the triennial maintenance, the combination that provides the maximum savings is that which has a solar absorptivity equal to 0.2 and an insulation thickness equal to 0.09. For this scenario of mixed energy used to condition the building, the optimum result is also obtained for the non-maintenance scenario with a total cost equal to 68.27 €/m², a total savings equal to 111.80 €/m², and a payback period of 4.05 years.

Table 13. Optimum results for cooling by pump and heating by gas.

Cool Roof Washing	Solar Absorptivity	Optimum Thickness (m)	Minimum Total Cost (€/m²)	Maximum total Savings (€/m²)	Payback Period (Years)
No	0.1	0.08	68.27	111.80	4.05
Decennial	0.1	0.08	69.81	110.26	4.48
Quinquennial	0.1	0.09	71.30	108.77	4.71
Triennial	0.2	0.08	74.99	105.08	4.97

In Figures 14–17, the insulation cost, the cool coating cost (including maintenance), the energy consumption cost, and the total cost are displayed with respect to the insulation thickness for the optimum absorptivities shown in Tables 12 and 13, for all the maintenance scenarios and for the two considered sources of energy.

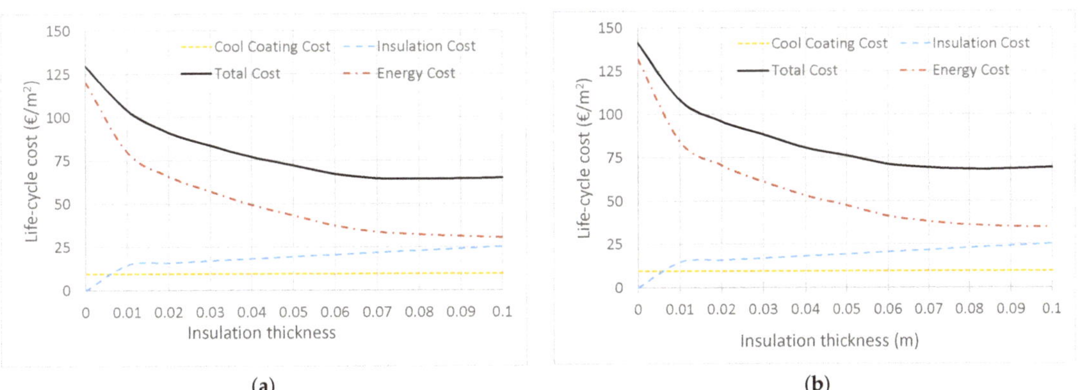

(a)

(b)

Figure 14. No washing life-cycle costs (GJ/m²): (**a**) cooling and heating by pump; (**b**) heating by gas and cooling by pump triennial washing.

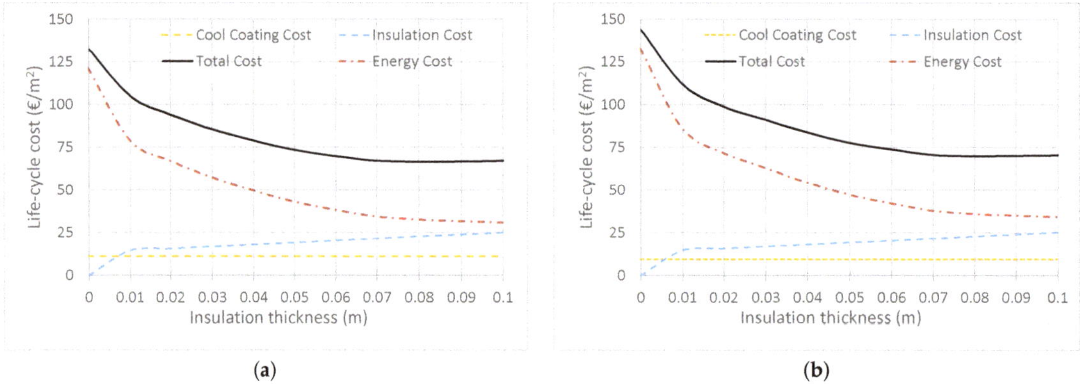

Figure 15. Decennial washing life-cycle costs (GJ/m^2): (**a**) cooling and heating by pump; (**b**) heating by gas and cooling by pump triennial washing.

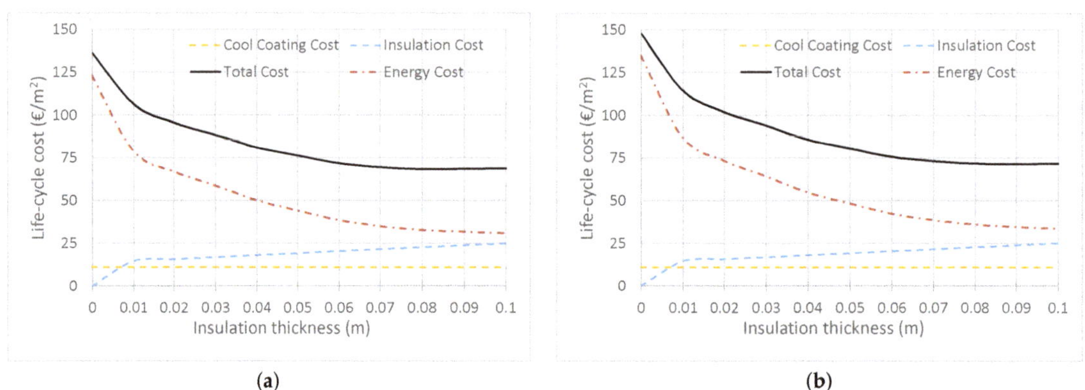

Figure 16. Quinquennial washing life-cycle costs (GJ/m^2): (**a**) cooling and heating by pump; (**b**) heating by gas and cooling by pump triennial washing.

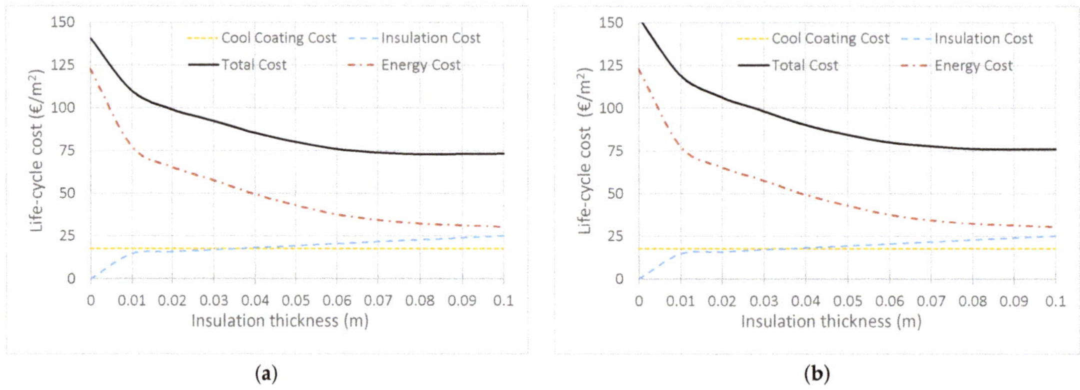

Figure 17. Triennial washing life-cycle costs (GJ/m^2): (**a**) cooling and heating by pump; (**b**) heating by gas and cooling by pump triennial washing.

As it can be seen from these figures, there is a rapid decrease in energy costs when the insulation thickness begins to increase from zero. This also results in a rapid decrease in total costs. However, for values of insulation thickness greater than 0.07 m, energy costs decrease at a much weaker rate according to the decrease in the rate of reduction of thermal loads. Thereby, the decrease in energy costs is not balanced by the increase in insulation costs, with this fact being much more evident for the higher frequency washing of the cool coating. This way, when using electricity as the only energy source for conditioning, the curve of total costs has a minimum for the insulation thickness equal to 0.08 m that corresponds to the optimum insulation thickness for all the cool roof maintenance cases; when using gas for heating and electricity for cooling, in the case of quinquennial cool roof maintenance, the optimum is reached for an insulation thickness equal to 0.09 m, whilst, for the cases of no-maintenance, decennial, and triennial cool roof maintenance, the curve of total costs exhibits a minimum for an insulation thickness equal to 0.08 m, even though, for the triennial maintenance, the minimum total cost is reached for a solar absorptivity equal to 0.2, as is shown in Table A5 in Appendix B.

As it can be observed in Tables A4 and A5 in Appendix B, even if, as it is reported in the energy analysis, the minimum energy consumption was obtained for triennial washing when absorptivity equals 0.1 and insulation thickness equals 0.1 m, energy cost savings are not enough to compensate the rise of cost derived of the increase in insulation thickness and of the cool roof maintenance. On the other hand, as it is deduced from those tables, all the studied ESRMs are cost-effective and produce positive savings. This result holds for all the different cases of maintenance of cool roofs and of machinery for conditioning.

Thus, from the performed economic analysis, it can be concluded that the optimum combination is that obtained for a cool coat solar absorptivity equal to 0.1, an insulation thickness equal to 0.08 m, no washing of the cool coat, and when using a pump for heating and cooling.

Finally, it is noteworthy to observe the limited scope of the cool roof maintenance through power washing as it can be deduced from the energy and economic analysis. This fact has been discussed formerly in the literature; this way, in Reference [50], it is pointed out that the differences induced by the washing of the cool roofs are very small for the LCA total period, which calls into question its suitability, especially when considering the possible damage to the cool coating during maintenance.

5. Conclusions

The thermal performance of social building roofs under Southern Spain climate and its refurbishment by means of combining a cool roof coat and a thermal insulation layer have been analyzed in the present paper. The thermal dynamic of the roofs was computed through a numerical model using a finite difference method in order to assess the impact of different combinations of cool coating reflectivities and insulation thickness when used to retrofit the roofs of social buildings built in the sixties of the last century at the city of Seville, Spain. The time-dependent impact of the aging effect on the roof coat reflectivity was also incorporated into the model.

Combinations of cool coating, for which values of solar absorptivity were equal to 0.1, 0.2, 0.3, 0.4, and 0.5, with insulation layer thickness ranging from 0 to 0.1 m, were analyzed, and the results obtained were compared to those reported for the initial reference roof belonging to the social housing. In order to perform the analysis under realistic conditions, the aging effect of the cool coatings was taken into account by considering a pattern for the aging effect and its impact on the solar reflectivity of the cool coat. This aging effect was estimated a monthly basis, and it was introduced in energy computations for the whole service lifetime of the refurbished roof to perform an LCCA in order to appraise the cost-effectiveness of the proposed combinations.

Then, a comprehensive study was conducted by taking into consideration different values for initial roof reflectivity and insulation thickness, different sources of energy for the conditioning of the building, and different maintenance patterns to restore the cool coating reflectivity.

The results obtained for the different combinations here studied point to significant savings in the operational energy. Such savings are larger as the absorptivity of the external coating of the retrofitted roof decreases and the insulation layer thickness increases. This way, a decrease in the annual total loads by 84.56% is found for the best case that is given by insulation thickness equal to 0.1, initial roof absorptivity equal to 0.1, and triennial washing.

Likewise, it has been concluded the limited impact of the cool roofs maintenance on the total loads during the LCCA period of time. This way, the values of total loads are similar for the considered maintenance scenarios, with the maximum difference of around 1.55% for each combination being considered. However, regarding the heating and cooling loads separately, some variations in the evolution of the loads can be observed for every maintenance scenario.

In order to evaluate the economical value of the different combinations studied for the refurbishment of the roofs belonging to the social housing considered, a 20-year LCCA was conducted. The LCCA pointed out that all the combinations of insulation layer thickness, roof solar reflectivity, and maintenance scenarios are cost-efficient. Again, the low impact of the maintenance on the total costs for the whole LCCA timespan is clear. However, the curves of costs result are affected by installation costs, so that the combination that produces the lowest operational cost is not the one with the lowest total cost for the entire period of time analyzed. Thus, it can be concluded from the LCCA performed that the optimum combination is that obtained for a cool coat solar absorptivity equal to 0.1, an insulation thickness equal to 0.08 m, no washing of the cool coat, and when using a pump for heating and cooling. Such a combination allows a total savings equal to 109.94 €/m^2, equivalent to the 63.10% of the reference costs, as well as a payback period of 4.54 years.

According to the results obtained in the present research, the use of an adequate combination of insulation and cool roof for the refurbishment of energy obsolete roofs is profitable in terms of energy consumption, economic costs, and environmental benefits. Because of all this, its use is recommended for the refurbishment of roofs belonging to the social housing in the climatic zone considered.

Author Contributions: Conceptualization, C.-A.D.-T., H.D.-T. and A.D.-D.; methodology, all the authors; software, all the authors; formal analysis, all the authors; investigation, all the authors; writing—original draft preparation, H.D.-T. and C.-A.D.-T.; writing—review and editing, all authors. All authors have read and agreed to the published version of the manuscript.

Funding: This research was funded by the Ministry of Economy and Competitiveness of the Spanish Government and the European Regional Development Fund through the research and development project "Parametric Optimization of Double Skin Facades in the Mediterranean Climate to Improve Energy Efficiency Under Climate Change Scenarios" (ref BIA2017-86383-R).

Conflicts of Interest: The authors declare no conflict of interest.

Appendix A. Thermal Loads Results

Table A1. Total loads (GJ/m^2) for the LC timespan.

	No Washing					Decennial Washing				
Insulation Thickness	Abs. = 0.1	Abs. = 0.2	Abs. = 0.3	Abs. = 0.4	Abs. = 0.5	Abs. = 0.1	Abs. = 0.2	Abs. = 0.3	Abs. = 0.4	Abs. = 0.5
0.0	5.035	2.8832	2.6243	2.3879	2.1733	3.1918	2.9121	2.6505	2.4117	2.1951
0.03	1.5207	1.3803	1.2511	1.1319	1.0225	1.5359	1.3941	1.2636	1.1432	1.0327
0.05	1.1469	1.0413	0.9443	0.8357	0.7480	1.1584	1.0517	0.9537	0.8441	0.7555
0.07	0.9190	0.8353	0.7583	0.6862	0.6221	0.9282	0.8436	0.7659	0.6931	0.6283
0.1	0.7183	0.6610	0.5968	0.5437	0.4890	0.7255	0.6676	0.6028	0.5491	0.4939
	Quinquennial Washing					Triennial Washing				
Insulation Thickness	Abs. = 0.1	Abs. = 0.2	Abs. = 0.3	Abs. = 0.4	Abs. = 0.5	Abs. = 0.1	Abs. = 0.2	Abs. = 0.3	Abs. = 0.4	Abs. = 0.5
0.0	3.2692	2.9703	2.6952	2.4479	2.1619	3.3182	3.0274	2.7555	2.5073	2.2820
0.03	1.5841	1.4378	1.3032	1.1791	1.0651	1.5968	1.4493	1.3136	1.1885	1.0736
0.05	1.1947	1.0847	0.9836	0.8705	0.7792	1.2042	1.0934	0.9915	0.8775	0.7854
0.07	0.9573	0.8701	0.7899	0.7148	0.6480	0.9649	0.8771	0.7962	0.7205	0.6532
0.1	0.7482	0.6885	0.6217	0.5664	0.5094	0.7542	0.6941	0.6266	0.5709	0.5135

Table A2. Heating loads (GJ/m^2) for the LC timespan.

	No Washing					Decennial Washing				
Insulation Thickness	Abs. = 0.1	Abs. = 0.2	Abs. = 0.3	Abs. = 0.4	Abs. = 0.5	Abs. = 0.1	Abs. = 0.2	Abs. = 0.3	Abs. = 0.4	Abs. = 0.5
0.0	3.1602	2.8832	2.6243	2.3879	2.1733	3.1918	2.9121	2.6505	2.4117	2.1951
0.03	1.5207	1.3803	1.2511	1.1319	1.0225	1.5359	1.3941	1.2636	1.1432	1.0327
0.05	1.1469	1.0413	0.9443	0.8357	0.7480	1.1584	1.0517	0.9537	0.8441	0.7555
0.07	0.9190	0.8353	0.7583	0.6862	0.6221	0.9282	0.8436	0.7659	0.6931	0.6283
0.1	0.7183	0.6610	0.5968	0.5437	0.4890	0.7255	0.6676	0.6028	0.5491	0.4939
	Quinquennial Washing					Triennial Washing				
Insulation Thickness	Abs. = 0.1	Abs. = 0.2	Abs. = 0.3	Abs. = 0.4	Abs. = 0.5	Abs. = 0.1	Abs. = 0.2	Abs. = 0.3	Abs. = 0.4	Abs. = 0.5
0.0	3.2692	2.9703	2.6952	2.4479	2.1619	3.3182	3.0274	2.7555	2.5073	2.2820
0.03	1.5841	1.4378	1.3032	1.1791	1.0651	1.5968	1.4493	1.3136	1.1885	1.0736
0.05	1.1947	1.0847	0.9836	0.8705	0.7792	1.2042	1.0934	0.9915	0.8775	0.7854
0.07	0.9573	0.8701	0.7899	0.7148	0.6480	0.9649	0.8771	0.7962	0.7205	0.6532
0.1	0.7482	0.6885	0.6217	0.5664	0.5094	0.7542	0.6941	0.6266	0.5709	0.5135

Table A3. Cooling loads (GJ/m^2) for the LC timespan.

	No Washing					Decennial Washing				
Insulation Thickness	**Abs. = 0.1**	**Abs. = 0.2**	**Abs. = 0.3**	**Abs. = 0.4**	**Abs. = 0.5**	**Abs. = 0.1**	**Abs. = 0.2**	**Abs. = 0.3**	**Abs. = 0.4**	**Abs. = 0.5**
0.0	1.8748	2.3057	2.7639	3.2457	3.7482	1.7626	2.1677	2.5986	3.0515	3.5240
0.03	0.7504	0.9353	1.1291	1.3305	1.5394	0.8098	0.9922	1.1841	1.3895	1.5783
0.05	0.5662	0.7045	0.8497	1.0069	1.1934	0.5886	0.6963	0.8392	1.0111	1.1608
0.07	0.4645	0.5768	0.6944	0.8150	0.9450	0.4467	0.5630	0.6801	0.8054	0.9304
0.1	0.3618	0.4697	0.5361	0.6308	0.7297	0.3445	0.4334	0.5254	0.6238	0.7196

	Quinquennial Washing					Triennial Washing				
Insulation Thickness	**Abs. = 0.1**	**Abs. = 0.2**	**Abs. = 0.3**	**Abs. = 0.4**	**Abs. = 0.5**	**Abs. = 0.1**	**Abs. = 0.2**	**Abs. = 0.3**	**Abs. = 0.4**	**Abs. = 0.5**
0.0	1.6985	2.0889	2.5041	2.9405	3.3958	1.6024	1.9707	2.3623	2.7741	3.2036
0.03	0.8043	0.9910	1.1818	1.3854	1.5764	0.7515	0.9474	1.1382	1.3369	1.5448
0.05	0.5855	0.6808	0.8192	0.9833	1.1370	0.5637	0.6522	0.7898	0.9469	1.1105
0.07	0.4262	0.5434	0.6608	0.7868	0.9138	0.4013	0.5172	0.6357	0.7600	0.8935
0.1	0.3280	0.4217	0.5133	0.6096	0.7074	0.3092	0.4030	0.4953	0.5919	0.6921

Appendix B. Life Cycle Total Costs

Table A4. Cooling and heating by pump. Total costs (€/m^2) for the LC timespan.

	No Washing					Decennial Washing				
Insulation Thickness	**Abs. = 0.1**	**Abs. = 0.2**	**Abs. = 0.3**	**Abs. = 0.4**	**Abs. = 0.5**	**Abs. = 0.1**	**Abs. = 0.2**	**Abs. = 0.3**	**Abs. = 0.4**	**Abs. = 0.5**
0.0	129.32	131.66	135.21	139.77	145.31	132.30	134.28	137.93	142.46	146.94
0.01	104.76	106.01	109.27	112.66	125.55	104.83	106.54	109.11	113.82	126.63
0.02	91.35	92.55	95.02	98.21	105.23	93.93	95.91	98.07	103.17	106.63
0.03	83.51	84.56	86.30	88.32	90.90	85.40	86.49	88.21	90.57	92.53
0.04	77.20	78.04	79.69	80.97	83.69	78.83	79.98	81.05	83.92	84.63
0.05	71.74	72.58	73.80	74.87	77.01	73.39	73.83	75.03	76.49	78.64
0.06	67.14	68.94	70.11	71.55	74.82	69.77	70.08	71.79	72.26	75.63
0.07	64.61	66.01	67.41	68.96	70.41	67.07	67.92	68.63	69.93	72.03
0.08	64.27	65.40	66.49	67.71	68.82	66.51	67.10	67.71	68.74	69.63
0.09	64.47	64.78	65.16	65.73	66.01	66.67	67.21	67.53	67.60	67.63
0.1	64.81	64.86	64.95	65.03	65.11	67.08	67.41	67.64	67.71	67.74

	Quinquennial Washing					Triennial Washing				
Insulation Thickness	**Abs. = 0.1**	**Abs. = 0.2**	**Abs. = 0.3**	**Abs. = 0.4**	**Abs. = 0.5**	**Abs. = 0.1**	**Abs. = 0.2**	**Abs. = 0.3**	**Abs. = 0.4**	**Abs. = 0.5**
0.0	135.76	138.16	141.76	146.01	150.50	140.60	143.02	146.13	149.32	153.54
0.01	106.21	109.31	112.66	116.03	120.76	109.76	112.83	116.21	119.96	124.15
0.02	95.38	96.85	98.22	100.92	103.76	98.82	100.23	101.97	104.08	106.49
0.03	88.20	89.77	90.18	92.37	94.88	92.16	93.55	95.19	97.28	98.68
0.04	80.86	82.10	83.56	84.98	86.52	85.15	85.20	86.62	88.21	89.91
0.05	76.15	76.69	77.81	79.02	80.31	79.96	80.64	81.75	82.73	84.07
0.06	71.74	72.39	74.10	75.22	76.81	75.79	76.60	77.59	78.39	79.11
0.07	69.37	70.10	71.22	72.46	73.74	73.74	74.43	75.37	76.24	77.04
0.08	68.33	68.39	68.46	68.78	68.99	72.82	73.19	73.58	73.90	74.38
0.09	68.40	68.55	68.61	68.83	69.11	72.89	73.26	73.81	74.21	74.61
0.1	68.72	68.71	68.89	69.38	69.57	73.11	73.39	74.07	74.48	74.71

Table A5. Cooling by pump and heating by gas. Total costs ($€/m^2$) for the LC timespan.

Insulation Thickness	No Washing					Decennial Washing				
	Abs. = 0.1	Abs. = 0.2	Abs. = 0.3	Abs. = 0.4	Abs. = 0.5	Abs. = 0.1	Abs. = 0.2	Abs. = 0.3	Abs. = 0.4	Abs. = 0.5
0.0	141.20	142.47	145.06	148.73	153.473	143.75	144.84	147.56	151.26	155.93
0.01	108.10	109.51	112.27	112.66	125.55	111.63	112.84	114.27	117.02	122.63
0.02	95.70	96.65	99.87	98.21	105.23	98.63	99.31	100.87	102.63	104.11
0.03	88.22	88.91	90.99	92.58	94.74	90.93	91.51	92.81	94.72	95.68
0.04	80.61	81.34	82.09	80.97	83.69	83.63	83.94	85.36	86.02	86.73
0.05	76.04	76.48	77.346	78.22	79.1	77.58	77.77	78.60	79.78	80.48
0.06	71.14	71.54	72.61	71.55	74.82	73.69	73.98	74.69	75.86	76.33
0.07	69.19	69.61	70.35	71.36	72.74	70.57	70.92	71.49	72.55	73.78
0.08	68.27	68.70	69.19	70.94	68.82	69.81	70.07	70.61	71.54	71.63
0.09	68.54	68.88	69.46	70.43	66.01	70.08	70.13	70.83	71.20	71.55
0.1	69.27	69.63	70.15	70.59	65.11	70.34	70.49	70.94	71.33	71.49
0.0	147.44	147.90	150.42	153.99	158.57	152.81	151.93	154.03	157.35	161.78
0.01	114.09	115.78	118.66	121.53	124.34	118.95	118.23	120.57	122.17	126.03
0.02	101.76	102.99	104.22	106.01	108.16	106.12	107.13	108.55	110.28	111.01
0.03	93.79	94.31	95.53	96.73	97.81	97.97	98.26	99.32	100.76	101.8
0.04	85.65	86.10	87.16	88.61	89.92	89.95	90.03	91.11	92.16	93.88
0.05	80.31	80.36	81.12	82.29	83.36	84.22	84.25	84.90	85.88	87.15
0.06	75.55	75.78	76.30	77.12	77.98	79.78	79.70	80.48	81.39	82.39
0.07	73.05	73.61	74.13	74.89	75.53	77.62	77.25	77.62	78.5	79.73
0.08	71.63	72.11	72.66	73.18	74.80	75.52	74.99	75.38	75.96	76.98
0.09	71.3	71.85	72.71	73.03	74.06	75.53	75.06	75.51	75.71	76.41
0.1	71.45	72.07	72.93	73.12	73.97	75.68	75.27	75.86	76.01	76.28

Appendix C. Comparison of Formulas to Compute the Payback Period

In this appendix, the formula given by Equation (13), used here to compute the payback period, is compared to the one used in former works as in Reference [31,63], where a constant value of the yearly energy consumption for the whole LCCA timespan is considered. The formula used in these reference to calculate the payback period (PB), when applied to the framework of the present work, is given by:

$$\begin{cases} PB = \dfrac{Ln\left[1 - (d - i)\dfrac{C_{I,ins} + P_2 C_{I,cc}}{A_s}\right]}{Ln\left(\dfrac{1+i}{1+d}\right)} & i \neq d \\[2em] PB = (1+i)\dfrac{C_{I,ins} + P_2 C_{I,cc}}{A_s} & i = d \end{cases} \tag{A1}$$

where A_s is the annual energy saving, taken constant, and $P_2 = 1 + P_1 M_s$, with M_s being a term that accounts for the present worth value of the cool coat maintenance, and P_1 is defined as

$$P_1 = \sum_{u=1}^{n} \frac{(1+i)^{(u-1)}}{(1+d)^u}.$$

More details about these formulas can been found in the cited works.

To perform the comparison between both formulas, the optimum combination of insulation layer and cool coat reflectivity found in the developed analysis, is considered. Because the formula given by Equation (A1) is only valid for constant values of yearly energy consumption, to compute A_s, the value of the energy consumption for the whole LCCA timespan of the optimal combination is divided by n, the number of years considered for the LCCA. Then, both formulas are applied for these values of energy consumption and for three different cases of i and d, the one used here, and two more combinations, in order to have some extra comparison values. The results of the comparison are shown in Table A6. In this table, the PB formula that uses constant values of yearly energy consumption is named Constant, and the presented here is named Variable.

Table A6. Comparison of formulas to compute the payback period.

PB Formula	$i = 2\%, d = 1.\%$	$i = 3\%, d = 1.5\%$	$i = 1.5\%, d = 3\%$
Constant	4.1834 [years]	4.1721 [years]	4.4442 [years]
Variable	4.1827 [years]	4.1710 [years]	4.4460 [years]

As it is observed in Table A6, the values provided by both formulas are very close. It is noteworthy to observe that the PB obtained for the optimal combination, when yearly energy consumption is considered constant, differs of the value of PB obtained when considering the energy consumption variable due to the aging effect, as in Table 12, which supports the stated need to take this effect into account when conducting the LCCA and the optimization analysis.

References

1. DB-HE2019. Basic Energy Saving Document. Spanish Royal Decrete 732/2019. Available online: https://www.codigotecnico.org/DocumentosCTE/AhorroEnergia.html (accessed on 15 June 2021).
2. European Commission. DG Energy. Available online: https://ec.europa.eu/clima/policies/strategies/2030-en (accessed on 2 July 2021).
3. Yu, J.; Tian, L.; Yang, C.; Xu, X.; Wang, J. Optimum insulation thickness of residential roof with respect to solareair degree-hours in hot summer and cold winter zone of China. *Energy Build.* **2011**, *43*, 2304–2313. [CrossRef]
4. Gentle, A.R.; Aguilar, J.L.C.; Smith, G.B. Optimized cool roofs: Integrating albedo and thermal emittance with R-value. *Sol. Energy Mater. Sol. Cells* **2011**, *95*, 3207–3215. [CrossRef]
5. Ramamurthy, P.; Sun, T.; Rule, K.; Bou-Zeid, E. The joint influence of albedo and insulation on roof performance: a modeling study. *Energy Build.* **2015**, *102*, 317–327. [CrossRef]
6. Arumugam, R.S.; Garg, V.; Ram, V.V.; Bhatia, A. Optimizing roof insulation for roofs with high albedo coating and radiant barriers in India. *J. Build. Eng.* **2015**, *2*, 52–58. [CrossRef]
7. Piselli, C.; Saffari, M.; Gracia, A.; Pisello, A.L.; Cotana, F.; Cabeza, L.F. Optimization of roof solar reflectance under different climateconditions, occupancy, building configuration and energy systems. *Energy Build.* **2017**, *151*, 81–97. http://dx.doi.org/10.1016/j.enbuild.2017.06.045. [CrossRef]
8. Shi, Z.; Zhang, X. Analyzing the effect of the longwave emissivity and solar reflectance of building envelopes on energy-saving in buildings in various climates. *Sol. Energy* **2011**, *85*, 28–37. [CrossRef]
9. Akbari, H.; Levinson, R.; Rainer, L. Monitoring the energy-use effects of cool roofs on California commercial buildings. *Energy Build.* **2005**, *37*, 1007–1016. [CrossRef]
10. Romeo, C.; Zinzi, M. Impact of a cool roof application on the energy and comfort performance in an existing non-residential building. A Sicilian case study. *Energy Build.* **2013**, *67*, 647–657. [CrossRef]
11. Akbari, H.; Bretz, S.; Kurn, D.M.; Hanford, J. Peak power and cooling energy savings of high-albedo roofs. *Energy Build.* **1997**, *25*, 117–126. [CrossRef]
12. Akbari, H.; Levinson, R. Evolution of cool-Roof standards in the US. *Adv. Build. Energy Res.* **2008**, *2*, 1–32. [CrossRef]
13. Akbari, H.; Konopacki, S. Calculating energy-saving potentials of heat-island reduction strategies. *Energy Policy* **2005**, *33*, 721–756. [CrossRef]
14. Akbari, H.; Matthews, H.D. Global cooling updates: Reflective roofs and pavements. *Energy Build.* **2012**, *55*, 2–6. [CrossRef]
15. Xu, T.; Sathaye, J.; Akbari, H.; Garg, V.; Tetali, S. Quantifying the direct benefits of cool roofs in an urban setting: Reduced cooling energy use and lowered greenhouse gas emissions. *Build. Environ.* **2012**, *48*, 1–6. [CrossRef]
16. Bozonnet, E.; Doya, M.; Allard, F. Cool roofs impact on building thermal response: A French case study. *Energy Build.* **2011**, *43*, 3006–3012. [CrossRef]
17. Santamouris, M. Cooling the cities—A review of reflective and green roof mitigation technologies to fight heat island and improve comfort in urban environments. *Sol. Energy* **2014**, *103*, 682–703. [CrossRef]
18. Seifhashem, M.; Capra, B.R.; Milller, W.; Bell, J. The potential for cool roofs to improve the energy efficiency of single storey warehouse-type retail buildings in Australia: A simulation case study. *Energy Build.* **2018**, *158*, 1393–1403. [CrossRef]
19. Boixo, S.; Diaz-Vicente, M.; Colmenar, A.; Castro, M.A. Potential energy savings from cool roofs in Spain and Andalusia. *Energy* **2012**, *38*, 425–438. [CrossRef]

20. Pisello, A.L.; Santamouris, M.; Cotana, F. Active cool roof effect: Impact of cool roofs on cooling system efficiency. *Adv. Build. Energy Res.* **2013**, *7*, 209–221. http://dx.doi.org/10.1080/17512549.2013.865560. [CrossRef]

21. Kolokotroni, M.; Gowreesunker, B.L.; Giridharan, R. Cool roof technology in London: An experimental and modelling study. *Energy Build.* **2013**, *67*, 658–667. [CrossRef]

22. De Masi, R.F.; Ruggiero, S.; Vanoli, G.P. Acrylic white paint of industrial sector for cool roofing application: Experimental investigation of summer behavior and aging problem under Mediterranean climate. *Sol. Energy* **2018**, *169*, 468–487. [CrossRef]

23. Kibert, C.J. *Sustainable Construction: Green Building Design and Delivery*; John Wiley and Sons: Hoboken, NJ, USA, 2016.

24. Vailati M.; Monti G. Earthquake-Resistant and Thermo-Insulating Infill Panel with Recycled-Plastic Joints. In *Earthquakes and Their Impact on Society*. *Springer Natural Hazards*; DÁmico, S., Ed.; Springer: Cham, Switzerland, 2016. [CrossRef]

25. Manfredi, V.; Masi, A. Seismic Strengthening and Energy Efficiency: Towards an Integrated Approach for the Rehabilitation of Existing RC Buildings. *Buildings* **2018**, *8*, 36. [CrossRef]

26. Vailati, M.; Monti, G.; Di Gangi, G. Earthquake-Safe and Energy-Efficient Infill Panels for Modern Buildings. In *International Conference on Earthquake Engineering and Structural Dynamics*; Rupakhety, R., Olafsson, S., Eds.; Springer: Cham, Switzerland, 2018; Volume 44. 62099-2-12. [CrossRef]

27. Zingre, K.T.; Kiran Kumar, D.E.V.S.; Wan, M.P.; Chao, C.Y.H. Effective R-value approach to comprehend the essence of integrated opaque passive substrate properties. *J. Build. Eng.* **2021**, *44*, 102865. [CrossRef]

28. Hernández-Pérez, I. Influence of Traditional and Solar Reflective Coatings on the Heat Transfer of Building Roofs in Mexico. *Appl. Sci.* **2021**, *11*, 3263. [CrossRef]

29. De Masi, R.F.; Gigante, A.; Ruggiero,S.; Vanoli, G.P. Impact of weather data and climate change projections in the refurbishment design of residential buildings in cooling dominated climate. *Appl. Energy* **2021**, *303*, 117584. [CrossRef]

30. Levinson, R.; Akbari, H. Potential benefits of cool roofs on commercialbuildings: conserving energy, saving money, and reducing emission ofgreenhouse gases and air pollutants. *Energy Effic.* **2010**, *3*, 53–109. http://dx.doi.org/10.1007/s12053-008-9038-2. [CrossRef]

31. Saafi, K.; Daouas, N. A life-cycle cost analysis for an optimum combination of cool coating and thermal insulation of residential building roofs in Tunisia. *Energy* **2018**, *152*, 925–938. [CrossRef]

32. Jo, J.H.; Carlson, J.D.; Golden, J.S.; Bryan, H. An integrated empirical and modeling methodology for analyzing solar reflective roof technologies on commercial buildings. *Build. Environ.* **2010**, *45*, 453–460. [CrossRef]

33. Sproul, J.; Wan, M.P.; Mandel, B.H.; Rosenfeld, A.H. Economic comparison of white, green, and black flat roofs in the United States. *Energy Build.* **2013**, *71*, 20–27. [CrossRef]

34. Zhang, Z.; Tong, S.; Yu, H. Life Cycle Analysis of Cool Roof in Tropical Areas. *Procedia Eng.* **2016**, *169*, 392–399. [CrossRef]

35. Yuan, J.; Farnham, C.; Emura, K.; Alam, M.A. Proposal for optimum combination of reflectivity and insulation thickness of building exterior walls for annual thermal load in Japan. *Build. Environ.* **2016**, *103*, 228–237. [CrossRef]

36. Spanish Royal Decrete 2429/1979. *Approving the Basic Building on Thermal Conditions in Buildings NBE-CT-79*; BOE Nro. 253; Norma Básoca de la Edificación: Barcelona, Spain, 1979; pp. 24524–24550.

37. Domínguez-Delgado, A.; Domínguez-Torres, C.A.; Domínguez-Torres, H. Energy and Economic Life Cycle Assessment of Cool Roofs Applied to the Refurbishment of Social Housing in Southern Spain. *Sustainability* **2020**, *12*, 5602. [CrossRef]

38. Hickson, R.I.; Barry, S.I.; Mercer, G.N.; Sidhu, H.S. Finite difference schemes for multilayer diffusion. *Math. Comput. Model.* **2011**, *54*, 210–220. [CrossRef]

39. Hagishima, A.; Tanimoto, J. Field measurements for estimating the convective heat transfer coefficient at building surfaces. *Build. Environ.* **2003**, *38*, 873–881. [CrossRef]

40. Walton, G.N. *Thermal Analysis Research Program Reference Manual*; NBSSIR 83-2655; National Bureau of Standards: Washington, DC, USA, 1983; p. 21.

41. Clark, G.; Allen, C. The Estimation of Atmospheric Radiation for Clear and Cloudy Skies. In Proceedings of the 2nd National Passive Solar Conference (AS/ISES), Philadelphia, PA, USA, 16–18 March 1978; pp. 675–678.

42. Kusuda, T. Heat transfer in buildings. In *Handbook of Heat Transfer Applications*; Rohsenow, W.M., Hartnett, J.P., Ganic, E.N., Eds.; McGraw-Hill: NewYork, NY, USA, 1985

43. *ASHRAE Fundamentals Handbook (S.I.)*; American Society of Heating, Refrigerating and Air Conditioning Engineers Inc.: Atlanta, GA, USA, 1997.

44. Sanjuan, C.; Suárez, M.J.; Blanco, E.; Heras, M.D.R. Development and experimental validation of a simulation model for open joint ventilated facades. *Energy Build.* **2011**, *43*, 3446–3456. [CrossRef]

45. Daouas, N. Impact of external longwave radiation on optimum insulation thickness in Tunisian building roofs based on a dynamic analytical model. *Appl. Energy* **2016**, *177*, 136–148. [CrossRef]

46. Mastrapostoli, E.; Santamouris, M.; Kolokotsa, D.; Vassilis, P.; Venieri, D.; Gompakis, K. On the ageing of cool roofs: Measure of the optical degradation, chemical and biological analysis and assessment of the energy impact. *Energy Build.* **2016**, *114*, 191–199. [CrossRef]

47. Eilert, P. *High Albedo (Cool) Roofs: Codes and Standards Enhancement (CASE) Study*; Pacific Gas and Electric Company: San Francisco, CA, USA, 2000.

48. Xue, X.; Yang, J.; Zhang, W.; Jiang, L.; Qua, J.; Xu, L.; Zhang, H.; Song, J.; Zhang, R.; Li, Y.; et al. The study of an energy efficient cool white roof coating based on styrene acrylate copolymer and cement for waterproofing purpose. Part I: Optical properties,

estimated cooling effect and relevant properties after dirt and accelerated exposures. *Constr. Build. Mat.* **2015**, *98*, 176–184. [CrossRef]

49.   Bretz, S.E.; Akbari, H. Long-term performance of high-albedo roof coatings. *Energy Build.* **1997**, *25*, 159–167. [CrossRef]
50.   Durability, Pausing for Reflection on Aging of the Cool Roof. *J. Archit. Coat.* **2008**, *63*, e5.
51.   Synnefa, A. Cool Roofs Standars and the ECRC Product Rating Program. European Cool Roofs Council. 2016. Available online: https://coolroofcouncil.eu/wp-content/uploads/2019/05/ECRC-Product-rating-manual_2017.pdf (accessed on 23 March 2021).
52.   Reglamento de Instalaciones Térmicas en los Edificios. Ministerio para la Transición Ecológica y el Reto Demográfico. Available online: https://energia.gob.es/desarrollo/EficienciaEnergetica/RITE/Paginas/InstalacionesTermicas.aspx (accessed on 2 March 2021).
53.   Rossi, M.; Rocco, V.M. External walls design: The role of periodic thermal transmittance and internal areal heat capacity. *Energy Build.* **2014**, *68*, 732–740. [CrossRef]
54.   Auliac, S.; Le Hyaric, A.; Morice, J.; Hecht, F.; Ohtsuka, K.; Pironneau, O. FreeFem++. Third Edition, Version 3.31-2, 2014. 2017. Available online: http:www.freefem.org (accessed on 5 June 2021).
55.   Agencia Estatal de Meteorología de España. Guía Resumida del Clima en España (1981–2010). 2010. Available online: http://www.aemet.es/es/conocermas/recursos_en_linea/publicaciones_y_estudios/publicaciones/detalles/guia_resumida_2010 (accessed on 2 June 2021).
56.   Domínguez Amarillo, S.; Sendra Salas, J.J.; Oteiza San José, I. *La Envolvente Térmica de la Vivienda Social. El Caso de Sevilla, 1939 a 1979*; 2016 Csic.; Editorial CSIC: Madrid, Spain, 2016; p. 174, ISBN 978-84-00-10124-4
57.   Cattarin, G.; Causone, F.; Kindinis, A.; Pagliano, L. Empirical and comparative validation of an original model to simulate the thermal behaviour of outdoor test cells. *Energy Build.* **2018**, *158*, 1711–1723. [CrossRef]
58.   *Código Técnico de la Edificación*; Ministerio de Fomento del Gobierno de España: Madrid, Spain, 2009.
59.   León, A.L.; Suárez, R.; Bustamante, P.; Campano, M.A.; Moreno, D. Design and performance of test cells as an energy evaluation model of facades in a mediterranean building area. *Energies* **2017**, *10*, 151–161. [CrossRef]
60.   US Department of Energy; Energy Efficiency and Renewable Energy Office; Building Technology Program. EnergyPlus 8.0.0. Available online: http://apps1.eere.energy.gov/buildings/energyplus/ (accessed on 10 June 2021).
61.   Duffie, J.A.; Beckman, W. *Solar Engineering of Thermal Processes*; John Wiley and Sons Inc.: Hoboken, NJ, USA, 2006.
62.   CYPE Engineers. CYPE Ingenieros Construction Price Generator. 2017. Available online: http://www.cype.es/ (accesed on 22 March 2021).
63.   Vincelas, F.F.C.; Ghislain, T.; Robert, T. Influence of the types of fuel and building material on energy savings into building in tropical region of Cameroon. *Appl. Therm. Eng.* **2017**, *122*, 806–819. [CrossRef]
64.   Available online: https://ec.europa.eu/eurostat/statistics-explained/index.php/Natural-gas-price-statistics (accesed on 2 March 2021).
65.   Nydahl, H.; Andersson, S.; Astrand, A.P.; Olofsson, T. Energy and Buildings Including future climate induced cost when assessing building refurbishment performance. *Energy Build.* **2019**, *203*, 109428. [CrossRef]

sustainability

MDPI

Article

A Research Methodology for Mitigating Climate Change in the Restoration of Buildings: Rehabilitation Strategies and Low-Impact Prefabrication in the "El Rodezno" Water Mill

Amadeo Ramos-Carranza, Rosa María Añón-Abajas and Gloria Rivero-Lamela *

Departamento de Proyectos Arquitectónicos, Escuela Técnica Superior de Arquitectura, Universidad de Sevilla, Av. Reina Mercedes 2, 41012 Seville, Spain; amadeo@us.es (A.R.-C.); rabajas@us.es (R.M.A.-A.)
* Correspondence: grivero@us.es

Abstract: New environmental challenges, coupled with the fact that 80% of the residential buildings that will exist in Europe in the year 2050 have already been built, mean that rehabilitation and restoration must be prioritised over new buildings. Construction is one of the largest generators of CO_2. Using prefabricated and industrialised products and systems can help to mitigate its harmful effects thanks to the greater control and environmental evaluation that can be carried out on these products from their manufacture until the end of their useful life (LCA). In the county of the Sierra de Cádiz (Andalusia, Spain), there are 85 water mills, many of which are derelict and in disuse, which, due to their location, size, and characteristics, are ideal for rehabilitation and restoration for residential use. Taking the "El Rodezno" mill as a case study, this paper proposes rehabilitation strategies using prefabricated industrialised elements that have a low environmental impact. The methodological discussion takes as its starting point the process of design and testing that Alvar Aalto applied in 1940 and from subsequent studies that have confirmed a research structure based on the project design and the built project with the appropriate field of study and confirmation of the applicable strategies and solutions. To this end, this article is written on the basis of the two main phases of Alvar Aalto's method, using the same terms that the Danish architect defined: *Scientific Observation*, for the study of preceding works and projects in light prefabrication and for the analysis of certain construction products and systems that, based on other research, have evaluated their LCA, and *Construction Period*, for the rehabilitation strategies of the "El Rodezno" mill, considering the studies and analyses of *Scientific Observation*. For the roof solution, we took as an example the rehabilitation of the roof carried out with the same methodology, construction criteria, and prefabricated products analysed in this article and used in the intervention strategies in "El Rodezno". The paper concludes with the validity of the methodology applied to test the starting hypotheses that lead to intervention strategies that confirm the environmental and economic advantages of industrialised prefabrication, the importance of the design and synergy that results from combining different construction systems, and technologies that improve the acceptance of prefabrication by the inhabitant and boost the circular economy.

Keywords: heritage regeneration; water mills; sustainable prefabrication; local industry; housing; climate change

check for
updates

Citation: Ramos-Carranza, A.; Añón-Abajas, R.M.; Rivero-Lamela, G. A Research Methodology for Mitigating Climate Change in the Restoration of Buildings: Rehabilitation Strategies and Low-Impact Prefabrication in the "El Rodezno" Water Mill. *Sustainability* **2021**, *13*, 8869. https://doi.org/10.3390/su13168869

Academic Editor: Pilar Mercader-Moyano

Received: 20 July 2021
Accepted: 2 August 2021
Published: 8 August 2021

Publisher's Note: MDPI stays neutral with regard to jurisdictional claims in published maps and institutional affiliations.

Copyright: © 2021 by the authors. Licensee MDPI, Basel, Switzerland. This article is an open access article distributed under the terms and conditions of the Creative Commons Attribution (CC BY) license (https://creativecommons.org/licenses/by/4.0/).

1. Introduction

Productive rural architecture refers to various types of buildings: granaries, silos, other grain stores, farmhouses, estates, or mills [1]. They are anonymous constructions that were built prior to the technological advances that resulted in industrial production. The move from craftsmanship to industrialisation entailed the loss of use and later abandonment of these buildings, which were built taking into account the resources available in the area, creating a system in balance with the environment. It must be recalled that these constructions, due to their size and the function they served, were architectural works

87

that required strategic positioning and recognition in an area to fulfil a very well-defined productive methodology for which there were few alternatives.

In the Sierra de Cádiz, there were 85 water mills on the banks of the rivers and streams that used to flow through the region [2]. The Sierra de Cádiz, with a pronounced orography, has a significant water network, supplied in a natural manner by the high rainfall in this region. Roads and paths, along with rivers and streams, established an infrastructure network superimposed on the territory that linked mills, riverbanks, crop areas, other diverse architecture, and settlements. Therefore, the mills were an essential part of the land, the topographical conditions that define it and its landscape, understood as identity values of the region (Figure 1). Its adaptation to the environment was also reflected in its relationship with the communication infrastructures that facilitated its accessibility and habitability. Additionally, along certain riverbanks, in mills and fields under cultivation, the water infrastructure was shared, creating a water recirculation system. Once the water had been used by mills and fields under cultivation, any that was left over returned to the river, recycling a natural resource as valuable as it was scarce [3].

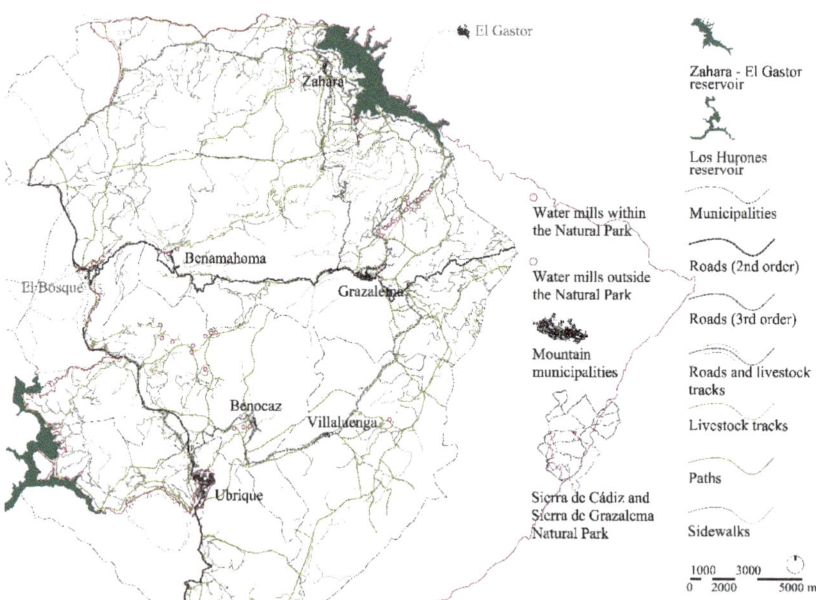

Figure 1. Current communication infrastructures of the Sierra de Grazalema Natural Park (Sierra de Cádiz). Source: prepared by the authors.

These mills are of a very basic type with a small and simple volumetry, which is usually very rational, without excessive pretensions. The inclusion in certain cases of the miller's own dwelling makes them more similar to the residential classification than the industrial classification, due to which transfers including the dwelling are believable. The knowledge that arises from traditional construction, the artisan trades that made it possible, results from the logical application of the materials [4] and the law of the "maximum internal economy" [5]. Despite their inherent values, these constructions had not been studied in the academic or institutional field until a few decades ago, and there has been even less reflection on the potential of their restoration and on the modus operandi in which their rehabilitation and re-habitation could be carried out.

It is important to bear in mind that 80% of the buildings that will be used for residential purposes in Europe by the year 2050 have already been built and, until that date, an annual increase of 1% is forecast. However, the restoration and rehabilitation of the residential housing stock will mean a larger percentage, which may reach 3% [6] and possibly increase, as the awareness of the need to recycle the existing architecture is greater and is supported by public institutions. Proposing the rehabilitation of existing architectures for residential use would also be consistent with the European Union's strategic guidelines and its policy of Comprehensive and Deep Renovation of residential buildings [7].

Construction is the cause of a third of CO_2 emissions, including generated waste, and 40% of energy consumption, which is why it is the goal of the Paris Agreement to minimise these two indicators. The use of industrialised prefabricated products makes it possible to better assess and control the environmental impacts that are generated, from their manufacture to the end of their useful life, following the cycle of recycling—manufacturing, function, and new recycling—that also boosts the circular economy.

Describing and contextualising a specific situation, this article starts from the hypothesis that these mills constitute an experimental field that serves to verify the validity of intervention strategies for the restoration and rehabilitation of these heritage architectures and, given their small size, their adaptation to family homes.

This article aims to explore the possible restoration of these architectures with new construction products and technologies that meet basic environmental criteria regarding recycling, lightness, flexibility, and modulation in the design of construction and structural systems that allow for self-assembly, facilitating the inclusion of unskilled labour, boosting the local economy, reducing greenhouse gases, and diminishing energy consumption. It also aims to verify the viability of certain strategies and solutions for the restoration, rehabilitation, and protection of these architectures which, due to the simplicity and viability of the strategies, can be transferred to the rehabilitation of dwellings of a family scale.

2. Methodology: Discussion and Rationale

In addition to research on the eco-efficiency of prefabricated products that must be used and the evaluation of the environmental impact throughout their life cycles, the design on which, to a large extent, the optimisation of materials depends, reducing waste or facilitating the total dismantling of what was built, plays a vital role. For this reason, the methodology that this study follows is based on the analysis and observation of the results of projects and built works. In the same vein, one of the most well-rounded methodologies was the one pioneered by Alvar Aalto during his time at MIT, Cambridge, USA, in 1940, in which he proposed a laboratory for researching prefabricated solutions for dwellings that were communal as well as individual [8] (pp. 173–186). Aalto distinguished three phases in his method: after the first one devoted to the construction of different types of houses that were the basis of the study, he continued with the phase that he called *Scientific Observation*, which was aimed at assessing the proper functioning and acceptability of these prototype built houses. The studies, analyses, and results derived from this phase should, in Aalto's opinion, be published "for academic and scientific use". To this end, he included the collaboration of universities and technological institutes with the intention of obtaining an objective assessment of what was designed and built. Aalto called the last phase the *Construction Period*, and its objective was to involve private companies.

The value of Aalto's method lies in its commitment to applied research that had to conclude, first, in a project design and then in its implementation. This means that the research that stems from the project or the architectural intervention requires its own methodology which, nonetheless, considers the advances on construction products or systems that contribute individual or one-off solutions to construction. Aalto's method also had the novelty of involving university research centres and faculties with technical and scientific capabilities to participate in the construction of the project and private companies with links to the construction sector, which would force one to consider, in the design phase, aspects such as profit and return on investment: a process that is, firstly, one of research and, subsequently, participatory (Figure 2).

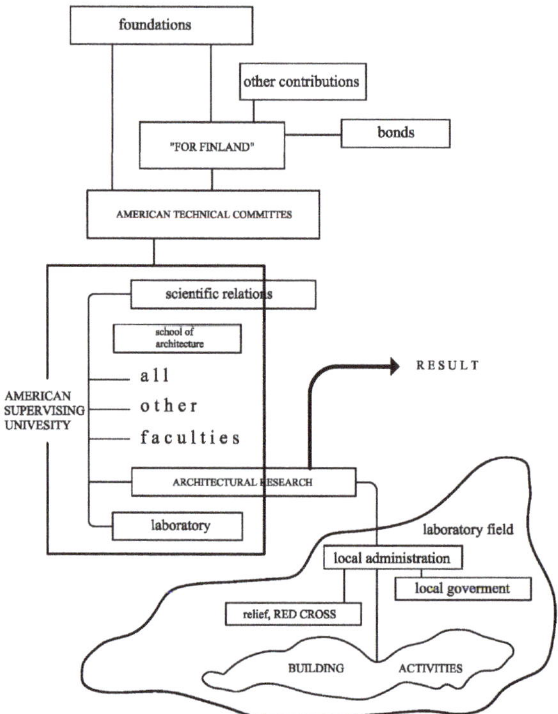

Figure 2. Organisational and participatory scheme of the project. Alvar Aalto 1940 [8] (p. 181).

In 1948, Richard Buckminster Fuller, during his time at Black Mountain College, applied a similar work method, albeit a much more simplified one. On this occasion, it was the construction of a geodesic dome with the assistance of students that turned out to be failure. He did achieve success, however, on his second attempt in 1949, thanks to the collaboration of a team of engineers who corrected the students' initial errors [9]. Aalto and Fuller's experiences emphasise a methodology in which the project is proposed as a theoretical testing model that validates or refutes the hypotheses that bring forth the proposal. The specificity of each project means that the conclusions are, in principle, exclusive to each one of them, although there are always generalisable results. What they have in common is that they consider that an experimental project may lead to a prototype or a model from which it is possible to define a series of main lines of intervention for similar cases. One understands, then, that some years later, Leonardo Benevolo would advocate a scientific research process in architecture based on the immense amount of experience built and accumulated over the years, which forced our research to consider the preceding projects and works of architecture, evaluating their results to create "a web of inductions and deductions, invention and calculation, and not a homogeneous succession of deductive operations" [10]. Benevolo stressed that "all contemporary experience must be systematically compared with that of the recent past in order to find the thread or threads of research capable of growth".

Research and advancement processes of this kind can be recognised even before Aalto's proposal, and, in the modern period, we can cite the experiences of the Bauhaus, especially the Haus am Horn, designed in 1923 by Professor Georg Muche in collaboration with Adolf Meyer and Walter Gropius: a laboratory test that set out to systematically verify the benefits of a new way of living. Perhaps these cases of research based on a project and built work methodology do not seem to conform to an empirical model as proposed by Bruce Archer [11], but these presented cases, based on a project action, express a fundamental and specific form of knowledge. Professor Christopher Frayling distinguishes three ways of conducting research through the project or the built work [12], and in what he calls "through art and design", he places in this type of methodology those works where the practice serves a purpose of the research: a project from which useful and necessary hypotheses can be made to solve a problem and, in the process, prove useful to the research itself. This is the methodology that this team of researchers has been developing for several years, with the conviction that "in architecture, to test ideas and knowledge, there is still a need for theory and practice enunciated and demonstrated in projects and in works" [13].

Professor Jorge Torres Cueco [14] positively rates "inferential" methodology, a term coined by Michael Baxandall in 1985 [15], also termed "generic critique", in which the construction work is and must be the starting point from which one justifies the relations that need to be established, inverting the research process, which would lead from the work to the background.

At present, especially as a result of the financial crisis of 2007, it is uncommon to complete the Aalto method, and the methodology of this article is closer to the "through art and design" approach of Frayling and the method of Alvar Aalto than to Baxandall's inferential method, since Frayling and Aalto's approaches allow architectural works from different periods to support a prior knowledge necessary for a practical implementation that considers the results of other previous projects and research. This ensures a greater interaction between basic and applied research.

Consequently, this article is written in two sections that borrow the terms *Scientific Observation* and *Construction Period*, which Aalto used in 1940, because they are well suited to both the contents and the methodology followed.

The first part, *Scientific Observation*, is intended to create a sufficient spectrum of architectures to support the purpose of the research. To this is added the analysis of certain products with which it is possible to create construction and structural solutions applicable to buildings on a domestic scale, *scientifically observed* from current environmental performance criteria that further strengthen the objective of affordable construction using local labour.

The second part, *Construction Period*, defines intervention strategies to restore a heritage site in poor condition that is reusable as a home and proposes a specific solution for covering the roof using, in both cases, materials with low environmental impact that help to mitigate the consequences of climate change and that have been analysed in this article.

3. Scientific Observation

3.1. Previous Architecture in Light Prefabrication: Values Applicable to the Proposed Intervention Strategies

The examples of architecture that we have studied with prefabricated systems are those that use light, prefabricated systems that comply with the conditions for recycling, the use of modules, and the standardisation of minimum dimensions that can be self-assembled. The result of this study has been sorted into a database that combines various information, considering not just the usual identification data that refer to its context (urban or rural location, climate, topography, access, etc.) but especially those that define its construction (data referring to the surroundings: materials, weight, dimensions, type of joints, implementation time or situation of the facilities, etc.) and its structural system (structural system, materials, dimensions, or bracing elements). The database also reports on whether there have been variants of the model, the degree of acceptance achieved, or its commercialisation, which are directly linked with its level of industrialisation (Figure 3).

GENERAL	CONTEXT	STRUCTURE	CONSTRUCTION	PLANIMETRY	IMAGES
PROYECT YEAR	URBAN TIPOLOGY	STRUCTURAL SYSTEM	MATERIAL OUTSIDE OF THE ENCLOSURES	LOCATION	LOCATION
CONSTRUCTION YEAR	LEVEL OF URBANIZATION	STRUCTURE MATERIALS	MATERIAL INSIDE THE ENCLOSURES	SOURCE : ARCHIVE / BIBLIOGRAPHY	SOURCE : ARCHIVE / BIBLIOGRAPHY
NAME	BOUNDARY CONDITIONS	STRUCTURAL DIMENSIONS	ENCLOSURE DIMENSIONS	ID. EXPEDIENTE	ID. EXPEDIENTE
AUTHOR	CLIMATE	BRACING ELEMENTS	WINDOW TYPOLOGIES	TYPOLOGY PLAN	BLACK AND WHITE / COLOUR
LOCATION	TOPOGRAPHY	APPLICATION REGULATIONS	INSULATION MATERIALS	DRAWING TYPE	
OTHERS LOCATIONS	GEOTECHNICAL DATA		ENERGY SYSTEM	SCALE	
CURRENT STATUS	SISMICITY		TYPE OF JOINTS BETWEEN MATERIALS	PLANE 1	IMAGE 1
TIPOLOGY			CONSTRUCTION TIME	PLANE 2	IMAGE 2
Nº OF FLOORS			TYPE OF ASSEMBLY	PLANE 3	IMAGE 3
Nº OF DWELLINGS			TYPES OF INSTALLATIONS	ETC.	ETC.
SURFACE AREA			CONSTRUCTION WEIGHT DATA		
MODEL VARIANTS					
DEGREE OF ACCEPTANCE					
DURABILITY					
ECONOMIC COST					
KEY WORDS					
BIBLIOGRAPHY					

Figure 3. Structure of contents in each record in the database. Source: the authors.

Based on these criteria, a selection was made of 78 examples of projects and construction works that begins with the Charles S. Ross House, by Frank Lloyd Wright, 1902, and ends with the Casa Garoza 10.1, by Juan Herreros, 2010. The structuring of the database enables the systematisation of solutions that are compatible in many cases with industrial production in the local area. Considering this condition, there is a greater probability of success both in the manner in which it is necessary to intervene as well as in the use of minimal modular elements that allow buildings to be habitable.

Following the methodology explained herein and based on the projects studied, it is useful to take note of some points that will later have bearing on the intervention strategies.

Firstly, and derived from the designs and projects for the Shingle Style carried out by Frank Lloyd Wright in the first decades of the 20th century, we must take note of the importance of the small-scale modulation employed in inward and outward-facing elements (porches and balconies) that were linked to more distinctive internal spaces and built with a single material. The Balloon Frame system used gives pride of place to the carpenter's craft as the basis of a highly adaptable construction system. In addition, in some houses, Wright used a floor grid that was modulated in accordance with the dimensions of the standardised wooden elements, such as in the Double Cottage for George Gerts (Whitehall, Michigan, 1902), using the "board and batten" technique (Figure 4) [16] under a 3-foot (91.44 cm) grid through which everything would be linked both in terms of space and construction. The use of a small base measurement restricts the maximum lengths of the materials to be used and establishes a proportional relationship among them.

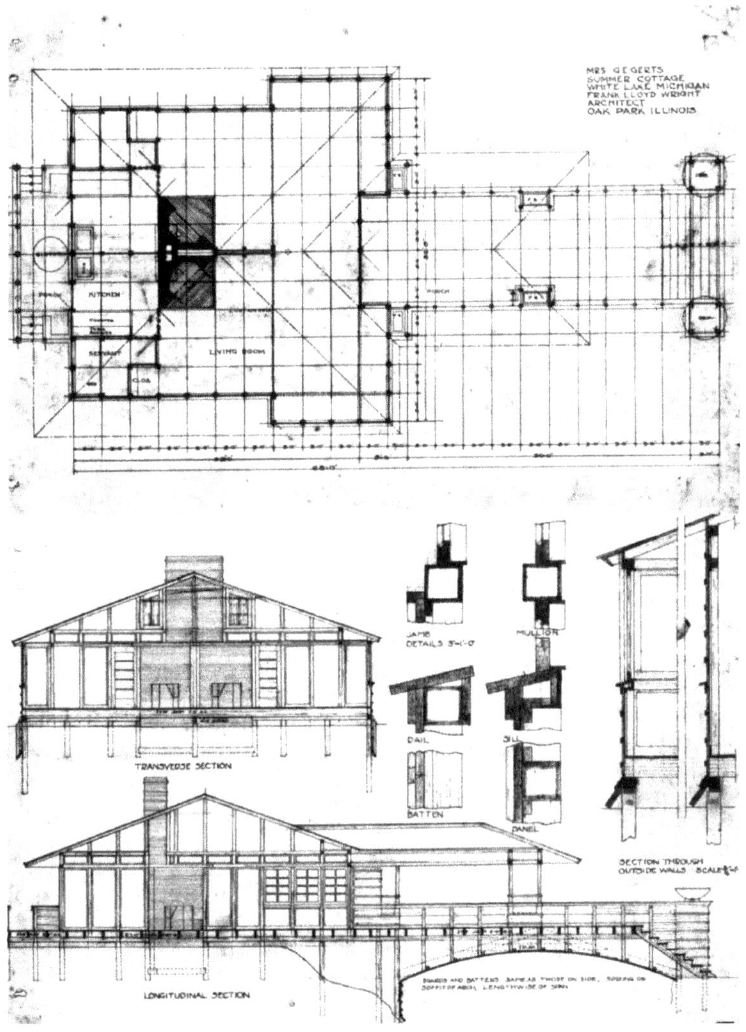

Figure 4. Doble Cottage for George Gerts (1902) Frank Lloyd Wright.

The grid as an instrument for control of the project has been widely used throughout the 20th century. Walter Gropius, in his houses numbers 16 and 17 built in the Weissenhof Siedlung (1927), used the dry assembly system—*Trockenmontage*—laying out a floor grid of 1.06 m × 1.06 m, the dimension of the standard door frame measurements. Additionally, well-known are the projects that the German architect developed in his American stage with the Packaged Houses.

Jean Prouvé also used a grid for his prefabricated houses, in this case a 1 m × 1 m grid. In addition to limiting the length of the prefabricated structural and construction elements, the condition was added that no item could exceed 80 kilos so as to ensure that all the components to be used could be handled by three or four workers. The aim was to facilitate self-assembly, a criterion that extended to numerous projects of his and was applied in various ways to all his designs. The engineer Léon Pétroff, who collaborated with Prouvé, patented an extendible system (1968) that consisted of a repeatable structural module that was adjusted to the dimensions of the grid (1 m × 1 m). Easy to transport due to its small dimensions and weight, it was also recoverable, as the joint between modules was bolted together. Prouvé used it in several of his works such as the Maison de Mme. Jaoul (Mainguerin, 1969) (Figure 5). This module is a braced box with which one can form a growing structure that works in a similar way to a spatial mesh, which enables one to conceive of other applications of this system, including as a linear and vertical structure in the same way as scaffolding, which would be easy to adapt to shore up unstable masonry walls. The prototypes with an axial doorway, sheet, and central nucleus that Prouvé designed were intended for housing but were also applied to schools with similar dimensions to the domestic scale. In some of them, such as that of Jules Ferry, built in the French town of Dieulouard (1952–1953), the prefabricated elements were combined with masonry walls, although it was still a new build [17], something that Prouvé had tried in the Meudon houses (1950–1952), in which, on masonry walls, a slab was deployed based on reticulated metal strips on which were mounted the prototype for the central doorway [18]. Prouvé designed the Maison Tropicale (1946–1949), also with an axial doorway, which, with a 2 m gallery and 1.2 m brise-soleil, he sought natural ventilation through an open gable that provided airflow, thus attempting to produce an architecture whose comfort and thermal control were based on incorporating passive elements [19]. In addition to providing a sustainable design, the combination of traditional construction techniques and prefabricated industrialised materials is feasible. This expands the possible intervention strategies in locations and architectures in which a mixed system is called for with apparently diverse building techniques. To put it another way, it also entails the inclusion of unskilled labour in the implementation of a project, without giving up the technology that the construction industry has to offer.

Figure 5. Maison para la Mme. Jaoul. Jean-Claude Drouin, architect; Jean Prouvé, consultant engineer; Léon Pétroff, engineer; Brisard, construction company (1969) [18]. Model of extendible module by Léon Petroff. Photography, authors.

The design of a module that could be repeated until it formed a larger whole has had very diverse variations, such as the Stadt Ragnitz designed by Günter Domenig & Eilfred Huth (1963–1969), the dimensions of which led to dense networks of prefabricated elements cable-stayed with diagonals. Additionally, the works of Eckhard Schulze-Fielitz are included, such as the Urban System, Research and Development (1965–1966), which was a cellular spatial system of 7.2 m × 7.2 m with four maximum elements in each node and screwed joints. He also proposed three types of climate control for the internal spaces: natural, semi-air-conditioned, and artificial, depending on the level of protection from rain, radiation, temperature, and wind [20] (Figure 6). Eckhard Schulze's proposal, which goes beyond the small scale of the home, alludes to a complex combination of and connection between residential units that would occupy each module measuring 7.20m × 7.20 m [21]. In this way, the rigidity that a grid apparently transfers, in this case in spatial terms, is converted into an adaptive and flexible system, depending on the number of modules this spatial grid will occupy. Each basic unit of the grid is understood as the container, and inside it, one conceives a free, clear space that allows for a diversity of uses or functions.

This idea would be of interest when aligning construction and structural systems that facilitate a flexible use of the interior when the external limits—the container—are rigid and unchangeable.

Figure 6. Urban System, Research and Development Company Ltd. Eckhard Schulze-Fielitz, 1965–1966 [22].

Lastly, it is interesting to cite the Naked House by Shigeru Ban (Kawagoe, Saitama, Japan, 2000). In addition to the wooden fibre-reinforced struts that support the trusses hidden by a false roof, the external casing is constructed with two panels of corrugated glass fibre-reinforced plastic [23]. The Ephemeral Pavilion for the Centro Andaluz de Arte Contemporáneo [Andalusian centre for contemporary art], constructed in Seville (2006) by architects Frank Mazzarella and Inma Donaire, is designed based on several walls that have an internal structure as a type of scaffold. The external casing is mini-wave panels of glass fibre-reinforced acrylic. These examples introduce new industrial materials and, above all, the combination of structural and enclosing systems of very different natures into the debate.

3.2. Prefabrication and the Construction Industry: Products and Systems Applicable to the Proposed Interventions

Prefabrication is a type of industrialisation of construction [24] that guarantees that the final product is the result of a process that is completely controlled and determined, in which it is possible to define the optimum conditions in which a construction element must be produced [25]. The exhaustive control of the dimensions, weight, mechanical characteristics, the possibility to choose the components, and the fabrication process distinguish it from other ways of manufacturing products that, without due control, may prove to be more environmentally harmful, contrary to the assessment of the impacts through the product life cycle and the circular economy [26].

Globalisation has meant a dislocation of industries and, with it, the atomisation of prefabricated components to form a final product. The case of cars is very illustrative, where the various mechanical components are made in various factories located in distant locations, ultimately to be assembled in a factory expressly dedicated to this function. In this international context, it is difficult to think of closed prefabrication, especially in construction, particularly when architecture is not well-suited to this idea of the final product, closed and repeatable at a quantity and with a fabrication and installation time similar to that used to produce a model of car. For architecture, the challenge consists of leveraging the advantages that industrialised prefabrication offers to generate different, innovative solutions [27] that, in addition to reducing environmental impacts, are energy-efficient and reduce building costs. Architectures can be created that comply with what Richard Llewelyn–Davies and John Weeks called in 1951 "socio-technological environments" [28]. The aim was that with prefabrication, one could define a way of building that would be accepted by its inhabitants, who normally rejected it. According to his theory, the image and prefabricated materials had to be visible in the new construction but compatible with the needs of the inhabitants in terms of function and space, spirit, and emotion. When this fails, the traditional architectural forms are imposed, and the reasons are forgotten as to why, in the 1960s, prefabrication and industrialisation were associated with a change in the paradigm of production in construction. At the end of the 1960s, Alexander Pike suggested some reasons why prefabricated architecture was rejected, reasons that still exist today [29].

- A greater presence of architects was needed in the production teams of companies involved in new materials and construction technologies.
- There was a lack of university education in the design of new technologies.
- There was a big difference in size and weight of the prefabricated products used in construction in comparison, for example, with what was happening in the automotive industry.
- The lifespan of a home was, at the time, approximately 65 years (today perhaps longer), comfortably exceeding the lifespan of a car or of any domestic appliance.
- A change in mentality was needed, both in the sectors and actors that participated in house building and in the people that were going to live in them.
- Products were needed that would be financially profitable and that could hold their own in the market for long enough to create companies and production plants engaged in prefabricated construction.

Prefabricated construction must be compatible with people's wish to be able to adapt accommodation to their needs. To the extent that the construction industry pays attention mostly to its sales rating, the way forward consists of designing systems that bring together both interests so that the construction industry fabricates products reducing energy consumption, minimising the carbon footprint, and driving the circular economy, to succeed in changing the current economic, social, and environmental model. Together with the choice of materials according to their life-cycle assessment, the design must achieve an optimisation of the materials and allow the space to be reconfigured so as to prolong the building's useful life until its complete disassembly for a new recycling. Nevertheless, it is difficult to foresee all the modifications that a building may undergo during its useful life, even more so in the case of single-family homes. For this reason, in 2003, some researchers proposed understanding the building as a container capable of adapting to different future scenarios, and so to assess its environmental impact, they proposed a dynamic life cycle assessment (DLCA) [30], studies that have been continued in other research; since 2013, LCA studies on the rehabilitation of single-family homes [6] have been more frequent, in which changes in the surrounding industrial and environmental systems have been considered [30]. This does not mean that prefabricated construction is not suitable, and, in fact, buildings built with conventional systems have been compared to those that have used prefabricated systems, concluding that prefabrication reduces environmental impacts by between 5% and 40%: it reduces greenhouse gas (GHG) emissions, energy consumption, resource depletion, and damage to ecosystems and is also 30% cheaper [31], although this figure varies depending on the geographic location, transportation, and labour costs.

Achieving widespread implementation of the industrial ecology continues to be a challenge [32], which translates the idea of natural ecosystems into construction: a continuous flow in which the waste from an activity, once exploited, is once again transformed and harmless to the ecosystem, enabling the cycle to begin again. The climate emergency has meant that industrial ecology seems current and innovative when, in reality, it dates back to the 1960s.

From all of this, we can obtain some keys and guidelines for our interventions that we can summarise as follows:

- The use of recycled materials or those that, after use, can be recycled.
- Considering the criteria for industrialised prefabrication, the number of different materials to be used must not be high.
- The commitment to an open system facilitates the substitution of elements and the adaptation of the construction for new uses.

- Implementing a circular economy in the construction process that returns to the environment near the project, using products, industries, and the workforce that are nearby.
- Encouraging the restoration and reactivation of pre-existing buildings as an engine for development, alternative to the global economic model.
- Based on these premises, the possibilities for using three prefabricated products are explored briefly. Due to their composite and environmental characteristics, as well as their modular nature and weight limitations, they would be suitable for our intervention strategies.

3.2.1. Laminated and Cross-Laminated Timber Panels

This product can be combined with others, also industrialised, creating construction systems with greater uses (sandwich panels of cross-laminated timber panels); therefore, it offers versatility in situations that require solutions suited to each requirement. Its widespread use also coincides with a suitable adaptation to the market due to its compatibility with other prefabricated products that follow a dry construction process, facilitating the disassembly and recycling of the products. It is important that it stays on the market due to the possibility of replacing it without causing obsolescence in the building.

The technical values confirm good thermal performance [33], with an insulating capacity up to three times better than that of a conventional enclosure, minimising energy consumption. In restoration, it would improve the function of the existing masonry enclosures, stone, bricks, etc., by lining with these products, increasing the insulating capacity of the whole resulting construction.

According to the study carried out by V. Tavares et al., in which the embodied, operational, and end-of-life energies have been compared, as well as the carbon emissions of products made with wood, these have the lowest impacts in all the categories, compared with the light steel structure or with conventional non-prefabricated solutions [31]. Given that it is possible to fabricate it using recycled materials, the waste generated is also reduced, nearly 60% in comparison to other traditional construction materials: reinforced concrete, ceramic, or metallic materials [34]. To this must be added the low effect of transport on the total impacts [35], as it is a product whose manufacture is widespread since it does not require high technology to produce it.

3.2.2. Industrialised Structural Steel

This section covers the structural profiles, bars and plates, and rolled steel parts, a wide range of elements that can be assembled in various ways using mechanical systems that enable their recovery, and therefore recycling, to a high percentage (90%).

To these elements is added extensive experience in the design of these construction and structural systems as evidenced in the examples selected and studied in the database of architectures referred to above in this article, which has also produced numerous patents for joining systems, anchors, and assemblies that increase the options for their use.

Its production system also enables it to be adjusted to any dimension without it requiring a modification in the production chain, i.e., a made-to-measure industrialised prefabrication. The current systems for controlling the dimensions through software and laser cutting increase the precision and expand the design possibilities of metal parts, regardless of the speed of production and their mechanical characteristics. On-demand production also enables the control of the dimensions and weight of the elements, aspects that are especially relevant for self-building.

In environmental terms, these structural steel products have also been analysed through an evaluation of their life cycle, above all considering the production stages of the product in a comprehensive way, from the extraction of the raw material needed to make it to the final deployment of the product for use in construction. As the study carried out by Pietro Renzulli et al. shows, the data vary not only between countries but also depend on the extraction and production systems of each factory. In any case, it is known that the steel production process is the most polluting part of its life cycle, due to the energy consumption required, to which we must add the waste it generates, although the slag produced by the manufacturing process can be reused as a fertilizer or for road surfacing [36].

Of the United Nations' 17 Sustainable Development Goals, it is important to guarantee the use of clean energy (goal 7), consumption patterns (goal 12), and climate action (goal 13), among others, for which controlling CO_2 emissions is required, goals that would also align with the European Union's Green Deal and the Paris Agreement. The production of these products with neutral steel would contribute to all of this.

3.2.3. Light Steel. Light Steel Framing

This system is the heir to the American *Balloon Frame* based on the construction of a network of elements that, due to the distance and deployment of their struts, take on a load-bearing function. Equally, there are numerous projects and construction works built with this system that give an example of its usefulness and the diversity that can be achieved.

Unlike the previous one, this system requires precise coordination between the elements and the other prefabricated products that complete the whole. In a certain way, as it is an open system in terms of components, its application is subject to the interdependence imposed by the framework of the profiles, but this can also produce an optimal result if the combination of these products is also based on environmental criteria. Some prototypes have been built and tested in accordance with this principle, such as the research carried out by Ornella Iuorio et al., which also compared the results with a wall built with traditional materials with the same thermal transmittance [37], with better results achieved by the prefabricated combination in almost all the categories in its LCA. Although each test shows that the results depend on the solution adopted in each project, there are mostly favourable conclusions in the construction systems combined with an LSF-based structural system, considering that the total restoration of the light structure is achieved at the end of its useful life. As these researchers indicate, the non-structural products of the ensemble can be replaced and reinserted into the life cycle. In addition to a better thermal behaviour, they would contribute to a greater implementation of the circular economy.

The distance between these profiles, which varies between 40 and 60 cm and reduces the dimensions of the elements to be used; the joining system between elements with recoverable systems; and its cold-rolled construction contribute to the lightness of the system, the ease of self-assembly, and its manageability.

4. Construction Period

4.1. Intervention Strategies in the El Rodezno Watermill

The 2030 Agenda and its Sustainable Development Goals, as well as the European challenge of the nearly zero energy buildings (NZEB), require the use of renewable energies: zero energy, in the operating systems of a building using renewable energies; zero emissions, using materials that do not emit substances that are harmful to the environment, are flexible and reusable or recyclable, and take into account the service life of the products used throughout the building's lifetime; and zero waste, in the use of construction solutions.

This approach was already included in the goals for *Horizon 2020* Cultural Heritage, Framework Programme, for the period 2014–2020 under the *Europe 2020* strategy, which is now in force in the new commitments that place the horizon in the year 2030 [38].

The goal consists of providing innovative solutions and knowledge, by means of strategies, methodologies, technologies, products, and adaptation and mitigation services, with a view to the conservation and management of the tangible cultural heritage of Europe that is exposed to risk due to climate change.

Of the 85 mills located, the scenarios are very diverse, and not all of them still have a sufficient wall structure to be recovered with light prefabricated systems, which would allow a habitability compatible with residential use. In fact, of the 37 mills that existed in the Sierra de Grazalema Natural Park, a region that is part of the Sierra de Cádiz, 13 of them still maintain a part of their construction that would make it possible to propose their restoration. Sixteen of them have already been converted into rural dwellings, without a clear intervention rationale: some have been completely replaced by a new-built dwelling, others with very extensive alterations that prevent their origin as a mill from being recognised, and in no case has an intervention been proposed based on sustainability principles. The remaining eight mills only present vestiges; some traces of their walls or part of the infrastructures that channelled the water to the mill remain [2] (pp. 507–508).

It is evident that any intervention in built heritage requires a deep formal, structural, and constructive study of its current state, so the strategies put forward here are merely suggestions that would require the subsequent verification and review of the proposed solutions. However, it must be considered that these mills are of a very common type, with one part devoted to the milling machinery that presents few variations in terms of construction and dimensions: a rectangular layout that ranges between 3.50 and 4.50 m, built with masonry supporting walls, using the materials from the area, with sand and lime mortar, usually around two feet thick. In less frequent cases, this base construction was extended with the miller's dwelling, giving rise to very variable combinations according to the family composition or the physical characteristics of the site, usually on a slope. The mills that included this house have been maintained better than those constructions intended exclusively for the production of flour, but they have also been the ones that have been transformed the most, precisely by including and maintaining residential use from the outset. In this research, the rehabilitation strategies for these rural architectures for residential use are applied only to the part that was devoted to milling because it is a replicable construction, with similar dimensions between different mills and, therefore, with a greater possibility of systematising the solutions that are proposed.

The "El Rodezno" watermill, located in Ubrique (Cádiz, Andalusia, Spain), is in a state of semi-ruin, retaining its structure of walls (Figures 7–9), so it lends itself well to a study of its rehabilitation with light prefabricated systems. This tests the adaptability of these industrialised materials and systems to the spatial and functional structures of existing buildings with or without heritage value. The symbiosis between tradition and innovation is also a field of exploration on the ability to comply with the general principles of a bioclimatic architecture, while the old walls that delimit the interior present an adequate thermal inertia. The new materials can provide an eco-efficient construction in addition to a rapid execution, reducing the energy consumption required during construction.

The mill has two floors, and its load-bearing walls are masonry constructed with stones from the area with a mortar of lime and sand, reaching a variable thickness between 60 and 65 cm. As in all mills, it had a gable roof, constructed with wooden elements and clad in Arabic tile: it is currently almost non-existent, as is the slab on the first floor (Figure 10). The floor does not have the regularity that it seems to show, which, apparently, could make the application of industrially prefabricated elements and systems difficult. The average dimensions of the bays are 4.30 and 3.30 m, dimensions that, however, would facilitate working with small, lightweight elements, avoiding the use of special machinery to assemble the prefabricated elements on the site.

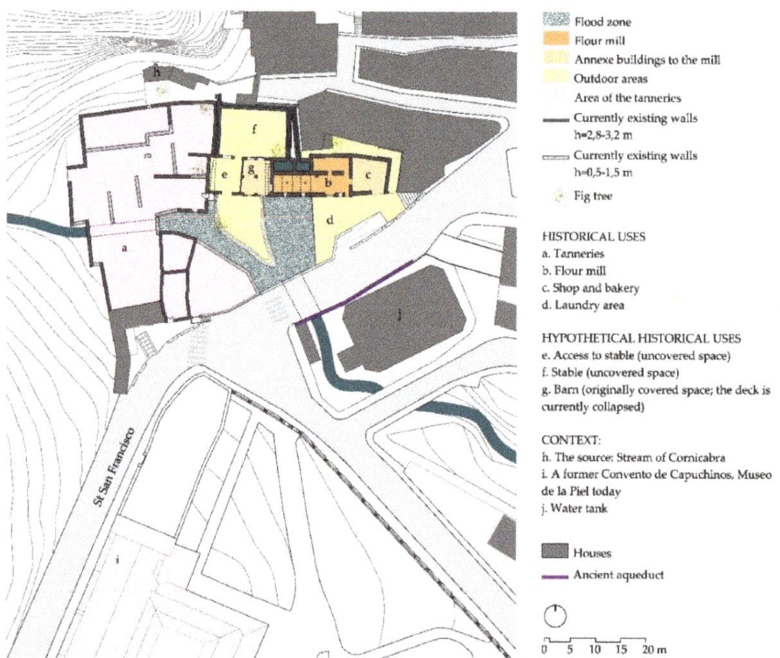

Figure 7. El Rodezno water mill. Current state. Ubrique, Cádiz, Spain. Plan of historic uses. Source: authors.

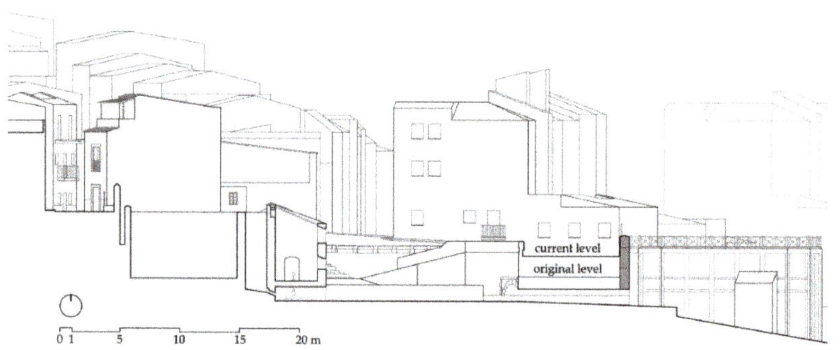

Figure 8. Cross section from Calle San Francisco. Source: authors.

Figure 9. Elevation from Calle San Francisco. Source: authors.

Figure 10. "El Rodezno" water mill, Ubrique, Cádiz, Spain. View from Calle San Francisco. Photography: authors, 2006.

The mill's current stability is due above all to the fact that it is a small construction and that the bonding between its different walls braces it as a whole. This way of building is of interest as it demonstrates the importance of the assembly between elements, necessary so that the overhauled structure as a whole can achieve stability and a light articulated structure that functions as a mechanism that can be inserted and that should form a solid structure with the existing walls, because, although these contribute by their own weight a certain structural stability, it is insufficient for the level of safety that is currently required in construction. This structural combination of walls and prefabricated structural elements would optimise the number and dimensions of the internal supports. An inspection of the pathologies observed in the masonry walls would also be necessary for their repair and/or consolidation and their stability or reinforcement, depending on whether what is required is an improvement in their base, in the wall itself, or in its crowning, following intervention criteria already analysed and studied in earthen walls in various conditions and of diverse natures [39,40].

There are various consolidation techniques for these types of walls, but their repair by composite reinforced coating applied on both their interior and exterior faces and in the corners guarantees stability and bonding at the weakest points without altering their external appearance. In addition, it also improves their ability to dissipate energy [41].

The improvement of the masonry walls can be an opportunity to rationalise their internal layout and thus achieve a better fit with prefabricated products. Something similar was done by Jean Prouvé in the Meudon houses (and in the Jules Ferry school) at the joint of the masonry wall with the sheet system whose roof protruded beyond the position of the wall. This solution avoided the encounter with a non-industrial or prefabricated construction element. Prouvé also built his houses on a reinforced concrete base executed on site, that is, a horizontal floor is needed for the assembly of the lightweight structural system. The reference dimension for the horizontal plane in the case of the "El Rodezno" mill would be the dimension +1.30 (Figure 9) where it is possible to resolve the accessibility from the outside.

Having said all this, the intervention avoids altering the external image without resulting in an excessive modification of the immediate environment, although according to article 20 of the Andalusian Historical Heritage Act of 2007 [42], any intervention on pre-existing architecture with heritage value must be acknowledged. In this regard, the internal intervention with the analysed materials respects this law and yet enables the recognition of the space occupied by the current semi-ruined status of the mill. The connection with the walls can be occasional given that the regular geometry of the new materials does not interfere with this approach, and, indeed, the slabs need not conform to the irregular contour they define, accepting the possibility of a physical and visible discontinuity between the walls and light structure.

The grid that must relate the different light elements must start from the dimensions of the industrialised products to be used. It is not a construction on a vacant surface where it is possible to start from a generic measurement, as Prouvé did with his 1 m $\times$ 1 m plot. The existence of a defined contour makes Gropius's strategy at the Weissenhof more appropriate, which started from the measurement of one of the construction elements (the door frames). In this way, several modulations can be proposed until the one that best fits between the outer walls of the mill is achieved. The smaller the grid dimension (as in Wright's houses), the greater the possibility of adjustment between construction systems as different as the artisanal and the industrial. The lower modulation extends the use of other possible prefabricated elements, which are also small and available on the market, to help resolve the encounters between different materials and different structural and construction systems. The panel that has been used repeatedly in the examples analysed in Section 3.1, and regardless of the dimension, shape, material used, or arrangement in the whole construction, indicates that it is an element that allows for adjustments and combinations and is the most suitable one for the basis of a modulation.

With this analysis, intervention strategies are proposed based on the use of prefabricated products that meet the criteria analysed in the previous sections, i.e.,

- Open systems, compatible with traditional construction.
- Frequently used products that guarantee availability in warehouses or rapid relocation.
- Products that correspond to production processes such as that of nearly zero energy schemes.
- Preference for local industrial production.
- Reduced number of materials.
- Dry assembly that enables the optimisation of implementation times, producing minimal or zero waste during the work.

- Systems that fabricate modular elements with small dimensions or that are easy to adapt to the construction site.
- Weight of prefabricated products limited to handling by two or three operators.
- Systems that enable the disassembly, reuse, and recycling of the prefabricated elements.

The systems and elements analysed in Section 3.2 may industrially produce linear products (beams and pillars), surface products (plates, boards, and panels that are self-load bearing or not), and spatial modules.

In the case of the mill, the use of spatial modules is not suitable due to the small dimensions, but the two first items were appropriate. The dimensions of the panels of the companies in the sector consulted (Thermochip for laminated wood panel and Egoin for cross-laminated wood panel) determine the structural modulation and the lengths of the linear elements. If we consider that the number of different materials to be used must be low, the options are wood or steel. In this case of restoration, wood seems the most suitable given the conditions of the environment in which the mill is located, its better environmental performance, and possible local production. As a result, to carry out the interior structural network, the following would be used:

- Laminated wood panels: *Thermochip* type. Dimensions: 550 mm × 2400 mm.
- Cross-laminated wood panels: *EGO_CLT* and *EGO_CLT MIX* panel type (with insulation). Usual dimensions (transport limit): 2400 × 10,000 mm.

A grid of wooden beams and pillars of 2400 mm × 1650 mm (Figure 11) sufficiently coincides with the dimensions of the panels chosen (Thermochip for the roof and EGO_CLT for the first-floor slab). A structure independent from the mill consists of 16 modules inserted between the pre-existing masonry walls. The connected modules would avoid the duplicity of pillars. To fill in the horizontal surfaces of the grid at "El Rodezno" (Figure 12), 48 Thermochip panels and 7 EGO_CLT panels of 2400 × 10,000 mm would be needed (each panel would be divided into three panels of 2400 mm × 3300 mm, filling in two modules).

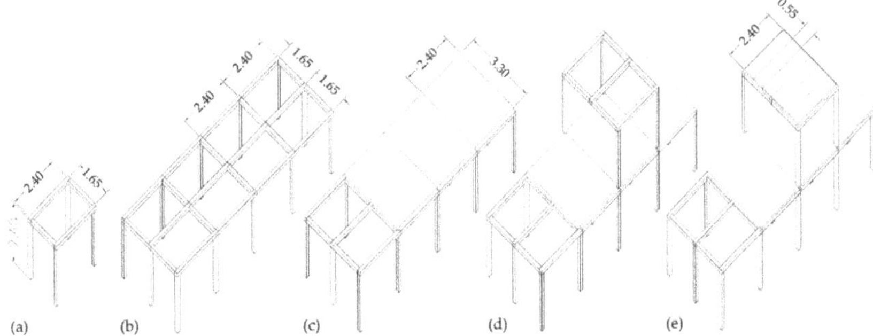

Figure 11. (**a**) Modular structure; (**b**) Addition of modules; (**c**) Placement of cross-laminated wood panel EGO_CLT panels; (**d**) Addition of modules at height; (**e**) Placement of Thermochip panels. Source: authors.

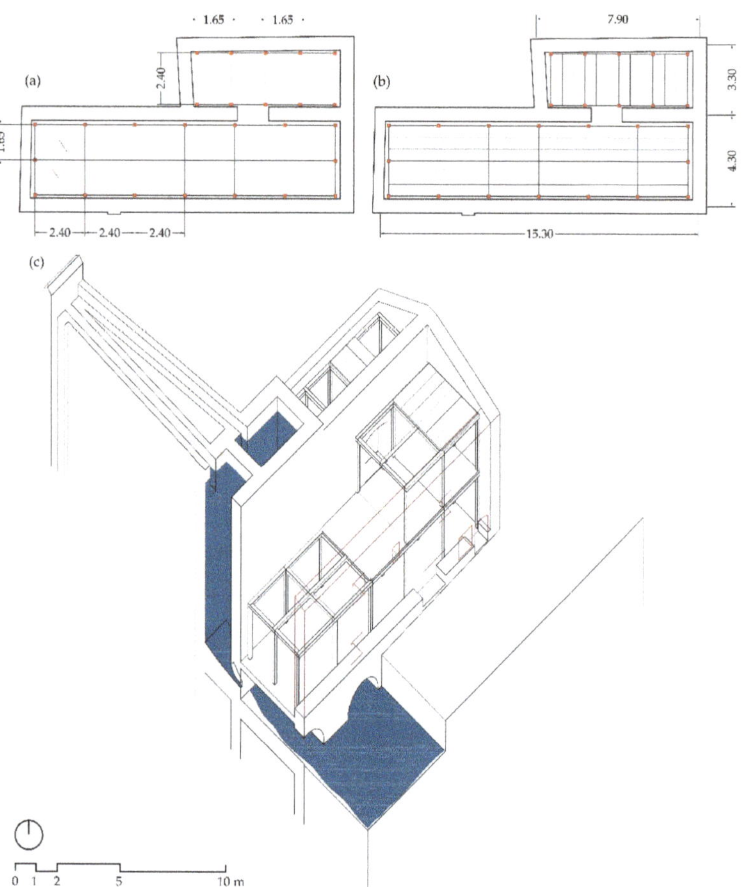

Figure 12. Insertion of the modular system at El Rodezno: (**a**) EGO_CLT panels; (**b**) Thermochip panels; (**c**) Floors and axonometry. Source: authors.

The columns and beams could also be made of industrialised structural steel, and the position of the columns in the contour of the mill could even be replaced by a framework of small profiles (light steel framing). In other words, the strategy of sizing from the panel not only allows the number of elements to be used to be optimised but is also highly compatible with the other two systems analysed in Section 3.2.

The small dimensions of the mill suggest a continuous and diaphanous interior with beams with a maximum dimension of 3300 mm. In this way, the pillars of the inserted lightweight structure are located next to the masonry walls, the slabs being resolved as a simple bay. The structural module delimited by beams and pillars is 2400 mm × 3300 mm in the main building and, in the smaller building, 3300 mm × 1650 mm, dimensions that are the result of subdivisions of the measurements of the wooden panels used. The final dimensions applied are suitable for the residential scheme. Thus, in a 2400 mm × 3300 mm (7.92 m^2) module, eliminating the slab panels, the staircase, and even a hoisting platform can be included to guarantee accessibility to all the floors of the house. The rest of the surface, five modules of 2400 mm × 3300 mm (39.60 m^2) to which would be added the surface of the second building, four modules of 3300mm × 1650 mm (21.78 m^2), allows different housing schemes to be developed. It is not a question of defining a solution or proposing various types of distributions but of facilitating a flexible and habitable space where the principles of Richard Llewelyn-Davies and John Weeks of the "socio-technological environments" are observed: the prefabricated light structure and its materials must be visible, and the designed spatial structure must be compatible with the functional adaptations, modifications, alterations, or divisions that its inhabitants wish to carry out over time. This is a frequent characteristic in the residential architectures that were included in the selective database of 78 works and projects in which continuous space and division with temporary or movable elements (furniture), built with different prefabricated materials and with very diverse shapes, forms, and volumes, are prioritised: Buckminster Fuller's Dymaxion House (1925); the Aluminaire House, by Albert Frey in collaboration with A. Lawrence Kocher (1930); the Canvas House by Albert Frey (1934); the Case Study House of Charles and Ray Eames (1949); the Zip-Up Enclosures of Team 4 (1968–1971); Shigeru Ban's Naked House (2000); the Maison Kerema by Lacaton & Vassal (2003–2005); or the Casa Garoza 10.1 by Juan Herreros (2010), among others.

Professor J. Terrados verifies that the progressive decrease in the size of houses is related to the way in which users assess the house's essential functions, prioritising the efficiency or the comfort of both the space and its construction [43] (p. 147), a view that fits with the approach of not carrying out a conventional distribution of the house.

It should be noted that the reduction of the surface areas also entails a proportional adjustment of the height of the interior space. Spatial continuity allows for many constructive resources, such as the one devised by Rudolf Schindler for the Schindler-Chace (1922), using a portico (*Schindler Frame*) that limited some elements to the height of 1.90 m so as not to interrupt the continuity of space, while the height of the house approached 244 mm, the standard dimension of the panels as sold on the market.

4.2. Construction of the Roof

One of the most important enclosing elements to be resolved is the roof. With laminated wood panels (Thermochip-type), they would maintain the same distribution as the wood panels used in the floors, optimising the number of construction elements. Its exterior location leads to the design of a coating that guarantees waterproofing and resolution with vertical walls. For this exterior cladding, using the prefabricated products studied in Section 3.2, the conditions of dry assembly, restoration of construction elements, lightness, and limitation of dimensions must be maintained to guarantee self-assembly and facilitate the transport of the products, and a simple solution that can be carried out by any kind of worker must be proposed.

Continuing the strategies analysed in the "El Rodezno" mill, the proposed solution is based on the cladding of a roof built in subsidised social housing where the methodology and construction criteria are those of the same research. In addition to the desired transfer between applied research and real construction, this solution is linked to the necessary rehabilitation of single-family homes that entail a functional and constructive improvement to increase comfort by reducing energy consumption and thus help to mitigate climate change.

The final finish by means of cold-formed light steel sheets makes it necessary to resize the external modulation based on the product's manufacturing limitations, also considering the design of the side edges of the sheet so that the union between elements is by overlapping and setting of materials, avoiding excessive handling, manufacturing, and the subsequent assembly processes [44] (Figure 13).

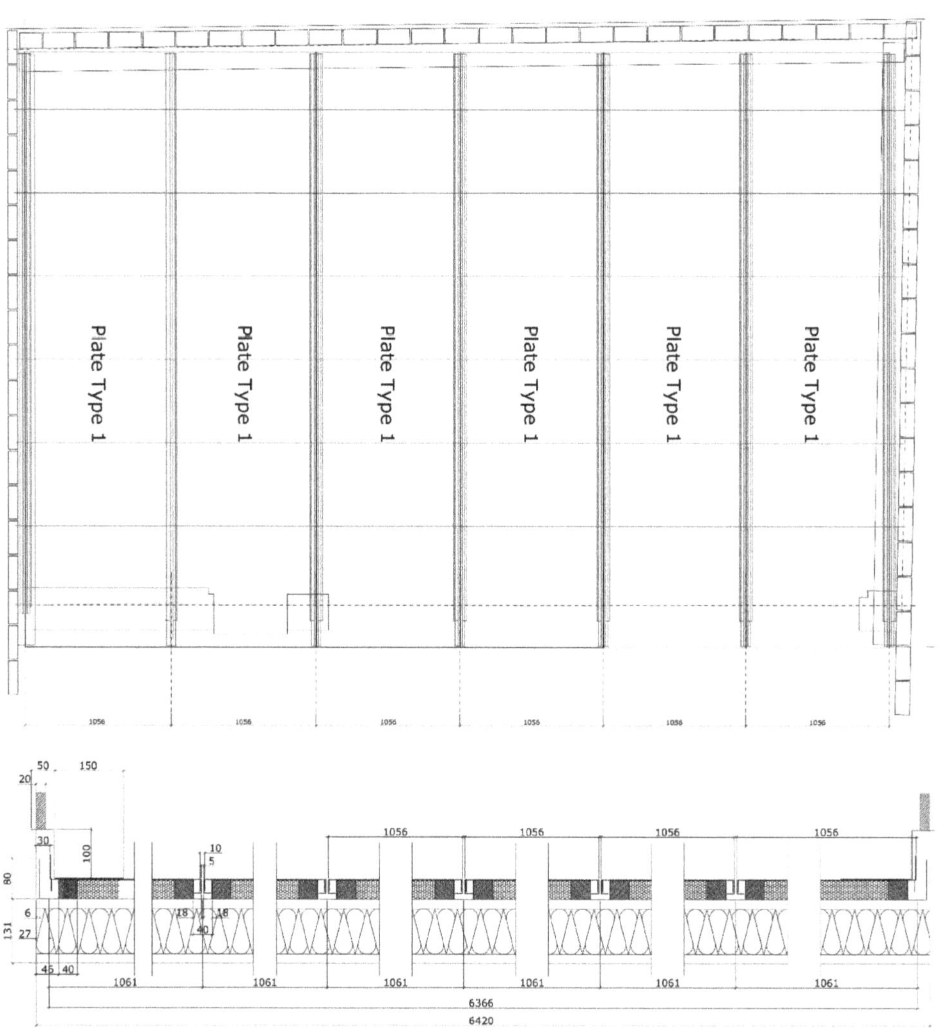

Figure 13. Roof plan and cross-section.: modulation of the guides and type of covering sheets.

Based on frequently used materials that are easily available on the market, such as pine wood slats and different profiles and light cold-formed steel sheets, a solution is designed where the union between materials is mechanical, by means of rivets, self-drilling screws, or guidelines where the outer sheets are fitted. These guidelines formed by the wooden slats and the cold-formed profiles are fixed to the Thermochip panels, and the gap that remains between the guidelines is filled with a light but rigid insulation so that the maintenance of the roof can be facilitated without having to pass through it, which would cause dents and deformations of the exterior cladding (Figure 14). The shape and dimensions of the steel sheets are digitally controlled, laser cut, and mechanically folded. Its ease of construction and the use of common materials that can be easily sourced make the generalisation of this solution viable.

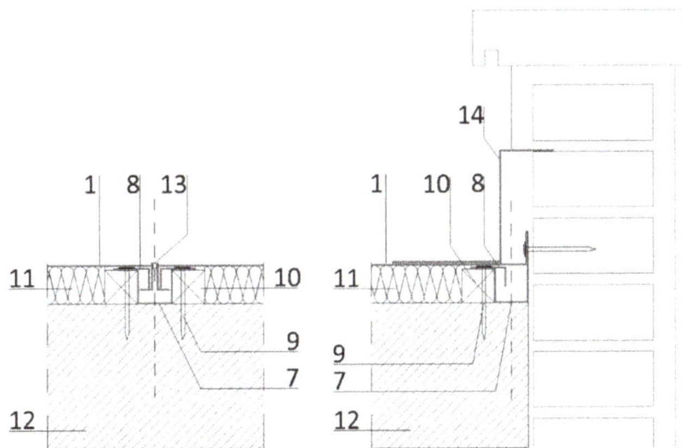

Figure 14. Detail. (**1**) Lacquered metal sheet thickness: 0.06 mm. Digital cut; (**7**) Profile W galvanized steel 40 × 40 × 40 mm; (**8**) Profile L galvanized steel 40 × 40 mm. Riveted joint with W profile; (**9**) Guide fixing screw to cover panel; (**10**) Pine wood strip 40 × 40 mm; (**11**) Lightweight insulating filler between guides. Thickness 40 mm; (**12**) Termochip slab panel; (**13**) Elastic sealing of the joints between sheets; (**14**) Lacquered sheet metal part. Thickness: 0.06 mm. For termination at the factory edge, sealed with the base plate.

The guidelines are hidden, achieving an external surface that is flat and consistent (Figure 15). This characteristic facilitates the solution of the front of the slab with the same material and types of profiles, as it is a continuous enclosure that can also continue along the outer side of the wall leaving the face of the slab concealed and insulated, where the joint occurs with the wall constructed with conventional resources. The solution can even be extended to the windows, also covering the darkening system, an area usually with significant energy losses (Figure 16).

Figure 15. Side edge and centre guides. Assembly process of the coating sheets. Factory-edged finish. Photographs: authors.

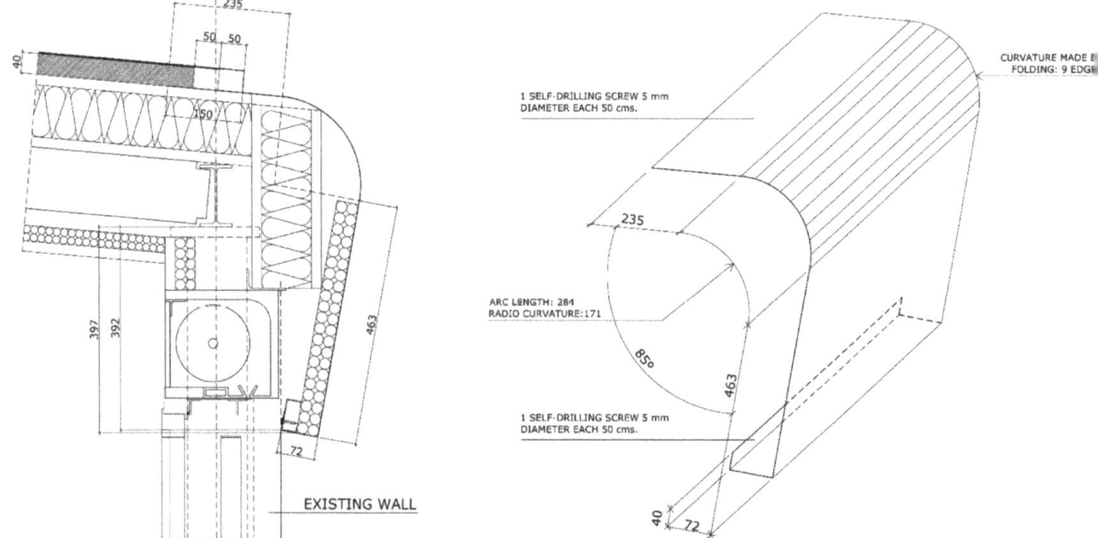

Figure 16. External casing of the front of the slabs and the darkening system. Source: authors.

5. Discussion of Results

The research methodology used, the intervention strategies in the "El Rodezno" mill, and the built roof model that, adjusted to the criteria of this research, are transferable to the studied mill architecture reaffirm the existing relationship between theory and practice that is a form of progress for architecture, thanks to the empirical value [45] provided by architectural designs and the critique that derives from architectural practice accumulated over decades. For this reason, the knowledge provided by the projects and the built works constitute a transfer between theory and practice that does not follow a chronological sequence, but rather constitutes a diachronic process of progress and improvement between theory and practice. From the built work, new solutions can be transferred to the strategies of a project, and vice versa, accepting that research in architecture is an open-ended process.

The research reveals the importance of small-scale renovations that need to be carried out with limited financial resources. This affects many homes, also considering that 80% of the housing stock that will exist in Europe in 2050 has already been built, necessitating both current processes for the comprehensive rehabilitation of buildings and urban environments, based on interventions that use non-polluting solutions and products [46].

The strategies proposed and the construction solutions implemented could be described as precision interventions in architectures with or without apparent value. These architectures are also transformed into an extensive field laboratory whose solutions lead to patents that can be marketed and, indeed, applied to constructions on a bigger scale.

We advocate compatibility between different construction techniques and technologies, even ones that would appear to be incompatible, such as industrial design and artisanal construction. This combination of opposing techniques would be along the lines of the "socio-technological environments" advocated by Richard Llewelyn-Davies and John Weeks. In this regard, the Frank Lloyd Wright buildings mentioned in Section 3.1 that were built in the early 20th century do not feature an industrialised exterior. Wood, in the different forms in which it is produced, prefabricated, industrialised, and widely marketed, conveys the idea of an artisanal construction linked to the carpenter's trade, and some researchers associate this circumstance with its greater acceptance [43] (p. 209). It also turns out that the environmental impact that derives from the evaluation of the life cycles of products made from wood are the lowest of those analysed in the literature and which have been mentioned in Section 3.2.

The dimensional and weight adjustment, which is a condition that characterises light prefabrication, is another factor that influences the acceptance of these construction systems, insofar as it allows the user to self-build and, in the process, to create a sense of identity and belonging in the home-dweller; "do-it-yourself" is the pleasure of doing it [47]. The sense of identity, of belonging of the dweller, is opposed to the idea of temporariness identified by A. Pike that is shown today, practically impossible for single-family homes, and more credible for those sectors of society that, due to their economic or employment situation, are more changeable, in addition to the humanitarian emergencies that occur frequently today for various reasons. Both situations convey a market need.

Modulating the house based on the basic products marketed by the construction industry results in a greater variability of interchangeable prefabricated components, enhancing the circular economy. This strategy is an old aspiration of prefabricated construction applied to housing, and it was already put forward by Walter Gropius in 1910 [48]. However, it did not succeed in gaining a foothold in the construction industry, which, in any case, was starting to focus its interest on reinforced concrete as a new building and structural system.

Prefabrication has generally been associated with innovation, creating new situations or solving problems with new forms. In housing, it has almost always generated interesting ideas, regardless of the different currents of thought with which it was associated or of certain circumstances that, during the 20th century (the lack of housing after the Second World War) or the beginning of the 21st century (natural disasters or migratory flows), have increased its demand and, with it, the interest of architects and the opening of numerous lines of research that address its study, optimisation, and consequences from the perspectives of various disciplines.

Regarding the roof solution, it should be noted that its implementation cost was EUR 18.49/m^2, including the auxiliary assembly elements, and is similar to the cost of covering a roof with traditional materials, which, under normal conditions, is around EUR 20.00/m^2 (valuation according to the Regional Government of Andalusia's construction price database, 2017). However, the labour yield of this lightweight cover is much better compared to a traditional one, taking 50% of the time it takes to complete a roof with traditional materials (compared to the time that was used in the implementation of the planned solution with the labour yields according to the construction price database of the Regional Government of Andalusia, 2017). It is not about defending the built solution but the validity of a research strategy aimed at looking for low-impact solutions combining materials that have been analysed in Section 3.2. To this is added profitability, thanks to its simplicity and the implementation by workers who do not have high qualifications. The economic competitiveness of construction without entailing a reduction in safety or comfort conditions is a factor worth considering for the acceptance by users of construction with prefabricated elements.

Finally, it should be noted that, together with evaluations of materials according to their life cycles and minimising environmental impacts, local climatic conditions are important, and, based on these conditions, incorporating passive solutions that are grounded in the experiences of the craftsmanship and knowledge that have proven to be useful in reducing energy consumption is equally as important. As Milagrosa Borrallo et al. state, "this allows us to decide on the type of actions in the field of architectural design, construction and management to include in the [evaluation] tool, based on the potential development of bioclimatic strategies in specific passive design systems" [49], a criterion which is recognisable in that the prefabricated light structure was implemented inside the masonry walls of the "El Rodezno" water mill.

6. Conclusions

Prefabrication can reduce impacts, the consumption of materials, and the generation of waste, promoting circularity within the construction sector. There are studies that quantify the reduction of environmental impact between 5% and 40%, in addition to economic savings around 30%, as opposed to traditional construction.

Assessments of the life cycles of industrialised products, which analyse and quantify environmental impacts, are essential studies to mitigate climate change and make necessary knowledge available to companies and architects. However, the intervention strategy in "El Rodezno" also demonstrates the importance of design, whereby the effects of climate change can be reduced by the choice and optimised fit of the products used. This factor, together with others of a social or cultural nature, should be considered in life cycle evaluation studies, so architectural projects need to follow a research methodology based on previous experiences on which they base their proposals.

In the rehabilitation of minor architectures located in rural environments, it is helpful to exploit those parts of the existing building made with manual means and artisanal techniques that can help reduce the energy consumption of the whole building. For the "El Rodezno" mill, it was also compatible with the objectives for the Cultural Heritage of Horizon 2020, now extended to 2030, and with Article 20 of the Andalusian Historical Heritage Act of 2007.

Wood-based prefabricated systems are those with the least environmental impact. Although they entail a greater increase in labour costs compared to other industrialised products, the user usually associates them with the manual and artisanal work of the carpenter. Solutions that hybridise prefabricated and traditional building systems, or use industrialised prefabricated products based on the use of wood, help to create a sense of belonging in the user and thus a better inclusion of industrialised prefabricated products in home renovation. Achieving a high degree of acceptance helps to change the mentality of users with regard to the use of this type of product and, in this way, to increase their demand, a factor that determines the profitability that justifies these products being available on the market for a sufficient time.

For the rehabilitation of minor architectures that can be adapted to residential use or the rehabilitation of single-family dwellings, the limitations of dimensions and weight increase the possibility of assembling, enlarging, piling, dismantling, or recycling all the structural and construction elements, with the resulting reduction in environmental impacts and a better evaluation of the LCAs. The use of products of these types that are usually available both in rural and urban environments boosts local employment and the circular economy, criteria that are compatible with the Sustainable Development Goals.

Author Contributions: Conceptualization, methodology, analysis, investigation, writing preparation, and visualisation: A.R.-C., R.M.A.-A. and G.R.-L. (33%/33%/33%). All authors have read and agreed to the published version of the manuscript.

Funding: This research was funded by two projects: "Grants for Specific Knowledge Transfer Actions (Mode A1). Year 2017. Project: 2020/00000423" and "Supports to the Research Group HUM-632: *PROJECT, PROGRESS, ARCHITECTURE*".

Conflicts of Interest: The authors declare no conflict of interest.

References

1. Olmedo Granados, F. La arquitectura agraria en Andalucía. In *Cortijos, Haciendas y Lagares en Andalucía De Cádiz*, 3rd ed.; Gil Pérez, M.D., Ed.; Junta de Andalucía, Consejería de Obras Públicas y Vivienda: Seville, Spain, 2007; pp. 13–19.
2. Rivero-Lamela, G. La Arquitectura de un Territorio Productivo: Los Molinos Hidráulicos de la Sierra De Cádiz. Ph.D. Thesis, Universidad de Sevilla, Sevilla, Spain, 2021.
3. Rivero-Lamela, G.; Ramos-Carranza, A. The Watermills of the Sierra de Cádiz (Spain). A Traditional Open Water Re-circulation System. *SPOOL* **2021**, *7*, 39–58. [CrossRef]

4.   Anasagasti, T.de. Discursos Leídos ante la Real Academia de Bellas Artes de San Fernando, en la Recepción Pública de Don Teodoro de Anasagasti. 1929. Available online: https://bibliotecadigital.jcyl.es/es/consulta/registro.cmd?id=1744 (accessed on 13 February 2021).

5.   Castellano Pulido, F.J. Bancales habitados: De la reutilización en la arquitectura tradicional al trabajo con el tiempo de César Manrique y Souto de Moura. *Proy. Prog. Arquit.* **2019**, *21*, 34–51. [CrossRef]

6.   Vilches, A.; García-Martínez, A.; Sánchez-Montañes, B. Life cycle assessment (LCA) of building refurbishment: A literature review. *Energy Build.* **2017**, *135*, 287. [CrossRef]

7.   European Commission. *Draft Horizon 2020 Work Programme 2014–2015 in the Area of Secure, Clean and Efficient Energy*; European Commission: Strasbourg, France, 2015.

8.   Aalto, A. Una ciudad americana en Finlandia. In *Alvar Aalto: De Palabra y Por; Escrito*; Schildt, G., Kapanen, E., García Ríos, I., Casabella, N., Eds.; El Croquis: Madrid, Spain, 2000; pp. 173–186.

9.   See Muñoz Jiménez, M.T. Verano de 1948. Buckminster Fuller en Black Mountain College. La Arquitectura como acontecimiento. *Proy. Prog. Arquit.* **2010**, *3*, 110–117. [CrossRef]

10.   Benevolo, L.; Melograni, C.; Giura Longo, T. *La Proyectación de la Ciudad Moderna*; Editorial Gustavo Gili, S.A: Barcelona, Spain, 2000; p. 9.

11.   Archer, B. The Nature of Research. *Co-Des. Interdiscip. J. Des.* **1995**, 6–13.

12.   Frayling, C.H. Research in Art and Design. *R. Coll. Art Res. Pap.* **1993**, *1*, 1–5.

13.   Ramos-Carranza, A. De la investigación, la enseñanza y el aprendizaje experimental de la Arquitectura. *Proy. Prog. Arquit.* **2015**, *12*, 14–17. [CrossRef]

14.   Torres Cueco, J. El proyecto de arquitectura como investigación académica. Una aproximación crítica. In *Colección Investigaciones IdPA_03*; RU: Sevilla, Spain, 2017; pp. 13–28.

15.   Baxandall, M. *Patterns of Intention: On the Historical Explanation of Pictures*; Yale University: New Haven, CT, USA, 1985.

16.   Futagawa, Y. (Ed.) *Brooks Pfeiffer, B. (tex). Frank Lloyd Wright. Monograph 1902–1906*; A.D.A. Edita: Tokyo, Japan, 1987; p. 42.

17.   Ramos-Carranza, A.; Bueno-Pozo, V. Design methodology for a school prototype: JeanProuvé's Jules Ferry School Group in Dieulouard, France, 1952–1953. *J. Archit.* **2019**, *24*, 385–407. [CrossRef]

18.   Sulzer, P. *Jean Prouvé: Oeuvre Complète = Complete Works. 4,s 1954–1984*; Birkhäuser: Basel, Switzerland, 2005.

19.   Cinqualbre, O.; Occeli, M.; Billé, R.; Archieri, J.-F. *El Universo De Jean Prouvé. Arquitectura/Industria/Mobiliario*; Fundación "la Caixa": Barcelona, Spain, 2021.

20.   Schulze-Fielitz, E. *Metasprache Des Raums = Metalanguage of Space*; Springer: Wien, Austria, 2010.

21.   Bezos Alonso, J.L. Dispositivos Abiertos: Habitares Open Sources. Ph.D. Thesis, University of Seville, Sevilla, Spain, 2017.

22.   Stadtbausysteme. Disponible en. Available online: http://www.nrw-architekturdatenbank.tu-dortmund.de/objekte_gross.php?gid=2222 (accessed on 1 June 2021).

23.   Mac Quaid, M. *Shigeru Ban*; Phaidon: New York, NY, USA, 2003; pp. 202–207.

24.   Salas, J. De los sistemas de prefabricación cerrada a la industrialización sutil de la edificación: Algunas claves del cambio tecnológico. *Inf. Constr.* **2008**, *60*, 19–34. [CrossRef]

25.   Staib, G.; Dörrhöfer, A.; Rosenthal, M. *Components and Systems: Modular Construction: Design Structure New Technologies*; Detail: Basel, Switzerland, 2008.

26.   Wadel, G.; Avellaneda, J.; Cuchí, A. La sostenibilidad en la arquitectura industrializada: Cerrando el ciclo de los materiales. *Inf. Constr.* **2010**, *62*, 37–51. [CrossRef]

27.   Ruiz-Larrea, C.; Prieto, E.; Gómez, A. Arquitectura, Industria y Sostenibilidad. *Inf. Constr.* **2008**, *60*, 35–45. [CrossRef]

28.   Llewelyn–Davies, R.; Weeks, J. Endless Architecture. *Archit. Assoc. J.* **1951**, 106–112.

29.   Pike, A. Deficiencias de la construcción industrializada aplicada a los problemas de la vivienda. *Cuad. Summa Nueva–Visión* **1968**, *13*, 3–7.

30.   Collinge, W.O.; Landis, A.E.; Jones, A.K.; Schaefer, L.A.; Bilec, M.M. Dynamic lifecycle assessment: Framework and application to an institutional building. *Int. J. Life Cycle Assess.* **2013**, *18*, 538–552. [CrossRef]

31.   Tavares, V.; Soares, N.; Raposo, N.; Marques, P.; Freire, F. Prefabricated versus conventional construction: Comparing life-cycle impacts of alternative structural materials. *J. Build. Eng.* **2021**, *41*, 2. [CrossRef]

32.   Cervantes Torre-Marín, G. Ecología industrial: Innovación y desarrollo sostenible en sistemas industriales. *Rev. Int. Sostenibilidad Tecnol. Humanismo* **2011**, *6*, 59–78.

33.   According to the Catalogue of Construction Elements of the CTE (Technical Building Code—Spain), the Thermal Conductivity of a Light Conifer is $\lambda = 0.13$ W/mK with a Thermal Transmittance of the Order of $1.58 < 0.59$ W/m^{2}K for Thicknesses from 60 to 200 mm. Available online: http://cte-web.iccl.es/materiales.php (accessed on 24 April 2020).

34.   Arrizabalaga Echeberría, J.J.; Rodríguez García, J. (Eds.) *Investigación Para Interpretar las Claves de los Diferentes Sistemas Constructivos Industrializables y su Posible Aplicación en la Vivienda de Protección Pública en el Ámbito de la CAPV/IDEFABRIK, Equipo de Investigación*; Departamento de Arquitectura, Universidad del País Vasco: San Sebastián, Spain, 2012; Available online: http://37.187.25.28/documentos/actualidad/doc_980.pdf (accessed on 13 January 2019).

35.   Vidal, R.; Sánchez-Pantoja, N.; Martínez, G. Life cycle assessment of a residential building with cross-laminated timber structure in Granada-Spain. *Inf. Constr.* **2019**, *71*, e289. [CrossRef]

36. Notarnicola, B.; Tassielli, G.; Renzulli, P.A. Industrial symbiosis in the Taranto industrial district: Current level, constraints and potential new synergies. *J. Clean. Prod.* **2016**, *122*, 133–143. [CrossRef]

37. Iuorio, O.; Napolano, L.; Fiorino, L.; Landolfo, R. The environmental impacts of an innovative modular lightweight steel system: The Elissa case. *J. Clean. Prod.* **2019**, *238*, 1–16. [CrossRef]

38. Horizonte. 2020. Available online: http://eshorizonte2020.cdti.es/index.asp?MP=87&MS=718&MN=2&TR=C&IDR=2043 (accessed on 18 February 2019).

39. Mileto, C.; Vegas López-Manzanares, F.; García-Soriano, L. La restauración de la tapia monumental: Pasado, presente y futuro. *Inf. Constr.* **2017**, *69*, 1–12. [CrossRef]

40. Bruno Quelhas, B.; Cantini, L.; Miranda Guedes, J.; Da Porto, F.; Almeida, C. Characterization and Reinforcement of Stone Masonry Walls. In *Structural Rehabilitation of Old Buildings*; Springer: Berlin/Heidelberg, Germany, 2013; pp. 131–155. [CrossRef]

41. Tomaževič, M.; Gams, M.; Berset, T.H. Strengthening of stone masonry walls with composite reinforced coating. *Bull. Earthq. Eng.* **2015**, *13*, 2003–2027. [CrossRef]

42. España, Ley 14/2007, de 26 de noviembre, del Patrimonio Histórico de Andalucía. Boletín Oficial del Estado, núm. 38, de 13 de febrero de 2008. Referencia BOE-A-2008-2494. 2007. Available online: https://www.boe.es/buscar/pdf/2008/BOE-A-2008-2494-consolidado.pdf (accessed on 2 December 2019).

43. Terrados, J. *Prefabricación Ligera De Viviendas. NUEVAS Premisas*; IUACC: Sevilla, Spain, 2012.

44. Ramos-Carranza, A.; Añón-Abajas, R.M. Patent Title: Light Covering For Construction Roofs. Application Number P201830264, 18 March 2019. Patent title with substantive examination. Publication No.: ES 2 725 129 B2 PCT/ES2019/070177. With Support from the Plan Propio de Investigación de la Universidad de Sevilla. Available online: https://www.sumobrain.com/patents/wipo/Light-covering-construction-roofs/WO2019180289A1.html (accessed on 15 February 2021).

45. Capitel, A. La arquitectura como modo de entender el mundo. Notas de un profesor veterano. *Proy. Prog. Arquit.* **2015**, *12*, 20. [CrossRef]

46. Moyano-Mercader, P.; Esquivias, P.M. Decarbonization and Circular Economy in the Sustainable Development and Renovation of Buildings and Neighbourhoods. *Sustainability* **2020**, *12*, 7914. [CrossRef]

47. Nash, G. *Do-it-Yourself. The Complete Handbook*; Sterling Publishing: New York, NY, USA, 1995; p. 8.

48. Herbert, G. *The Dream of the Factory-Made House. Walter Gropius and Konrad Wachsman*; The MIT Press: Cambridge, MA, USA, 1984; p. 33.

49. Borrallo-Jiménez, M.; Lopez de Asiain, M.; Herrera-Limones, R.; Lumbreras Arcos, M. Towards a Circular Economy for the City of Seville: The Method for Developing a Guide for a More Sustainable Architecture and Urbanism (GAUS). *Sustainability* **2020**, *12*, 7421. [CrossRef]

Article

An Inclusive Model for Assessing Age-Friendly Urban Environments in Vulnerable Areas

Raquel Agost-Felip [1,2], María José Ruá [2,3,*] and Fatiha Kouidmi [1]

1 Departament of Developmental, Educational and Social Psychology and Methodology, Universitat Jaume I, 12071 Castelló de la Plana, Spain; ragost@uji.es (R.A.-F.); al243181@uji.es (F.K.)
2 Local Development Interuniversity Institute (IIDL), Universitat Jaume I, 12071 Castelló de la Plana, Spain
3 Departament of Mechanical Engineering and Construction, Universitat Jaume I, 12071 Castelló de la Plana, Spain
* Correspondence: rua@uji.es

Abstract: Population aging is becoming a major challenge in many countries. This paper deals with the elderly's specific needs in the public open space as it can play a significant role in their social inclusion and could be especially relevant in deprived areas. The main goal is to build a model to evaluate the vulnerability of the public space by focusing on the elderly's needs, using indicators. A previous analysis of the scientific and policy-oriented literature and of the technical standards and regulations linked with accessibility and social aspects that affect the elderly in urban areas was performed to identify the main dimensions for evaluation. The interjudge agreement technique was applied to validate the indicators with a panel of experts in technical and social disciplines. The model was applied to a vulnerable area in Castellón (East Spain), based on indicators adapted to the specific context features. The agreement level reached by experts was used to weight the indicators. The application of the model permitted the vulnerability in the suggested dimensions to be estimated and a global integrated index of vulnerability in the area to be calculated. It could assist in urban planning decision making toward age-friendly and, therefore, inclusive cities.

Keywords: neighborhood regeneration; urban realm; accessibility; social inclusion; active aging; social services

Citation: Agost-Felip, R.; Ruá, M.J.; Kouidmi, F. An Inclusive Model for Assessing Age-Friendly Urban Environments in Vulnerable Areas. *Sustainability* **2021**, *13*, 8352. https://doi.org/10.3390/su13158352

Academic Editor: Pilar Mercader-Moyano

Received: 2 July 2021
Accepted: 20 July 2021
Published: 27 July 2021

Publisher's Note: MDPI stays neutral with regard to jurisdictional claims in published maps and institutional affiliations.

Copyright: © 2021 by the authors. Licensee MDPI, Basel, Switzerland. This article is an open access article distributed under the terms and conditions of the Creative Commons Attribution (CC BY) license (https://creativecommons.org/licenses/by/4.0/).

1. Introduction

According to the International Longevity Forum, in 2018 there were more people older than 65 years than children under the age of 5 for the first time in history. Besides, and predictably, between 1950 and 2050, people aged over 80 will increase from 14 to 379 million. Consequently, the elderly's needs are becoming progressively more relevant. Many countries face major challenges to ensure that their health and social systems are ready to make the most of this demographic shift. The World Health Organization (WHO) has been warning about aging since the 1980s. In 1982, Vienna held the first World Assembly on Aging, where countries adopted the International Plan of Action with a variety of health and nutrition, employment and income security, education, social welfare housing, and environment initiatives. It was continued in 2002 at the Madrid Second World Assembly on Aging, where the key challenge was "building a society for all ages".

According to European Union (EU) data, in 2060, about 30% of people will be aged older than 65, and 12% will be 80 years or older. Regarding the urban context, governments, authorities, politicians, and economists must change their approach in relation to cities' development and management, especially public spaces [1].

In line with sustainable development goals (SDG) (UNO, 2015) and the New Urban Agenda (2016), architecture and urbanism should ensure people-centered and inclusive built environments. It is an acceptable notion that senior citizens very often encounter serious problems caused by insufficient mobility and less socialization than other inhabitants.

Therefore, accessibility to urban spaces is crucial for welfare in urban communities [2], and the elderly can be considered a vulnerable population.

This paper deals with the elderly's specific needs in the public open spaces of built environments given increasing population aging and the exclusion and isolation from society that this population may suffer due to deficiencies of built environments, which may be particularly serious in deprived areas. The main goal is to build a model to evaluate the vulnerability of public spaces by focusing on old people and using key influential aspects in urban spaces to promote their inclusion in city life, particularly in deprived neighborhoods where residential and social vulnerabilities usually concur.

To this end, our analysis is twofold: on the one hand, the technical perspective is examined, especially those factors linked mainly with accessibility and livability in the public space; on the other hand, the social perspective that considers the actual demographic situation and the specific needs and available services for the elderly in the urban context. The practical implementation of measures to improve the elderly's welfare needs to acquire profound knowledge of their specific needs and the suitability of today's existing instruments to ensure their quality of life. For vulnerable populations, the Social Welfare Services provided by Administrations are vital because they have first-hand information [3,4]. As a novelty, this work analyses and incorporates social indicators from Social Welfare Services' potential information by considering that vulnerable areas should be prioritized to undertake urban interventions that increase social inclusion. The first point of view is linked with the physical context, which is the urban fabric, and the second one is connected to the co-existing social fabric. The combination of both perspectives is a must in a people-centered urban model.

A previous theoretical framework was constructed based on the scientific and policy-oriented literature and also on the technical standards and regulations connected to accessibility and the social aspects that affect the elderly in urban areas. From this review, the main dimensions to build up a standardizable evaluation model were proposed. Then, the evaluation model was applied to a previously identified vulnerable area in the city of Castellón (East Spain). The model was based on indicators, which were selected after considering the technical and social perspectives and were adapted to the specific context's features. To validate the indicators, a panel of experts in technical and social disciplines selected the appropriate indicators in the study area to confer them proper weights in the vulnerability evaluation The local application was twofold: on the one hand, the accurate diagnosis and practical implementation required a microscale analysis in order to detect the actual needs of the elderly people in the area and to design ad-hoc solutions. On the other hand, the model was intended for deprived areas, and the city's recent Urban Land Plan identified the vulnerable neighborhoods. It would be possible to extrapolate the dimensions of analysis proposed after the desk review, in Section 3, to other cities, although an analysis of the local features should be undertaken thereafter in order to select appropriate indicators for the suburban area. The selection of the specific neighborhood in the city was made considering the highest aging ratio, but it would also be applicable to other areas in the city.

2. Materials and Methods

Work Stages and Structure

Figure 1 presents the stages followed in this work, together with the qualitative and quantitative methodologies and the main results obtained from each stage.

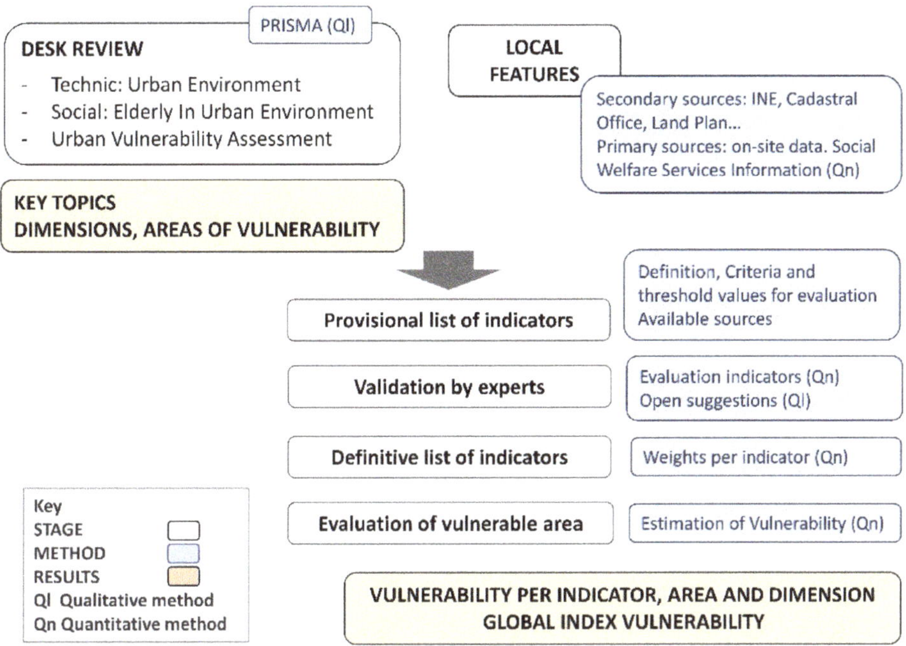

Figure 1. Stages, methodology, and results.

A theoretical framework was built from the analysis of the literature by identifying the key factors and requirements to obtain a model to evaluate inclusiveness in open spaces in urban environments by focusing on senior citizens' needs. The main key topics, identified from the literature review, are presented in Section 3. Sections 3.1 and 3.2 after considering the technical view and the social view. Key factors, such as accessibility in built environments and technical requirements, were examined together with the analyses of the elderly's needs, including the WHO's age-friendly city (AFC) concept and social sustainability from SDG. In addition, Section 3.2.3 introduces the relevance of the implementation of policies to materialize the inclusion of the elderly and how local administrations should use the resources to this end. Finally, previous studies have focused on vulnerability evaluations also being examined to prioritize interventions in deprived areas, presented in Section 3.2.4. From this multidisciplinary review, implementing the Preferred Reporting Items for Systematic Reviews and Meta-Analysis methodology (PRISMA), a standardizable model of dimensions and areas was suggested to evaluate vulnerability. Next, the selected area and its particular characteristics are presented in Section 4.1 to find the appropriate indicators that adapted to both site characteristics and available information. Section 4.2 includes the indicators for the evaluation; first, a provisional ad hoc list of indicators is presented, and the indicators were validated by the interjudge agreement technique. Some experts in the multifaceted topic, with profound knowledge about the selected area, analyzed and evaluated the provisional indicators. From their feedback, a final list of indicators was drawn up and applied to the selected area. Section 5 introduces the main work issues for discussion, which leaves room for further research in this field and summarizes the main reached conclusions.

The following section presents the literature review on the main technical and social factors. From this starting point, the initial list of indicators for the evaluation model was built up and reviewed by some experts to decide on the definitive list. The model was applied to a selected vulnerable area by drawing some conclusions to suggest interventions in urban open spaces that could improve the quality of life of citizens in general and of the elderly in particular.

3. Literature Review

Figure 2 illustrates the identification of the influential variables obtained from a literature review and the theoretical framework connected to the technical and social perspectives using the PRISMA methodology and by searching for references from different databases. The source was the Scopus database. The selection of the updated references (2005–2021), with the proper keywords and the subject area limited the number of references to a group, and two reviewers checked both titles and abstracts. They selected some final studies that were directly linked to the scope of this work, which were reviewed in depth. From the technical perspective, keywords "Urban OR Building" AND "Accessibility OR Design" AND "Elderly OR OLD-AGED" and the subject area "Engineering" yielded 24 references. Likewise, from the social perspective, keywords "Active Ageing" AND "Urban OR Planning" and the subject area "Engineering" produced 32 references.

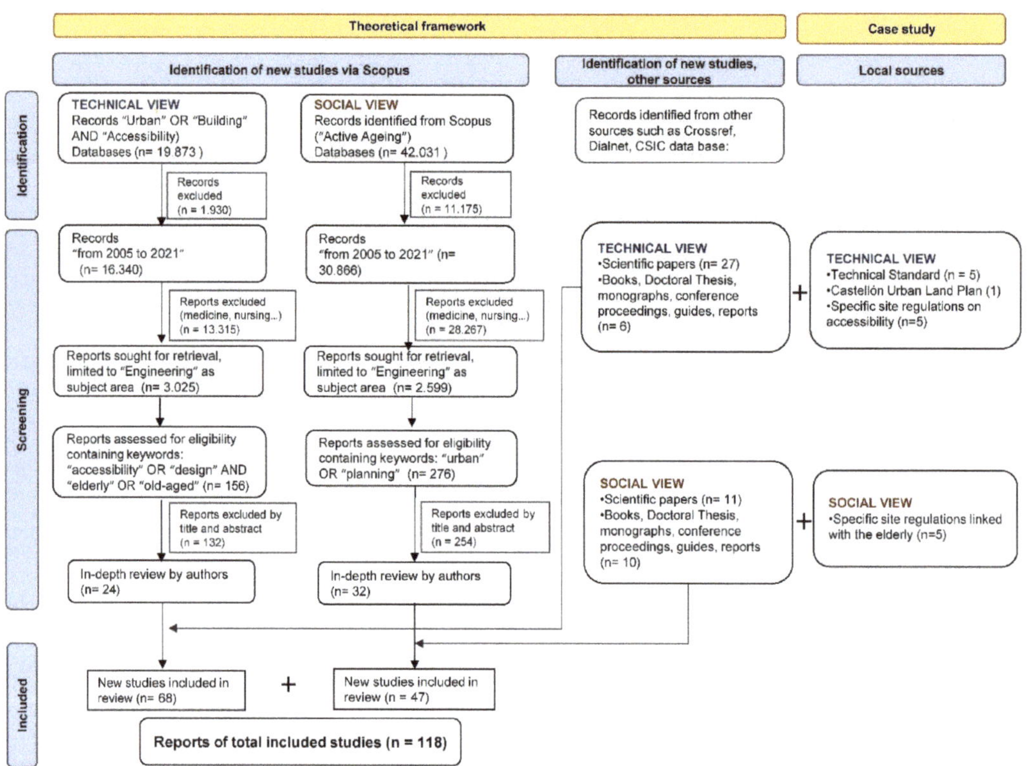

Figure 2. Illustration of the PRISMA methodology (adapted from [5]).

Other sources such as Dialnet, Cross Ref or the CSIC database were used, and not only scientific references, but also other sources applicable to this study were found, such as reports, guidelines, handbooks, etc.

In the physical urban context, a review of the Technical Standards connected to accessibility in built environments was also done. The regulations currently in force and linked with urban issues and accessibility in the study area were examined to identify the key factors to be included in the model.

Some WHO recommendations for the Active Aging (AA) population and AFC were reviewed to obtain some key ideas. Similarly, the targets included in the United Nations' SDG related to the elderly were analyzed. Some specific-context policies were examined by focusing on the case study to apply the model. Finally, the role of the Social Welfare Services available for the elderly and for the vulnerable populations in the area was also observed to identify the instruments available to support them in urban areas. The information and aid from this part of Local Administrations were crucial to gain a complete picture of the elderly's situation at the local scale.

3.1. Technical Conditions of Public Spaces

The analysis of the literature showed that accessibility is vital to ensure the inclusion of citizens with mobility issues. Accessibility has evolved in the public space from an inexistent consideration to become a universal design concept in every product, including urban spaces. The main conclusions of the review varied and are summarized as follows:

The relationships between the city's sociality and spatiality are highlighted by many authors [6–8]. The universal design concept inevitably arises [9–15], and urban universal accessibility is specifically analyzed by some authors [16,17].

The latter is connected to everyone's right to mobility regardless of disability, age or gender. Therefore, some authors connect this to social sustainability and social justice concepts [18–24]. In line with this, some authors point out some factors, such as density, accessibility, mobility, integration (connections and street networks), diversity of services, mixed use, environmental quality, safety, and social capital (sense of belonging, participation) [25], while others highlight that accessibility and quality of social life, conservation of resources, quality of built environments, protecting disadvantaged groups, and commercial and economic opportunities are observed in urban renewal practices [26]. Some factors underlying social factors have been suggested, such as health and physical comfort, accessibility, integration, economy, and participation [27]. Ahrentzen and Tural reviewed 37 research articles and identified six built environment characteristics: barriers, supports and features that fit; spatial organization and layout; environmental cues; ambient qualities; assistive technologies; and gardens and outdoor spaces. They concluded that accessibility-oriented features dominated the studies [28].

La Rosa et al. considered demands and preferences for different social groups' accessibility, e.g., children and the elderly [29], and some studies focus on the elderly's specific accessibility needs [30–34].

Other authors centered on specific infrastructures; for example, Basbas et al. (2010) analyzed the facilitation of pedestrian trips in urban areas by pointing out that the elderly usually face more difficulties during trips on foot than other age groups [35]. Wen et al. (2018) identified some landscape features desired by old-aged people, such as natural, esthetic, comprehensible, and diverse, with accessible and well-maintained infrastructure and facilities [36]. Shan et al. (2020) examined the acoustic environment in the elderly's public activity space to design a new elderly healthy urban park environment [37], which they associated with a healthy and active way of living [38,39]. Tao and Cheng (2019) analyzed the elderly's spatial accessibility to healthcare services in Beijing [40]. Brake (2008) identified appropriate options for delivering urban transportation to older people [41].

In order to examine the technical perspective, some technical standards containing requirements for an accessible urban and built environment were reviewed. UNE is Spain's only Standardization Organization, designated by the Spanish Government. Although standards are voluntarily applied, they are an excellent guide to detect the key topics that achieve complete accessibility in urban spaces and are basic guides to legislate in specific technical topics. Six documents were reviewed: two standards on Universal Accessibility (UNE 170001-1, Universal accessibility. Part 1: MGLC criteria to facilitate accessibility to the environment and Universal accessibility and UNE 170001-2 Part 2: Accessibility management system), and four more standards that focus on built environments (UNE 41510 Accessibility in urbanism. UNE 41524 Accessibility in building. General design rules for spaces and elements in buildings. Links, equipment and use. UNE-ISO 21542 Building construction. Accessibility and usability of built environments).

Some policy-oriented references were also examined, such as reports, guides or mandatory regulations for the practical implementation of accessible spaces. As the case study is located in Spain, complete national, regional, and local regulations are linked to accessibility in built environments [42], which were identified and analyzed. In Spain, the main reference is the Technical Code for Buildings (CTE, Código Técnico de la Edificación, in Spanish), specifically Basic Document SUA9 on Accessibility to buildings and nearby urban spaces (SUA, standing for Seguridad de Utilización y Accesibilidad, in Spanish, meaning Secure Use and Accessibility). Regionally in Spain, as competences on urban matters are delegated to the Autonomous Communities, the main reference in the Valencian Region, where Castellón is located, is Decree 65/19, which regulates accessibility in buildings and public spaces (Decreto 65/2019, de 26 de abril, del Consell, de regulación de la accesibilidad en la edificación y en los espacios públicos). Locally, Local Administrations are in charge of managing urban spaces. The Municipal Ordinance on Accessibility in Castellón was adapted to the city's specificities (Ordenanza Municipal de Accesibilidad, 26 April 2007).

3.2. Senior Citizens' Needs in Public Spaces

The social view completed the analyses and considered some people-centered factors. When talking about the elderly in society, the AA and AFC concepts arise. They have been promoted by the WHO since the early 1990s and have been largely analyzed in the literature. Moreover, the available tools that Local Administrations have to support the elderly and to take care of their needs were analyzed. In this work, Social Welfare Services were considered to play a crucial role with vulnerable populations [43]. They face the difficulties that the most vulnerable populations encounter in the city and provide exact information to adapt the social specificities of the case study. This section summarizes all these factors.

3.2.1. Active Aging

Active aging (AA) is defined as "the process of optimizing opportunities for optimizing opportunities for health, participation and security in order to enhance quality of life as people age" (WHO, 2002). Six key factors are considered to influence the promotion of AA: health and social services; healthy way of life; biology and genetics; physic environment; social environment; economic situation.

The importance of living healthy aging is highlighted by many authors. AA is the logical consequence of the demographic, social, economic, and political changes that have been taking place in societies worldwide [44]. Some authors indicate that the objective of AA is to maintain the elderly's physical activity and their productivity, as well as life expectancy [45]. Others define healthy aging as the process of developing and maintaining the (mental and physical) functional capacity for well-being in older age [46], while some confirm that AA is built on four main pillars: health, participation of the elderly, safety to improve the lives of the elderly as they get older, and lifelong learning [47,48].

Afacan (2013) conducted a study in the city of Ankara and indicated that an inclusive open environment not only allowed elderly people to feel safer, but promoted more regular uses of urban spaces. It highlighted accessibility and plain simple signage as key factors to increase the aging population's social participation [49]. Elsawahli et al. (2017) concluded that in the public realm, older adults are sensitive to poor lighting and inadequate walkway conditions, and regular maintenance is required to promote their physical activity [50]. Focusing on mobility, Musselwhite et al. (2015) pointed out that connection to communities and social networks enables older people to contribute to and connect with society, and mobility is associated with positive mental and physical health by facilitating physical activity and independence, while reducing social isolation [51]. Regarding housing needs, Gharaveis (2020) indicated that design interventions can increase physical functioning both inside and outside long-term residential facilities [52]. Some authors point out that aging should be seen as a life opportunity, a challenge and an enjoyment for elderly people who wish to become active members of society [53–55].

The AA Index (AAI), created in 2012 by the United Nations Economic Commission for Europe (UNECE), together with the European Commission Directorate General for Employment, Social Affairs and Inclusion and the European Centre for Social Welfare Policy and Research, measures the extent to which older people live independent lives, participate in paid employment and social activities, and their capacity to actively age. The index is calculated from 22 indicators grouped into four domains: Employment; Participation in Society; Independent, healthy and secure living; Capacity of enabling environment for AA. From the results, differences appear among countries, with Sweden at the top, followed closely by Denmark. The UK, Italy, Portugal, Spain, and Malta are middle-ranked countries, with Greece and many Central European countries at the bottom, which highlights having to make more policy efforts in the latter countries [56]. Thalassinos et al. (2019) conducted research to assess how the AAI correlated with economic and labor market credentials and how it would impact EU Member States' economic development to find important dissimilarities among them [57].

3.2.2. Age-Friendly City

The Vancouver Protocol (2006) pointed out the relationship between aging demographics and urban processes. In the urban environment, the AFC is gaining more relevance because the elderly population represents a high and growing percentage of the population in society, hence the need to respond adequately to their daily and basic needs. Eight domains should be observed in an AFC: Outdoor Spaces and Buildings; Transportation; Housing; Respect and Inclusion; Social Participation; Civic Participation and Employment; Communication and Information; and Community Support and Health Services [58,59].

De Oliveira et al. (2019) found a huge number of references in the scientific literature from 2007 to 2017 about the implantation and evaluation of AFCs in the world and highlighted international concern about age pyramid change and longevity [60]. Rémillard-Boilard et al. (2021) compared the experience of eleven cities located in different countries. Their study explored the key goals, achievements, and challenges faced by local age-friendly programs and identified four priorities: changing the perception of older age, involving key actors in age-friendly efforts, responding to the diverse needs of older people, and improving the planning and delivery of age-friendly programs [61]. Sengers and Peina (2021) focused on age-friendly housing options by considering private spaces in the city [62], while Marquet el al. (2017) explored the effects of neighborhood morphology and walkability in active travel patterns of aging older adults [63].

Regarding social aspects, some authors point out that the elderly living in unsafe places with architectonic barriers are more likely to suffer isolation and depression, and the social network should be maintained to keep quality of life high. Some point out that age is not very often explicitly integrated into urban socio–spatial inequality analyses [64], while others highlight that older adults, especially those living alone with no family support, are particularly sensitive to the local environment [65]. Iamtrakul et al. (2019) conducted research to understand the health-related problems of age groups between different life styles and neighborhood characteristics in Thailand and recommended encouraging AA by optimizing health, participation, and security opportunities to enhance quality of life [66]. Fung (2020) indicated that the community support of AA becomes a crucial part of the urban neighborhood's collective resilience [67]. Some authors introduced the gender perspective and found that old age punishes women more than men because they live longer and under worse conditions [68,69]. Flores et al. (2019) evaluated an AFC by analyzing its relationship to life satisfaction by considering the elderly's age cohort variables and whether they live alone or with someone else. Despite differences being found in the various old-aged people groups, they were all domains of outdoor spaces and buildings, and community support and health services were significantly related to life satisfaction [70]. Yung (2016) worked with eight focus groups in elderly community centers in two urban renewal districts of Hong Kong and perceived the most important needs, as well as social and physical activities, community life facilities and services, the social network, and a clean pleasant environment [34].

In line with the WHO, the UN SDG aim to develop friendly environments for the elderly by reaching AFC and communities and to reinforce long-duration care [71,72]. When looking at the varied perspectives of SDG, practically all of them present specific targets connected to people's well-being in general, and, in some cases, by focusing on vulnerable populations, including the elderly; for example, targets 2 and 7 of SDG 11 for sustainable cities and communities are set out as "Provide access to safe, affordable, accessible and sustainable transport systems for all, improving road safety, notably by expanding public transport, with special attention to the needs of those in vulnerable situations, women, children, persons with disabilities and older persons" and "Provide universal access to safe, inclusive and accessible, green and public spaces, in particular for women and children, older persons and persons with disabilities", respectively.

Deprived areas and vulnerable neighborhoods face more difficult challenges because of their greater deficiencies, which are implicit given their vulnerability condition [73]. Bayar and Türkoğlu (2021) compared the experiences of older adults living in two neighborhoods of Istanbul (Turkey). They concluded that the city's engagement and sociability level are connected to higher incomes [74]. The study by Carroll et al. (2020) in a deprived neighborhood in Copenhagen found that targeting the appropriate kind of age-friendly programs might enhance empowerment through physical spaces as a starting point for social interaction [75]. Curl and Mason (2019) investigated how the environment influences the walking and well-being of the older adults living in disadvantaged urban areas [76].

Some relevant studies have centered specifically on the elderly in built environments and are summarized in Table 1.

Table 1. Relevant studies about elderly people's interaction with urban environments.

Research Paper	Topic	Main Findings
[77]	Environment influences older adults' health and activity participation	Some limitations in the literature. Key topics to consider in future research: climate, pollution levels, street lighting, traffic conditions, accessibility and appropriateness of services and facilities, socio–economic conditions, esthetics, pedestrian infrastructure, community life, exposure to antisocial behavior, social network participation, environmental degradation, urbanism level, exposure to natural settings, familiarity with the local environment, among others.
[78]	Determinants of an age-friendly community in Melaka (Malaysia)	Housing, social participation, respect, civic participation and employment, health services provided for the elderly, and outdoor spaces and environment for the elderly to perform physical activities; only respect, social participation, and outdoor spaces were significant.
[79]	Perceived suitability of urban and housing environments for aging populations in Spain and Mexico	35 variables for each scale, neighborhood-public space and buildings-dwellings in Spain and Mexico. In both cases, indicators are organized as seven dimensions: design, accessibility & mobility, comfort, maintenance, security and health, use and control, and stimulus.
[80]	Older adults' experiences with outdoor space and buildings	They found crucial accessibility and appropriate infrastructure.
[81]	Quantifiable spatial indicators to assess local lived environments according to AFC domains (Australia)	Remarkable indicators: outdoor spaces and buildings: walkability for transport; access to public open space within 400 m; intersections serviced with pedestrian crossings; access to public seating; access to public toilets; accessible buildings. social participation domain: access to neighborhood houses/community centers; recreational services that cater for older people. The respect and social inclusion domain: access to social clubs/senior citizens clubs. The civic participation and employment dimension: proportion of the population aged 60+ years regularly volunteering; proportion of the population working beyond the official retirement age.

Table 1. *Cont.*

Research Paper	Topic	Main Findings
[82]	Measurement scale	Five domains: personal characteristics, place-related, socio–economic environment, governance, and health-related; 15 criteria and 99 indicators. Some noteworthy indicators: the place-related dimension and public open space criteria: street lighting, the area to open spaces ratio, public recreation and open spaces, quietness, maintenance. The socio–economic dimension: quality of life, social interaction, happiness, social inclusion, social inequalities, social participation, social support. The health dimension: social life or sense of community.
[83]	Index of quality of life for the elderly; local territorial context; neighborhood level	The index was calculated from objective and subjective indicators. It considers five broad areas, namely, quality of life, business and labor, services and environments, population, and leisure, included in two main streams: elderly quality of life and key urban environment features.
[84]	Focuses on public open spaces	The problems which the elderly frequently encounter in public open spaces: pavements and roads, pollution, safety, insufficient maintenance and management, traffic and socio–cultural problems.
[85]	Focuses on green public spaces	It highlights the effect of green public spaces on the average urban quality of life and stresses that levels of safety, maintenance, accessibility, and availability of equipment are key factors of well-being.
[86]	Focuses on public open spaces	Mobility, accessibility, and availability of open spaces connected to the elderly's satisfaction with the out-of-home environment, encouraging the availability of public open spaces and pedestrian-friendly neighborhood environments.

Certain conclusions can be drawn from the review of the studies in Table 1. Firstly, while some of the studies rely on some or all of the AFC domains [78,80,81,86] (6, 8, 2 and 2 domains, respectively), others propose a new domain structure containing different variables or indicators [79,82,83]. Some authors draw conclusions from a desk review on the topic, for example, in [77], 83 studies were examined, concluding that not only should environmental aspects be included when analyzing influences on older adult health, but also personal circumstances. The latter is also present in other studies, where the subjective factors are considered, for example, in [79], where variables such as "self-perceived quality of life" or "place attachment", or in [83], where 13 out of the 29 indicators are "Subjective Indicators". On the other hand, the desk review of other studies is focused on physical attributes, for example, mobility, accessibility, and availability of open spaces [84] or green spaces [85]. Reference [80] provides a different approach connecting AFC to environmental justice. The authors consider that the work is a useful framework for social workers to systematically document and evaluate their age-friendly community efforts. This is aligned with our work, which proposes the involvement of Social Services in the diagnosis, considering its application to vulnerable areas. The scale of application also differs depending on the study. In some target villages or cities [78,79,81,86], others are applied at a suburban scale [83–85]. As for the implications of the studies, some could be implemented in local, national, and, sometimes, international policy making and planning [78–83], while others have a more regional or local focus [84–86].

3.2.3. Policies That Address the Elderly's Social Inclusion

Population aging is a challenge for governments to comply with international commitments to human rights and for the State because it must act to guarantee these social rights by protecting and safeguarding them with public policies [87]. Aging European countries have widespread implications for current and future social and economic policies all across Europe [88]. Kurek et al. (2011) point out that the EU response to the aging population challenge by the 'AA' policy aims to increase the employment rate of elderly workers and to raise their retirement age and also refers to their participation in social, economic, cultural, spiritual, and civic affairs [89].

In the EU, each Member State undertakes its national strategy by attending to its particularities. Eurostat data show that the prospected demographic trend in Spain increases for the general population until 2050, but decreases after this period. When we look at the projected old-rate dependency ratio, Spain is one of the Member States with the highest ratio in 2100, which doubles from 29.3 to 60.1. Currently, Spain has a National Strategy for Older People's AA and for their Good Treatment for 2018–2021. It contains five main strategic lines: 1. improve the rights of older workers and extend working life; 2. participation and decision making in society; 3. promote healthy independent living in suitable and safe environments; 4. non discrimination, equal opportunities, and pay attention to situations of fragility and more vulnerability; 5. avoid mistreating and abusing the elderly. Each line of action has specific targets.

Under the national strategy umbrella, all Spanish Autonomous Communities develop their regional strategy. In the Valencian Region, the intention of the Valencian Plan for Inclusion and Social Cohesion (VICS, 2017) is for social inclusion to reach all citizens. The work of Garrido and Jaraíz (2017) studies the influence of territorial–urban policies on social inclusion [90]. According to Caballer et al., (2017), who analyzed the Regional Valencian Government policy, by focusing on old-aged people, it aims to improve their lives by better health services, accessibility to buildings and dwellings, public transport, promoting education, etc. [91].

The implementation of this plan relies, among other factors, on the important role of Social Welfare Services to support vulnerable populations. Law 3/2019, on Inclusive Social Services of the Valencian Community, is the reference regulation. Its professional performance includes protecting the elderly in vulnerable situations and also supporting care givers (Article 36). It also highlights the importance of vulnerable spaces that, due to their urban/residential, social, labor or economic characteristics, require integrated action (Article 25). Article 25.3 points out the deficit of community or socio–cultural equipment or resources, the existence of substandard housing or difficulties for urban mobility, as some aspects that indicate vulnerable spaces.

It is important to consider that municipalities are the last step on the scale to ensure old-aged people's care. The Social Welfare Service Departments in municipalities are responsible for guaranteeing that this population stratum is included. The city of Castellón joined the Network of Cities and Communities Friendly to the Elderly at the end of 2014. The diagnostic stage was completed in two years. Through a plenary agreement dated November 30, 2017, the Action Plan of the project "Castelló, a friendly city for the elderly" for the 2017–2019 period was approved and included the AFC domains to be improved.

3.2.4. Vulnerability Evaluation

The scope of this work ultimately centered on evaluating vulnerability by considering technical and social perspectives. On the one hand, the analysis focused on a vulnerable urban area with many deficiencies in the urban and built spheres; on the other hand, senior citizens' needs are placed in the spotlight by assuming old age to be a life cycle stage that can be potentially reached by every inhabitant.

Table 2 presents some remarkable works on this topic in the Spanish context. They all fall within the sustainability framework, including economic and social features and characteristics connected to the built environment. Some look at social issues in more depth, while others are more urban-centered.

Table 2. Summary of studies on vulnerability in urban environments.

Reference Study	Scale	Dimensions and Examples of Indicators
[92]	Regional	Nine areas: demography; health; education; employment; housing; urbanism; social relations; participation; perceived and projected vulnerability.
[93]	National	Economic subsystem: Demographic: population evolution, dependence index, etc. Economic: income, unemployment, etc. Social subsystem: Resources: institutional, communitarian, etc. Social cover: beneficiaries of subsidies; support in the area, etc. Urban subsystem: Location: isolation; proximity to infrastructures, etc. Infrastructures: education; health; urban facilities (pavement, lighting, etc.)
[94]	National	Social level: Percentage of unqualified population; percentage of illiterate peoples; female unemployment, etc. Demographic and family situations: percentage of the population > 65 years; dependence index; household composition, etc. Living conditions: average dwelling area per inhabitant; maintenance of buildings, etc.
[95]	National	Socio-demographic vulnerability: percentage of population > 64 years; over-aging index; immigrants index; percentage of single-parent family, etc. Socio-economic vulnerability: unemployment; illiterate population, etc. Residential vulnerability: percentage of dwellings <31 m^2; percentage of badly conserved dwellings, etc. Subjective vulnerability: percentage of dwellings affected by nearby green areas; percentage of dwellings affected by inefficient communications, etc.
[96]	Regional	Residential vulnerability: poor living conditions; evictions, etc. Economic vulnerability: unemployment, reduction in family income, etc. Social vulnerability: welfare cuts, health cuts, etc.
[97,98]	National	Urban vulnerability: green areas; accessibility to health services, etc. Building vulnerability: constructive quality, accessibility in residential buildings, etc. Economic vulnerability: location of proximity services; socioeconomic stratification, etc. Socio-demographic vulnerability: population older than 65 years; level of education, etc.
[99]	Regional	Residential vulnerability: accessibility in residential buildings, average area/inhabitant, etc. Socio-demographic vulnerability: dependence index with gender perspective; immigrants, etc. Socio-economic vulnerability: illiterate population; income; unemployment, etc.
[100,101]	Local	Urban vulnerability: urban density; proximity to public transport, etc. Building vulnerability: energy efficiency; accessibility, etc. Socio-demographic vulnerability: population older than 65 years; immigrants, etc. Socio-economic vulnerability: social subsidies; level of education, etc.

In this work, we pay attention to the role of Social Welfare Services in intervening in deprived urban areas. They possess plenty of valuable information because they manage the resources that the Local Administrations provide to the most vulnerable population. On a regional scale, Giménez-Bertomeu et al. (2020) prepared a research report for the Regional Valencian Government [92]. It intended to incorporate vulnerability indicators from Social Services' information, which is aligned with the scope of this work. They included nine areas: demography; health; education; employment; housing; urbanism; social relations; participation; perceived and projected vulnerability. They concluded that socio–demographic and socio–economic dimensions present only a few dimensions, while physical variables are better defined. They also emphasized the lack of subjective variables and pointed out that information should be disaggregated into specific population groups (the elderly, traders, experts, etc.) when, for example, it comes to difficulties in urban mobility, deficiencies in lighting or sanitary infrastructures, among others. One very important conclusion was reached: the complete lack of sources of information on an intraterritorial scale and in relation to environmental aspects of material (equipment, infrastructures) and immaterial (perceptions, opinions, feelings about the territory) natures.

4. Results

From the analysis of the above references, the key topics to evaluate the accessibility of public open spaces were proposed by separately considering the technical and social perspectives. They were grouped and formed three dimensions that presented different areas to organize the several features connected to the use of public spaces:

1. Dimension Urban and Building Vulnerability (UBV)

    - Area Urban environment (UE): to measure aspects linked with the urban realm
    - Area Infrastructures (I): to detect lack of important infrastructures connected to accessibility and safety in public spaces
    - Area Facilities (F): to evaluate the presence or absence of basic services for the elderly (e.g., health)
    - Area Buildings (B): to evaluate access to buildings

2. Socio-Demographic Vulnerability (SDV): this dimension was divided in two areas

    - Demographic (D), to include the demographic features of the study area
    - Social (SO), to evaluate social aspects in the area, such as support from local administrations or the elderly's participation in the area

3. Socio-Economic Vulnerability (SEV): it reflects the economic features in the area

4.1. Case Study

Castellón is located in the Valencian Region (East Spain). It is a medium-sized city with 174,264 inhabitants, (Spanish National Statistics Office, INE, 2020).

The city is divided into nine districts that are, in turn, divided into 110 census sections. There are six administrative structure areas: North, South, East, West, Center, and Grao. The highest percentage of aged-people is identified in the Center administrative district.

On the local scale, García et, al. analyzed and wrote a report to identify vulnerable areas in the city of Castellón according to the city's Land Urban Plan [100]. This was taken as the basis to select the vulnerable area, which was the case study herein analyzed. The particular features, available information, and knowledge about the area allowed us to draw up a tailored list of indicators to be included in the evaluation model.

As seen in Figure 3, the selected case study is a neighborhood located in the Centre district formed by census sections 05005 and 05013. This area was identified as being vulnerable by the recent Urban Land Plan. The coincidence of the neighborhood borders with the census section borders made it easier to find the disaggregated information at the desired level. The neighborhood is characterized by the presence of the Bullring and is limited to the north by the Ribalta Park. The total area covers 58,553.64 m^2. According to the Urban Land Plan, some urban features were vulnerable; for example, urban density was 176.93 dwellings/ha, which exceeds the threshold value of 100 dwellings/ha. The nearby park conferred the surroundings green areas, but vegetation is scarce in the neighborhood. Constructions in the building stock are old and obsolete: 77.78% of the buildings present bad energy performance, 10% have deficient conservation, and 44% display poor building quality.

All these features were confirmed during the visit to the neighborhood, when some deficiencies were identified (see Figure 4). Although the Ribalta Park is close to the area, green areas and vegetation are scarce in the neighborhood, and the presence of trees should be increased to create shaded areas and to reduce the heat island effect. Moreover, existing trees have inappropriate tree pits. Some problems linked with accessibility in pedestrian areas also appear: in many spots, pedestrian routes are occupied by commerce or parked cars, most sidewalks are narrow, and the public space generally lacks maintenance. Some buildings are abandoned, and many commercial and residential buildings are not accessible and need wider accesses or ramps to ensure vertical circulation. A high percentage of the building stock is old and presents general degradation due to lack of maintenance. The general conclusion reached is that most of the elements considered for urban dimension are better maintained in the main avenues surrounding the neighborhood. However, in the inner parts of the study area, all the urban elements are nonexistent or improperly or badly maintained.

Figure 3. Location of the study area.

Some processes in which citizens participate were also taken into account to devise the Land Urban Plan. The focus groups performed with old-aged people were especially connected to this work because they are considered a vulnerable population group. The main findings of the concerns voiced by the participants were: lack of maintenance and cleaning in public spaces in certain areas, non-accessible transport, need for bike lines, lack of lighting in some spots, etc. [101] (p.102).

According to the 2019 INE data, census section 5005 and census section 5013 had 27% and 27.9%, respectively, of the population aged older than 65 years, both of which were higher than the city's average value of 18.9%.

As logically assumed, the population in their 80s faces more mobility difficulties than those in their 60 s. The older populations in the census sections are illustrated in Figure 5 by considering the age ranges of 65–69, 70–74, 75–79, 80–84, and older than 85 years (2019 INE data). Note that senior citizens older than 80 years represent more than 20% of the population aged older than 65 years in census sections 5005 and 5013.

Figure 4. Some problems in the public spaces in the area.

Regarding economic features, the Urban Land Plan report for the identification of vulnerable areas highlighted that Social Welfare Services were required more often than the average and social subsidies, while the dependence aid in this area presented higher values than average. Some economic data, such as income or unemployment in the area, were not directly linked with the elderly, but are good indicators that reflect economic vulnerability. The average income per household in the city is €30,929, with €26,813 and €24,964 in census sections 5005 and 5013, respectively (INE, 2019).

An important indicator to measure social dynamism has to do with civic associations. According to the Municipality, the city of Castellón has 776 registered associations. The general associations per administrative area in the city of Castellón are as follows: 205 are located in the Center District, followed by the East, North, West, South, and Grao Districts, with 158, 148, 117, 80, and 58, respectively. Only 33 are intended for old-aged people, which represent 4.2%. According to Mollar et al., (2020), 18% of the elderly in Castellón voluntarily participate in associations. There are 20 associations registered in a

600-m radius from the neighborhood's geometric center. Seven are intended for old people: four are retiree associations, two are day care centers, and one is one of the three C.E.A.Ms. in the city (standing for *Centro Especial de Atención a Mayores* in Spanish, meaning Elderly Care Center).

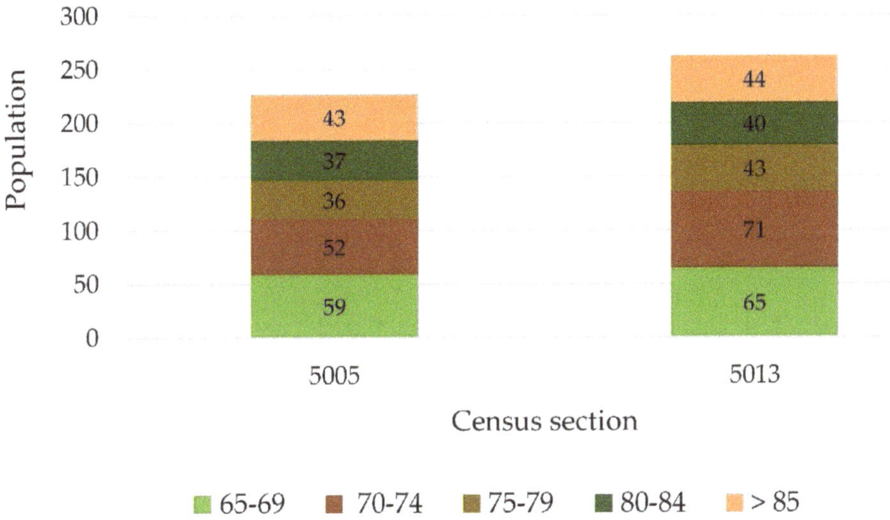

Figure 5. Distribution of old people in age intervals in the study area.

The Municipality is active in promoting AA. Specific subsidies are offered for AA (http://www.castello.es/archivos/387/Anuncio_Concesion_Subvenciones_2020.pdf, last accessed on 31 May 2021). In addition, services for the elderly are provided, such as psychological and legal assessments.

4.2. List of Indicators for the Case Study

Having collected data from the case study, a provisional list of 24 indicators, and arranged in relation to dimensions and areas, is presented in Table 3.

In order to validate the indicators, some experts were asked to evaluate them and to make suggestions to improve the model. Interjudge agreement was implemented [102]. This technique is based on the degree of agreement reached by some people of guaranteed experience and professional solvency. The main goal is to reach the maximum consensus.

Selection of experts is based on the following criteria: independence; years of experience, and expertise in accessibility in the urban open space or in social sustainability, and vulnerability. They must all have a similar educational and cultural background [103]. Seven experts were selected: four women and three men, as can be seen in Table 4. In April 2021, the experts were requested to complete an on-line survey (through Google forms) to evaluate the relevance of the proposed indicators on a scale from 1 to 5, where 1 denotes not important (NI), 2 means slightly important (SI), 3 suggests not important nor unimportant (NIN), 4 means important (I), and 5 is very important (VI). The survey was completed with two final open questions with which the experts were required to suggest new indicators that were not on the provisional list and to make open suggestions about the model. In June 2021, in a second round, the experts were informed about the results and were asked to evaluate the newly included indicators (by e-mail). By doing so, the level of agreement could be used as an estimation of the weight to be applied to each indicator by using weighted average values. Therefore, the definitive model included 27 indicators.

Table 3. Provisional list of indicators.

Dimension	Area	Indicators
UBV	UE	1. Green areas
		2. Daytime acoustic comfort
		3. Nighttime acoustic comfort
		4. Accessible paving
		5. Ramps
		6. Pavement maintenance
		7. Pedestrian routes
	I	8. Accessible public transport
		9. Lighting
		10. Adapted traffic lights
	F	11. Health facilities
		12. Elderly care facilities
		13. Commercial facilities
	B	14. Buildings' age
		15. Accessibility in residential buildings
		16. Accessibility in non residential buildings
SDV	D	17. Population older than 64 years
		18. Immigrants
		19. Aging ratio
	SO	20. Social services attention to dependency
		21. Social subsidies
		22. Participation in the community
SEV	SE	23. Promoting AA
		24. Household income

Table 4. Experts for the interjudge technique.

Judge	Area	Current Position	Years	1st Round	2nd Round
1	Architecture	President of the Professional Association of Architects Castellón province	21	20 April 2021	30 June 2021
2	Architecture	Professor of Building Engineering at the Universitat Jaume I in Castellón	29	14 April 2021	30 June 2021
3	Urbanism	Technician of Urbanism Department in the Castellón Municipality	19	14 April 2021	1 July 2021
4	Architecture	Professor at the School of Architecture in Zaragoza	10	22 April 2021	2 July 2021
5	Social Services	Head of the Social Welfare Department in the Castellón Municipality	35	22 April 2021	2 July 2021
6	Economy	Head of the Interuniversity Institute of Local Development in the Valencian Region	33	20 April 2021	30 June 2021
7	Social Services	Coordinator in the Elderly People Unit in the Castellón Municipality	29	22 April 2021	30 June 2021

After completing the survey, the agreement degree was calculated by the percentage according to Cohen's kappa coefficient using the scale 0%, 0–20%, 21–40%, 41–60%, 61–80%, 81–99%, and 100%, meaning a poor, slight, fair, moderate, substantial, almost perfect, and perfect agreement, respectively. The average value, $\tilde{x}$, of each indicator was calculated to obtain an overview for every single indicator. We considered the upper scale part, with values 4 and 5, I and VI, as a criterion to include the indicator whenever at least 61% of the agreement was reached (substantial, almost perfect or perfect), and rejected the indicator otherwise. A excel sheet was used for calculation of the level of agreement and the statistical basic descriptors (average, minimum, maximum values).

Table 5 summarizes all the results and shows the number of responses per mark in the middle columns and the level of agreement as a percentage in the columns on the right. The last column shows the percentage when considering categories I and VI together. They were almost all higher than 61%, except for the indicator "immigrants" with 57% agreement. This meant rejecting only one of the 24 indicators, which is presented in the shaded cell in Table 4. The maximum value was reached in 10 of the 24 indicators, such as "green areas" and "pavements", with perfect agreement (100%).

Table 5. Evolution of the proposed list through a desk review and interjudge agreement validation.

Indicator	Evaluation (nr Answers)						Level of Agreement (%)					
	1	2	3	4	5	x	NI	SI	NIN	I	VI	I + VI
UBV.UE.1. Green areas	0	0	0	3	4	4.6	0	0	0	43	57	100
UBV.UE.2. Daytime acoustic comfort	0	0	1	4	2	4.1	0	0	14	57	29	86
UBV.UE.3. Nighttime acoustic comfort	0	0	1	1	5	4.6	0	0	14	14	71	86
UBV.UE.4. Accessible Paving	0	0	0	3	4	4.6	0	0	0	43	57	100
UBV.UE.5. Ramps	0	0	0	2	5	4.7	0	0	0	29	71	100
UBV.UE.6. Pavements	0	0	0	2	5	4.7	0	0	0	29	71	100
UBV.UE.7. Pedestrian route	0	0	0	3	4	4.6	0	0	14	57	29	86
UBV.UE.8. Cleanliness	0	1	1	1	4	4.1	0	14	14	14	57	71
UBV.UE.9. Adapted urban furniture	0	0	1	3	3	4.3	0	0	14	43	43	86
UBV.I.1. Accessible public transport	0	0	1	2	4	4.4	0	0	14	29	57	86
UBV.I.2. Lighting	0	1	1	2	2	3.8	0	17	17	33	33	67
UBV.I.3. Adapted traffic lights	1	0	2	1	4	4.4	14	0	29	14	57	71
UBV.F.1. Health facilities	0	0	1	3	3	4.3	0	0	14	43	43	86
UBV.F.2. Elderly care facilities	0	0	1	3	3	4.3	0	0	14	43	43	86
UBV.F.3. Commercial facilities	0	0	0	3	4	4.6	0	0	0	43	57	100
UBV.B.1. Buildings' age	0	0	2	3	2	4.0	0	0	29	43	29	71
UBV.B.2. Accessibility in residential build.	0	0	1	2	4	4.4	0	0	14	29	57	86
UBV.B.3. Accessibility in non residential bu.	0	0	2	1	4	4.3	0	0	29	14	57	71
SDV.D.1. Population over 64 years	0	1	0	2	4	4.3	0	14	0	29	57	86
SDV.D.2. Population over 79 years	1	0	1	2	3	3.9	14	0	14	29	43	71
Immigrants	0	2	1	2	2	3.6	0	29	14	29	29	57
SDV.D.3. Aging ratio	0	0	0	3	4	4.6	0	0	0	43	57	100
SDV.SO.1. Social services attention to dep.	0	0	0	4	3	4.4	0	0	0	57	43	100
SDV.SO.2. Social subsidies	0	0	2	2	3	4.1	0	0	29	29	43	71

Table 5. *Cont.*

Indicator	Evaluation (nr Answers)							Level of Agreement (%)					
	1	2	3	4	5	x	NI	SI	NIN	I	VI	I + VI
SDV.SO.3. Participation in civic community	0	0	0	5	2	4.3	0	0	0	71	29	100
SDV.SO.4. Participation in Governance	0	0	2	3	2	4.0	0	0	29	43	29	71
SEV.SE.1. Promoting AA	0	0	2	1	4	4.3	0	0	29	14	57	71
SEV.SE.2. Household income	0	0	1	2	3	3.7	0	0	14	29	43	71

With the open questions in the survey, where the experts were asked to make suggestions, some indicators were proposed to complete the list: degree of cleanliness and urban furniture adapted to the old-aged people in the urban category. Thus, two new indicators were included: UBV.UE.8 and UBV.UE.9. They were added to Table 4 and are depicted in bold.

About the socio–demographic category, one suggestion was to include the number of people older than 74: SDV.D.2. One of the experts suggested splitting the indicator "Participation in Community" into two subindicators: one for civic voluntary participation and the other for participation in government organizations and decision making. Old-aged people's voluntary participation can be measured with the associations registered in the city. However, the second indicator was linked with governance. In fact from 2018, Castellón has had the Council of Old-Aged People, in which the elderly are periodically informed and can assess those issues concerning them (http://www.castello.es/web30/pages/generico_web10.php?cod1=25&cod2=1426, last accessed on 31 May 2021). It was added to the initial list as SDV.SO.4 in bold. This indicator, together with SEV.SE.1, affects the whole city equally, and it is not possible to disaggregate it by census section. However, it was included because its existence sheds light on some actions performed by Local Administrations to promote elder citizens' well-being.

With the socio–economic category, the judges suggested including unemployment for people older than 60 years. However, this disaggregated information was not found. Unemployment data exist for the whole population in the area (source: Municipality) or are divided into ages (under 25, 25–45, and older than 45 years), but no data are disaggregated into census sections (source: INE).

4.3. Application to the Case Study

In order to apply the model, the indicators were measured in areas according to different criteria and on a three normalized values scale, where 0 means no vulnerability, 0.5 means medium vulnerability, and 1 high vulnerability. There is also a threshold value used for evaluation, which depends on the indicator.

Table 6 provides detailed information about the measurements of indicators, along with a brief definition, the source from where data were obtained, and the measurement criteria according to a fixed threshold value. The last column depicts the evaluation in the analyzed area.

As the number of indicators (n) may vary from one category to another and to assign the same weight to each category, the vulnerability index for each dimension and area (Ivulj) was calculated using Equation (1):

$$I_j = \text{Sum of } I_i / n \tag{1}$$

where I_j indicates the vulnerability index for dimension or area j, I_i is the estimated value for indicator i, and n = number of indicators in dimension or area j.

A global index can also be obtained by calculating the mean value of the indices of all the dimensions from detecting vulnerability from an integrated view. The equation for the global index is the sum of all the categories as defined by Equation (2):

$$IGvul = Sum\ Ij/3 \tag{2}$$

where IGvul is the global vulnerability index and Ij is the vulnerability index for dimension j.

Figure 6 shows the vulnerability index values obtained for dimensions and areas, as well as the global index.

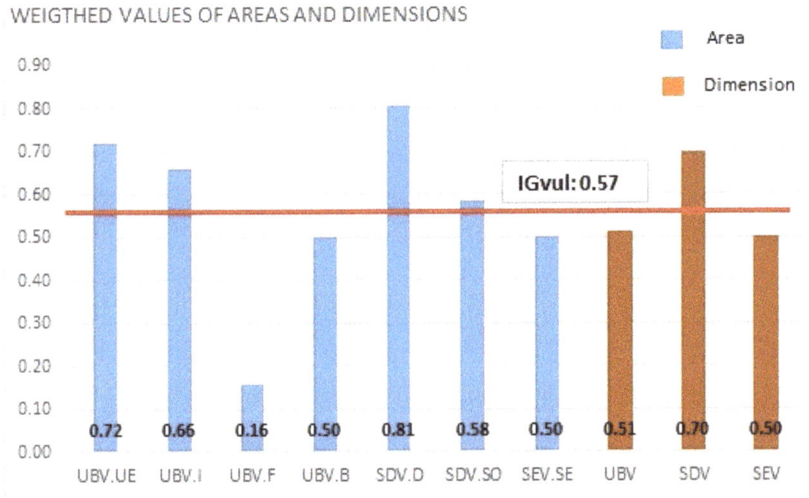

Figure 6. Vulnerability indices per category and dimension and the global index.

Many urban indicators were checked in situ. When calculating indices per area, they all exceeded 0.5 except for the Facilities area.

Urban vulnerability was highest in the Urban Environment area (UBV.UE), reaching 0.72, followed by the Infrastructure area, 0.66, but it decreased to 0.16 in the Facilities area, resulting in a global dimension vulnerability UBV of 0.51. Socio-demographic vulnerability, for demographic area (SDV.SD), was 0.81, which combined with the value 0.58 of the Social area (SDV.SO) resulted in a value for the dimension SDV of 0.70. Socioeconomic dimension had the lowest value of 0.5.

The global index integrated the three dimensions, resulting in a value of 0.57.

5. Discussion and Conclusions

Population aging is becoming increasingly more important in policies. When it comes to urban planning, the needs and problems that older people encounter should be carefully considered. The biggest challenges for the WHO's AFC concept are found in obsolete and vulnerable areas, which are normally inhabited by elderly people who tend to have lower incomes. The inclusive public space promotes informal relationships and becomes a neighborhood meeting place, where social networks are strengthened and social capital increases. These relationships have a positive effect on the health and quality of life of the elderly.

Table 6. Details of the evaluation in the area.

Indicator	Definition (Source)	Threshold Values. Criteria	Evaluation Scale	Evaluation in the Area	Weight	Vulnerability
UBV.UE.1. Green areas	Green area per inhabitant (m^2/inhabitant.) (1)	Threshold value 10 m^2/inhabitant. Minimum value according to World Health Organization (WHO) criteria and the Valencian Region Urban Planning and Landscape Law	<10 m^2/inh.: 1 10–15 m^2/inh.: 0.5 >15 m^2/inh.: 0	1	1.00	1.00
UBV.UE.2. Daytime acoustic comfort	Percentage of streets with noise level $\geq$ 55 dBA, 8–22 h (1)	Level of noise over 55 dBA in the daytime WHO criteria	>55 dBA: 1 50–55 dBA: 0.5 <50 dBA: 0	1	0.86	0.86
UBV.UE.3. Nighttime acoustic comfort	Percentage of streets with noise level > 45 dBA, 22–8 h (1)	Level of noise over 45 dBA, during night WHO criteria	>45 dBA: 1 40–45 dBA: 0.5 <40 dBA: 0	1	0.86	0.86
UBV.UE.4. Accessible Paving	Percentage of pavements over 1.5 m width (1–2)	(Meters of pavement width >1.5 m/Total meters of pavements in the area) × 100	<50%: 1 50–75%: 0.5 >75%: 0	0.5	1.00	0.50
UBV.UE.5. Ramps	Percentage of slopes under 8% on pavements (1–2)	(Meters of ramps on pavements with slopes <8%/Total meters of pavement in the area) × 100	<50%: 1 50–75%: 0.5 >75%: 0	0.5	1.00	0.50
UBV.UE.6. Pavements	Percentage of maintained pavements (2)	(Meters of well-maintained pavement m/Total meters of pavement in the area) × 100	<50%: 1 50–75%: 0.5 >75%: 0	0.5	1.00	0.50
UBV.UE.7. Pedestrian itinerary	Percentage of available pedestrian routes (1–2)	(Meters of available pedestrian routes/Total meters of routes in the area) × 100	<25%: 0 25–50%: 0.5 >50%: 1	0.5	1.00	0.50
UBV.UE.8. Cleanliness	Percentage of cleanliness observed in the area (2)	(Meters of streets presenting proper cleanliness/Total meters of streets in the area) × 100	No cleanliness: 1 Medium: 0.5 Cleanliness: 0	0.5	0.71	0.36
UBV.UE.9. Adapted urban furniture	Existence of adapted urban furniture (2)	Presence of adapted urban furniture	Scarce: 1 Medium: 0.5 Important: 0	1	0.86	0.86

Table 6. *Cont.*

Indicator	Definition (Source)	Threshold Values. Criteria	Evaluation Scale	Evaluation in the Area	Weight	Vulnerability
UBV.UE						**0.72**
UBV.I.1. Accessible public transportation	Percentage of accessible public transport stops (1)	(Number of accessible stops/total number of accessible stops) × 100	<50%: 1 50–75%: 0.5 >75%: 0	0.5	0.86	0.43
UBV.I.2. Lighting	Percentage of illuminated street (1)	Minimum values: <35 lux (road traffic) y <20 lux (pedestrian streets, inner courtyards)	<50%: 1 50–75%: 0.5 >75%: 0	0.5	0.67	0.33
UBV.I.3. Adapted traffic-lights	Percentage of adapted traffic lights (1–2)	(Adapted traffic lights/total traffic lights in the area) × 100	<50%: 1 50–75%: 0.5 >75%: 0	1	0.71	0.71
UBV.I						**0.66**
UBV.F.1. Health Facilities	Proximity to health facilities (<600 m) (1–2)	Distance from the geometric center of the neighborhood < 600 m	<600 m: 0 [600–1000] m: 0.5 >1000 m: 1	0	0.86	0.00
UBV.F.2. Elderly care Facilities	Proximity to elderly care facilities (<300 m) (1–2)	Distance from the neighborhood's geometric center < 300 m	<300 m: 0 [300–600] m: 0.5 >600 m: 1	0.5	0.86	0.43
UBV.F.3. Commercial Facilities	Proximity to commercial facilities (<300 m) (1–2)	Distance from the neighborhood's geometric center < 300 m	<300 m: 0 [300–600] m: 0.5 >600 m: 1	0	1.00	0.00
UBV.F						**0.16**
UBV.B.1. Age buildings	Percentage of buildings over 50 years old (3)	(Number of buildings >50 years/total number of buildings in the area) × 100.	>75%: 1 50–75%: 0.5 <50%: 0	1	0.71	0.71
UBV.B.2. Accessibility in residential buildings	Percentage of inaccessible residential b. (3–5 floors no elevator) (3)	(Number of inaccessible buildings/total number of buildings in the area) × 100	>25%: 1 10–25%: 0.5 <10%: 0	0.5	0.86	0.43
UBV.B.3. Accessibility in non-residential build.	Percentage of inaccessible non residential buildings (2, 3)	(Number of inaccessible buildings/total number of buildings in the area) × 100	>25%: 1 10–25%: 0.5 <10%: 0	0	0.71	0.00

Table 6. *Cont.*

Indicator	Definition (Source)	Threshold Values. Criteria	Evaluation Scale	Evaluation in the Area	Weight	Vulnerability
UBV.B						**0.50**
UBV						**0.51**
SDV.D.1. Population over 64 years	Percentage of population over 64 years (4)	(Number of people aged >64 in the area/total population in the area) × 100. ±10% Mean value in the city $\tilde{x}$: 18.4%	>1.1$\tilde{x}$: 1 [0.9$\tilde{x}$, 1.1$\tilde{x}$]: 0.5 <0.9$\tilde{x}$: 0	1	0.86	0.86
SDV.D.2. Population over 79 years	Percentage of population over 79 years (4)	(Number of persons aged >79 in the area/total population in the area) × 100. ±10% Mean value in the city $\tilde{x}$: 5.4%	>1.1$\tilde{x}$: 1 [0.9$\tilde{x}$, 1.1$\tilde{x}$]: 0.5 <0.9$\tilde{x}$: 0	1	0.71	0.71
SDV.D.3. Aging ratio	Persons aged $\geq$ 65 in relation to persons aged 15–64	(Number persons aged $\geq$ 65/number of persons aged 15–64) × 100 ±10% Mean value in the city $\tilde{x}$: 27.73%	>1.1$\tilde{x}$: 1 [0.9$\tilde{x}$, 1.1$\tilde{x}$]: 0.5 <0.9$\tilde{x}$: 0	0.5	0.57	0.29
SDV.D						**0.87**
SDV.SO.1. Social Services attention to dep.	Percentage of interventions for dependency in the area (5)	(Number of interventions for dependency in the area/population in the area) × 100. ±10% Mean value in the city $\tilde{x}$: 0.98	>1.1$\tilde{x}$: 1 [0.9$\tilde{x}$, 1.1$\tilde{x}$]: 0.5 <0.9$\tilde{x}$: 0	1	1.00	1.00
SDV.SO.2. Social subsidies	Percentage of social subsidies in the area (5)	(Number of social subsidies in the area/population in the area) × 100 ±10% Mean value in the city $\tilde{x}$: 0.6	>1.1$\tilde{x}$: 1 [0.9$\tilde{x}$, 1.1$\tilde{x}$]: 0.5 <0.9$\tilde{x}$: 0	1	1.00	1.00
SDV.SO.3. Civic voluntary participation	Percentage of civic associations in the area for the elderly	(Number of associations for the elderly/total number of civic associations in the area) × 100 ±10% Mean value in the city $\tilde{x}$: 4.2%	>1.1$\tilde{x}$: 1 [0.9$\tilde{x}$, 1.1$\tilde{x}$]: 0.5 <0.9$\tilde{x}$: 0	0	0.71	0.00
SDV.SO.4. Governance	Existence of mechanisms for old-aged people to participate in the local government	Identification of mechanisms of old-aged people to participate in the local government	Non existence: 1 Existence: 0	0	1.00	0.00

Table 6. *Cont.*

Indicator	Definition (Source)	Threshold Values. Criteria	Evaluation Scale	Evaluation in the Area	Weight	Vulnerability
SDV.SO						**0.54**
SDV						**0.70**
SEV.SE.1. Promotion of Active Ageing	Existence of Administration initiatives for the elderly (5)	Identification of mechanisms to promote AE by the local government	Non existence: 1 Existence: 0	0	0.71	0.00
SEV.SE.2. Household Income	Average annual net income per household (4)	Annual net income per household in euros $\pm 10\%$ Mean value in the city $\widetilde{x}$: €30,929	$>1.1\widetilde{x}$: 0 $[0.9\widetilde{x}, 1.1\widetilde{x}]$: 0.5 $<0.9\widetilde{x}$: 1	1	0.71	0.71
SEV.SE						**0.5**
SEV						**0.5**

Sources: (1) Land Use Plan; (2) In situ measurements; (3) Cadastral Office; (4) National Statistics Office; (5) Municipality.

In this work, we propose a model for evaluating vulnerability in the public space of an urban area of the city, from the perspective of older people. The model targets vulnerable areas, assuming that they should be prioritized for urban interventions and considering that urban plans can have great potential to improve the quality of life of citizens. In particular, older people could be considered to be a vulnerable population because they face a higher risk of isolation and exclusion as they tend to encounter greater problems in terms of mobility, accessibility, lower income, and so on.

Previous literature on the interaction between the urban space and the elderly (Table 1) and on vulnerability assessment (Table 2) provides a valuable starting point from which to develop a proposal. All the studies on vulnerability consider aspects related to the physical attributes of the urban environment (buildings, infrastructure, urban features . . .), together with population features (demography, social participation, economic status . . .).

Some of the above work is applicable at the local, regional or even national level. In this model, we propose the suburban level because the model has a bottom-up approach, where some on-site measures are needed, together with the availability of social information. Regarding the latter, we propose the inclusion of information from the Social Welfare Services of the local Administration. It is a valuable source that provides a very accurate idea of the vulnerability of older people given that these services manage local resources to promote the well-being of the elderly, making it possible to detect and correct the strengths and weaknesses of policies.

Site- specific features and the availability of information are analyzed in order to propose ad hoc indicators. The level of agreement reached by a panel of experts is used to validate the indicators and to improve the model. Vulnerability is obtained by indicator, area or dimension. A global index can be obtained to compare the vulnerability of areas within a city. For the case study, all the vulnerability indices by dimensions and areas exceed 0.5 in a scale from 0 to 1, except for the Facilities area, which is 0.16, probably due to the fact that the neighborhood is located in the city center. By areas, the highest value is 0.81 in Socio-demographic vulnerability, due to the high rate of aging in the area, followed by 0.72 in Urban Environment area, due to the deficiencies in accessibility, lighting, and maintenance, mainly encountered in the inner parts of the neighborhood. The values for the proposed dimensions are 0.50, 0.51, and 0.70 for UBV, SEV, and SDV, respectively. The global index is 0.57, which confirms the vulnerability of the area. A first look at the results suggests interventions and better maintenance in the urban space, for example, in the renovation of buildings, planting more trees and vegetation, or reducing the noise level in the avenues, among other necessities. The information from the Social Services indicates a high ratio of subsidies demanded in the area, compared to the average ratio for the city. This denotes a high degree of social vulnerability, which is consistent with the data on household income. On the other hand, participation in social life is not an issue in the area and the latest active-aging policies indicate good practices in the city that should be maintained.

The results presented are valid for the local case study outlined in this paper. A different urban area would need to adapt the indicators to its characteristics and starting conditions and to the urban planning requirements set down in the regulations. This makes the model dependent on the availability of reliable and disaggregated information. On the other hand, some of the information is obtained on site, from primary sources, which entails some limitations; on the one hand, for it to be achievable, the scale of application should not exceed the neighborhood scale; and on the other hand, information must be available from the Social Services. The model considers the technical input of experts, as recommended by other studies. The validation of the indicators has been made through the interjudge agreement technique, but other techniques with a panel of experts would be applicable, the main limitation being the selection of appropriate specialists.

The vulnerability values reached by indicators, areas, and dimensions could be useful to allocate funds to mitigate detected weaknesses in the area. The proposed model could be useful to the local public administration in order to facilitate and optimize urban planning and management, prioritizing interventions in vulnerable areas. It could be useful to introduce key factors at the regulatory level in order to support urban planners when they are redesigning neighborhoods and to strengthen the sense of community by promoting more age-friendly spaces. In addition, it could be a useful tool for social workers, who could systematically document and evaluate their work intended to develop an age-friendly community and to detect social deficiencies in order to implement appropriate social policies and to intervene at the public policy level.

Author Contributions: Conceptualization, R.A.-F. and M.J.R.; methodology, M.J.R.; software, F.K.; validation, all authors; formal analysis, all authors; investigation, all authors; resources, all authors; data curation, F.K.; writing—original draft preparation, R.A.-F. and M.J.R.; writing—review and editing, professional proofreading service; visualization, R.A.-F. and M.J.R.; supervision, R.A.-F. and M.J.R.; project administration, R.A.-F.; funding acquisition, R.A.-F. All authors have read and agreed to the published version of the manuscript.

Funding: This research was funded by the Regional Valencian Government (Vicepresidencia y Conselleria de Igualdad y Políticas Inclusivas de la Generalitat Valenciana), through Project Fomento de la Investigación en Servicios Sociales de la Comunitat Valenciana [21I158].

Institutional Review Board Statement: Not applicable.

Informed Consent Statement: Not applicable.

Acknowledgments: We thank the panel of experts for their valuable contribution to this work. We also thank the Municipality of Castellón that provided many of the data for research purposes.

Conflicts of Interest: The authors declare no conflict of interest.

References

1. Jablonska, J.; Trocka-Leszczynska, E. Public Space for Active Senior. *Adv. Intell. Syst. Comput.* **2020**, *1214*, 81–88. [CrossRef]
2. Lotfi, S.; Manouchehri, A. Evaluating urban service accessibility in the medium sized cities of Iran. *Theor. Empirica* **2015**, *10*, 77–95.
3. Fantova, F. Crisis de los cuidados y servicios sociales. *Zerbitzuan* **2015**, *60*, 47–62. [CrossRef]
4. Rocha, H.B. Social work practices and the ecological sustainability of socially vulnerable communities. *Sustainability* **2018**, *10*, 1312. [CrossRef]
5. Page, M.J.; Bossuyt, P.M.; Boutron, I.; Hoffman, T.C.; Mulrow, C.D. The PRISMA 2020 statement: An updated guideline for reporting systematic reviews. *BMJ* **2021**, *372*, 1312. [CrossRef]
6. Colleoni, M. A social science approach to the study of mobility: An introduction. In *Understanding Mobilities for Designing Contemporary Cities. Research for Development*; Pucci, P., Colleoni, M., Eds.; Springer: Cham, Switzerland, 2016. [CrossRef]
7. Pafka, E.; Dovey, K.; Aschwanden, G.D.P.A. Limits of space syntax for urban design: Axiality, scale and sinuosity. *Environ. Plan B Urban Anal. City Sci.* **2020**, *47*, 508–522. [CrossRef]
8. Tang, B.S.; Wong, K.K.H.; Tang, K.S.S.; Wai Wong, S. Walking accessibility to neighbourhood open space in a multi-level urban environment of Hong Kong. *Environ. Plan B Urban Anal. City Sci.* **2020**, *48*, 1340–1356. [CrossRef]
9. Alonso López, F. La Accesibilidad en Evolución: La Adaptación Persona-Entorno y su Aplicación al Medio Residencial en España y Europa. Doctoral Thesis, Universidad Autónoma de Barcelona, Barcelona, Spain, July 2016. Available online: https://ddd.uab.cat/record/166087 (accessed on 20 May 2021).
10. Arjona, G. La accesibilidad y el diseño universal entendido por todos. De cómo Stephen Hawking viajó por el espacio. In *Democratizando la Accesibilidad*; La Ciudad Accesible: Granada, Spain, 2015; Volume 4, Available online: http://riberdis.cedd.net/handle/11181/4655 (accessed on 2 June 2021).
11. Bianco, L. Universal design: From design philosophy to applied science. *J. Access. Des. All.* **2020**, *10*, 70–97. [CrossRef]
12. Bonilla, A.N.; Alcívar, D.E.; García, J.E.; Carrillo, A.J. Movilidad y accesibilidad universal en la arquitectura. Caso Universidad San Gregorio de Portoviejo. Ecuador. *Revistarquis* **2019**, *8*, 24–36. [CrossRef]
13. De Asís, R.; Aiello, A.L.; Bariffi, F.; Campoy, I.; Palacios, A. La accesibilidad universal en el marco constitucional español. *Revista Derechos y Libertades* **2007**, *16*, 57–82. Available online: http://hdl.handle.net/10016/7130 (accessed on 2 June 2021).
14. Dropkin, D. Accessibility for all. *Build. Eng.* **2008**, *83*, 12–13.
15. Olaru, D.; Smith, N.; Ton, T. Activities, Accessibility and Mobility for Individuals and Households. WIT Transactions on the Built Environment. In Proceedings of the 11th International Conference on Urban Transport and the Environment in the 21st Century, Online. 16–18 June 2021; Brebbia, C.A., Wadhwa, L., Eds.; WIT Press: Algarve, Portugal, 2005; Volume 77, pp. 373–383. Available online: https://www.witpress.com/Secure/elibrary/papers/UT05/UT05037FU.pdf (accessed on 21 July 2021).
16. Calonge-Reillo, F. Recursos de movilidad y accesibilidad urbana en los municipios del sur del área metropolitana de Guadalajara, México. *Urbano* **2018**, *21*, 48–57. [CrossRef]

17. Cocco, F.; Alonso-López, F. Ajustes razonables en la rehabilitación de polígonos de viviendas: Aplicación al Barrio Montserrat de Terrassa (Barcelona). *Archit. City Environ.* **2015**, *10*, 31–58. [CrossRef]
18. Elkouss, E. La accesibilidad: Hacia la plena integración social del discapacitado en el entorno urbano y natural. *Ci[ur]]* **2006**, *46*, 3–87. Available online: http://polired.upm.es/index.php/ciur/article/view/261/256 (accessed on 2 June 2021).
19. Faura-Martínez, U.; Lafuente-Lechuga, M.; García-Luque, O. Riesgo de pobreza o exclusión social: Evolución durante la crisis y perspectiva territorial. *Rev. Esp. Invest. Sociol.* **2016**, *156*, 59–76. [CrossRef]
20. Hong, J.; Shing, S.S. A study on infrastructure-centered publicness in urban public space through a look at Dutch architectural policies and practices. *J. Asian Archit. Build. Eng.* **2016**, *15*, 33–40. [CrossRef]
21. Kyttä, M.; Broberg, A.; Haybatollahi, M.; Schmidt-Thomé, K. Urban happiness: Context-sensitive study of the social sustainability of urban settings. *Environ. Plann. B Plann. Des.* **2016**, *43*, 34–57. [CrossRef]
22. Pitarch-Garrido, M.D.; Salom, J.; Fajardo, F. Detección de barrios vulnerables a partir de la accesibilidad a los servicios públicos de proximidad. El caso de la ciudad de Valencia. *An. Geogr. Univ. Complut.* **2018**, *38*, 61–85. [CrossRef]
23. Wu, J.; He, Q.; Chen, Y.; Lin, J.; Wang, S. Dismantling the fence for social justice? Evidence based on the inequity of urban green space accessibility in the central urban area of Beijing. *Environ. Plan. B Urban. Anal. City Sci.* **2018**, *31*, 913–930. [CrossRef]
24. Wu, K.C.; Song, L.Y. A case for inclusive design: Analyzing the needs of those who frequent Taiwan's urban parks. *Appl. Ergon.* **2017**, *58*, 254–264. [CrossRef]
25. Alipour, S.M.H.; Galal Ahmed, A. Assessing the effect of urban form on social sustainability: A proposed 'Integrated Measuring Tools Method' for urban neighborhoods in Dubai. *City Territ. Archit.* **2021**, *8*, 254–264. [CrossRef]
26. Yildiz, S.; Kivrak, S.; Gültekin, A.B.; Arslan, G. Identifying the key factors in construction projects that affect neighbourhood social sustainability. *Sustain. Cities Soc.* **2020**, *60*, 102173. [CrossRef]
27. Almahmoud, E.; Doloi, H.K. Built environment design—Social sustainability relation in urban renewal. *Facilities* **2020**, *38*, 765–782. [CrossRef]
28. Ahrentzen, S.; Tural, E. The role of building design and interiors in ageing actively at home. *Build. Res. Inf.* **2015**, *43*, 582–601. [CrossRef]
29. La Rosa, D.; Takatori, C.; Shimizu, H.; Privitera, R. A planning framework to evaluate demands and preferences by different social groups for accessibility to urban greenspaces. *Sustain. Cities Soc.* **2010**, *36*, 346–362. [CrossRef]
30. Fernández, J.L.; Parapar, C.; Ruiz, M. El envejecimiento de la población. *Cuadernos de la Fundación General CSIC. LYCHNOS* **2018**, *2*, 6–11. Available online: http://www.fgcsic.es/lychnos/es_es/articulos/envejecimiento_poblacion (accessed on 2 June 2021).
31. Iwarsson, S.; Wilson, G. Environmental barriers, functional limitations, and housing satisfaction among older people in Sweden: A longitudinal perspective on housing accessibility. *Technol Disabil.* **2006**, *18*, 57–66. [CrossRef]
32. Pacheco, A. Espacio Público y Envejecimiento Activo en los Barrios Bardegueral y Los Llanos. *Territorios en Formación.* **2017**, *11*, 101–119. [CrossRef]
33. Puyuelo, M.; Gual Ortí, J. Diseño prospectivo y elementos de uso en parques urbanos a partir de la experiencia de las personas mayores. *Medio Ambiente y Comportamiento Humano* **2009**, *10*, 137–160. Available online: http://repositori.uji.es/xmlui/handle/10234/22811 (accessed on 12 October 2020).
34. Yung, E.H.K.; Conejos, S.; Chan, E.H.W. Social needs of the elderly and active aging in public open spaces in urban renewal. *Cities* **2016**, *52*, 114–122. [CrossRef]
35. Basbas, S.; Konstantinidou, C.; Gogou, N. Pedestrians' needs in the urban environment: The case of the city of Trikala, Greece. *WIT Trans. Built Environ.* **2010**, *111*, 15–52. [CrossRef]
36. Wen, C.; Albert, C.; Von Haaren, C. The elderly in green spaces: Exploring requirements and preferences concerning nature-based recreation. *Sustain. Cities Soc.* **2018**, *38*, 582–593. [CrossRef]
37. Shan, W.; Xiu, C.; Ji, R. Creating a Healthy Environment for Elderly People in Urban Public Activity Space. *Int. J. Environ. Res. Public Health* **2020**, *17*, 7301. [CrossRef] [PubMed]
38. Chandrasiri, O.; Arifwododo, S. Inequality in Active Public Park: A Case Study of Benjakitti Park in Bangkok, Thailand. *Int. J. Procedia Eng.* **2017**, *198*, 193–199. [CrossRef]
39. Lee, C.; Moudon, A.V. Neighbourhood design and physical activity. *Build. Res. Inf.* **2008**, *36*, 395–411. [CrossRef]
40. Tao, Z.; Cheng, Y. Modelling the spatial accessibility of the elderly to healthcare services in Beijing, China. *Environ. Plan. B Urban Anal. City Sci.* **2019**, *46*, 1132–1147. [CrossRef]
41. Brake, J.F. Identifying appropriate options for delivering urban transportation to older people. *WIT Trans. Built Environ.* **2008**, *101*, 57–66. [CrossRef]
42. Espínola, A. *Comparativa Sobre Normativa de Accesibilidad en Urbanismo y Edificación en España*; La Ciudad Accesible: Granada, Spain, 2018; Available online: http://hdl.handle.net/11181/5367 (accessed on 2 June 2021).
43. Casado, D.; Fantova, F. Los sistemas de bienestar en España: Evolución y naturaleza. *Doc. Soc.* **2017**, *186*, 55–80.
44. Urrutia, A. Envejecimiento activo: Un paradigma para comprender y gobernar/Active ageing: A paradigm for understanding and governing. *Aula Abierta* **2018**, *47*, 45–54. [CrossRef]
45. López Martínez, A. Análisis de las Relaciones Sociales y la Fragilidad en Mayores de 75 Años Residentes en Castellón de la Plana. Ph.D. Doctoral Thesis, Universitat Jaume I, Castellón, Spain, July 2017. Available online: https://www.tdx.cat/handle/10803/481957#page=1 (accessed on 20 May 2021).

46. Leiton, Z.E. El envejecimiento saludable y el bienestar: Un desafío y una oportunidad para enfermería / Healthy ageing and well-being: A challenge, but also an opportunity for nursing. *Enferm. Univ.* **2016**, *13*, 139–141. [CrossRef]
47. Limón, M.R. Envejecimiento activo: Un cambio de paradigma sobre el envejecimiento y la vejez / Active Aging: A change of paradigm on aging and old age. *Aula Abierta* **2018**, *47*, 45–54. [CrossRef]
48. Pérez, J.; Abellán, A.; Aceituno, P.; Ramiro, D. Un Perfil de Las Personas Mayores en España, 2020. Indicadores Estadísticos Básicos. Informes Envejecimiento en Red. Numero, 25. 2020. Available online: http://envejecimiento.csic.es/documentos/documentos/enred-indicadoresbasicos2020.pdf (accessed on 20 May 2021).
49. Afacan, Y. Elderly-friendly inclusive urban environments: Learning from Ankara. *Open House Int.* **2013**, *38*, 52–63. [CrossRef]
50. Elsawahli, H.; Ahmad, F.; Ali, A.S. A qualitative approach to understanding the neighborhood environmental influences on active aging. *J. Des. Built Environ.* **2017**, *17*, 16–26. [CrossRef]
51. Musselwhite, C.; Holland, C.A.; Walker, I. The role of transport and mobility in the health of older people. *J. Transp. Health* **2015**, *2*, 1–4. [CrossRef]
52. Gharaveis, A. A systematic framework for understanding environmental design influences on physical activity in the elderly population: A review of literature. *Facilities* **2020**, *38*, 625–649. [CrossRef]
53. Parra Rizo, M.A. Envejecimiento Activo y Calidad de Vida: Análisis de la Actividad Física y Satisfacción Vital en Personas Mayores de 60 Años. Ph.D. Doctoral Thesis, Universidad Miguel Hernández, Elche, Spain, May 2018. Available online: http://dspace.umh.es/bitstream/11000/4457/1/TD%20Parra%20Rizo%2c%20Maria%20Antonia.pdf (accessed on 8 June 2021).
54. Van Hoof, J.; Kazak, J.K.; Perek-Bialas, J.M.; Peek, S.T.M. The Challenges of Urban Ageing: Making Cities Age-Friendly in Europe. *Int. J. Environ. Res. Public Health* **2018**, *15*, 2473. [CrossRef]
55. Van Hoof, J.; Kazak, J.K. Urban ageing. *Indoor Built Environ.* **2018**, *27*, 583–586. [CrossRef]
56. Zaidi, A.; Howse, K. The Policy Discourse of Active Ageing: Some Reflections. *J. Popul. Ageing* **2017**, *10*, 1–10. [CrossRef]
57. Thalassinos, E.; Cristea, M.; Noja, G.G. Measuring acive ageing within the European Union: Implications on economic development. Equilibrium. *Q. J. Econ.* **2019**, *14*, 591–609. [CrossRef]
58. Montañés, R. Castellón: Una Ciudad Amigable con las Personas Mayores. Bachelor's Thesis, Jaume I University, Castellón, Spain, May 2017. Available online: http://hdl.handle.net/10234/169038 (accessed on 8 June 2021).
59. Van Hoof, J.; Marston, H.R. Age-friendly cities and communities: State of the art and future perspectives. *Int. J. Environ. Res. Public Health* **2021**, *18*, 1644. [CrossRef] [PubMed]
60. De Oliveira, S.M.L.; Pessa, S.L.R.; Schenatto, F.J.; de Lourdes Bernartt, M. Cities and Population Aging: A Literature Review. *Adv. Intell. Syst. Comput.* **2019**, *824*, 1644. [CrossRef]
61. Rémillard-Boilard, S.; Bufel, T.; Phillipson, C. Developing Age-Friendly Cities and Communities: Eleven Case Studies from around the World. *Int. J. Environ. Res. Public Health* **2021**, *18*, 133. [CrossRef]
62. Sengers, F.; Peine, A. Innovation pathways for age-friendly homes in Europe. *Int. J. Environ. Res. Public Health* **2021**, *18*, 1139. [CrossRef] [PubMed]
63. Marquet, O.; Hipp, A.; Miralles-Guasch, C. Neighborhood walkability and active ageing: A difference in differences assessment of active transportation over ten years. *J. Transp. Health* **2017**, *7*, 190–201. [CrossRef]
64. Hochstenbach, C. The age dimensions of urban socio-spatial change. *Popul. Space Place* **2019**, *25*, e2220. [CrossRef]
65. Zhang, C.J.; Barnett, A.; Johnston, J.M.; Lai, P.C.; Lee, R.S.; Sit, C.H.; Cerin, E. Objectively measured neighbourhood attributes as correlates and moderators of quality of life in older adults with different living arrangements: The ALECS cross-sectional study. *Int. J. Environ. Res. Public Health* **2019**, *16*, 876. [CrossRef]
66. Iamtrakul, P.; Chayphong, S.; Klaylee, J. The study on age-friendly environments for an improvement of quality of life for elderly, Asian mega city, Thailand. *Lowl. Technol. Int.* **2019**, *21*, 123–133.
67. Fung, J.C. Place Familiarity and Community Ageing-with-Place in Urban Neighbourhoods. In *Advances in 21st Century Human Settlements*; Leong, C.H., Malone-Lee, L.C., Eds.; Springer: Singapore, 2020. [CrossRef]
68. Del Barrio, E.; Pinzón, S.; Marsillas, S.; Garrido, F. Physical Environment vs. Social Environment: What Factors of Age-Friendliness Predict Subjective Well-Being in Men and Women? *Int. J. Environ. Res. Public Health* **2021**, *18*, 798. [CrossRef]
69. Cordero del Castillo, P. European year of active aging and the intergenerational solidarity. *Humanismo y Trabajo Social* **2012**, *11*, 101–117. Available online: http://hdl.handle.net/10612/3430 (accessed on 8 June 2021).
70. Flores, R.; Caballer, A.; Alarcón, A. Evaluation of an age-friendly city and its effect on life satisfaction: A two-stage study. *Int. J. Environ. Res. Public Health* **2019**, *16*, 5073. [CrossRef]
71. Huenchuan, S. (Ed.) *Envejecimiento, Personas Mayores y Agenda 2030 Para el Desarrollo Sostenible: Perspectiva Regional y de Derechos Humanos*; Libros de la CEPAL, 154 (LC/PUB.2018/24-P); Comisión Económica para América Latina y el Caribe (CEPAL): Santiago, Chile, 2018; Available online: https://repositorio.cepal.org/bitstream/handle/11362/44369/1/S1800629_es.pdf (accessed on 8 June 2021).
72. Huenchuan, S.; Rivera, E. (Eds.) *Experiencias y Prioridades Para Incluir a las Personas Mayores en la Implementación y Seguimiento de la Agenda 2030 para el Desarrollo Sostenible*; (LC/MEX/SEM.245/1); Comisión Económica para América Latina y el Caribe (CEPAL): Santiago, Chile, 2019; p. 151. Available online: https://repositorio.cepal.org/handle/11362/44600 (accessed on 8 June 2021).
73. Cucó-Giner, J. Un barrio marginado no es un barrio marginal. A propósito de Nazaret (Valencia). *Revista de Dialectología y Tradiciones Populares* **2016**, *71*, 151–171. [CrossRef]

74. Bayar, R.; Türkoğlu, H. The relationship between living environment and daily life routines of older adults. *A/Z ITU J. Fac. Archit.* **2021**, *18*, 29–43. [CrossRef]
75. Carroll, S.; Jesperen, A.P.; Troelsen, J. Going along with older people: Exploring age-friendly neighbourhood design through their lens. *J. Hous. Built Environ.* **2020**, *35*, 555–572. [CrossRef]
76. Curl, A.; Mason, P. Neighbourhood perceptions and older adults' wellbeing: Does walking explain the relationship in deprived urban communities? *Transp. Res. Part A Policy Pract.* **2019**, *123*, 119–129. [CrossRef]
77. Annear, M.; Keeling, S.; Wilkinson, T.; Cushman, G.; Gidlow, B.; Hopkins, H. Environmental influences on healthy and active ageing: A systematic review. *Ageing Soc.* **2014**, *34*, 590–622. [CrossRef]
78. Lai, M.M.; Lein, S.Y.; Lau, S.H.; Lai, M.L. Determinants of age-friendly communities. *Gerontechnology* **2014**, *13*, 228. [CrossRef]
79. Mercader-Moyano, P.; Flores-García, M.; Serrano-Jiménez, A. Housing and neighbourhood diagnosis for ageing in place: Multidimensional Assessment System of the Built Environment (MASBE). *Sustain. Cities Soc.* **2020**, *62*, 102422. [CrossRef]
80. Ravi, K.E.; Fields, N.L.; Dabelko-Schoeny, H. Outdoor spaces and buildings, transportation, and environmental justice: A qualitative interpretive meta-synthesis of two age-friendly domains. *J. Transp. Health* **2021**, *20*, 100977. [CrossRef]
81. Davern, M.; Winterton, R.; Brasher, K.; Woolcock, G. How Can the Lived Environment Support Healthy Ageing? A Spatial Indicators Framework for the Assessment of Age-Friendly Communities. *Int. J. Environ. Res. Public Health* **2020**, *17*, 7685. [CrossRef]
82. Lak, A.; Rashidghalam, P.; Amiri, S.N.; Myint, P.K.; Baradaran, H.R. An ecological approach to the development of an active aging measurement in urban areas (AAMU). *BMC Public Health* **2021**, *21*, 4. [CrossRef]
83. Astiaso Garcia, D.; Cumo, F.; Pennacchia, E.; Stefanini-Pennucci, V.; Piras, G.; De Notti, V.; Roversi, R. Assessment of a urban sustainability and life quality index for elderly. *Int. J. Sustain. Dev. Plan.* **2017**, *12*, 908–921. [CrossRef]
84. Sonmez Turel, H.; Malkoc Yigit, E.; Altug, I. Evaluation of elderly people's requirements in public open spaces: A case study in Bornova District (Izmir, Turkey). *Build. Environ.* **2007**, *42*, 2035–2045. [CrossRef]
85. Bosia, D.; Montacchini, E.; Savio, L.; Tedesco, S. Aging-People Accessibility to Urban Garden: A Case Study in Turin. *Adv. Intell. Syst. Comput.* **2020**, *1205*, 327–334. [CrossRef]
86. Rehal, P.; Chani, P.S.; Atreya, S.; Sehgal, V. Ageing-Friendly Neighbourhoods: A Study of Mobility and Out-of-Home Activity. *Lect. Notes Civ. Eng.* **2021**, *113*, 267–280. [CrossRef]
87. Triviño, M.; Oyarzo, V.; Brito, E.; Vega, P.; Ojeda, M.; Rojas, A.; Ivanissevich, L. Una vejez, una ciudad y un vacío. Una visión in situ de la acogida a la ancianidad que ofrece Rio gallegos, Pcia. de Sta. Cruz, 2015. *Informes Científicos Técnicos UNPA* **2016**, *8*, 256–272. [CrossRef]
88. Walker, A.; Maltby, T. Active ageing: A strategic policy solution to demographic ageing in the European Union. *Int. J. Soc. Welf.* **2012**, *21*, S117–S130. [CrossRef]
89. Kurek, S.; Rachwal, T. Development of entrepreneurship in ageing populations of the European Union. *Procedia Soc. Behav. Sci. Welf.* **2011**, *19*, 397–405. [CrossRef]
90. Garrido, M.; Jaraíz, G. Políticas inclusivas en barrios urbanos vulnerables. *Áreas: Revista Internacional de Ciencias Sociales* **2017**, *36*, 141–151. Available online: https://dialnet.unirioja.es/servlet/articulo?codigo=6246390 (accessed on 21 July 2021).
91. Caballer, A.; Castillo, A.; Martínez, M.A.; Flores, R.; Alarcón, A.H.; Agost-Felip, R.; Mulet, M.; Díaz, M.J. *Estudio del Envejecimiento Activo en la Ciudad de Castellón Desde el Paradigma de la Organización Mundial de la Salud*; Publisher Fundación Dávalos-Fletcher: Castellón de la Plana, Spain, 2017.
92. Giménez-Bertomeu, V.M.; Acebal Fernández, A.; Ferrer-Aracil, J.; Cortés-Florín, E.M.; De Alfonseti Hartmann, N.; Mira-Perceval Pastor, M.T.; Domenech-López, Y. *Vulnerabilidad Territorial: Indicadores Para su Medición Desde los Servicios Sociales*; Limencop, S.L.: Alicante, Spain, 2020.
93. Sorribes i Monrabal, J.; Perelló Oliver, S. Hacia un sistema de indicadores de vulnerabilidad urbana. *Barataria. Revista Castellano Manchega De Ciencias Sociales* **2006**, *6*, 87–103. [CrossRef]
94. Egea, C.; Nieto, J.A.; Domínguez, J.; González, R.A. Vulnerabilidad del tejido social de los barrios desfavorecidos en Andalucía. Análisis y potencialidades. *Cuadernos Geográficos* **2008**, *45*, 325–327. [CrossRef]
95. Alguacil, J.; Camacho, J.; Hernández-Ajá, A. La vulnerabilidad urbana en España. Identificación y evolución de los barrios vulnerables. *Empiria. Revista de Metodología de Ciencias Sociales* **2013**, *27*, 73–94. [CrossRef]
96. Piñeira-Mantiñán, M.; González-Pérez, J.M.; Lois-González, R.C. Vulnerabilidad urbana y exclusión. La fragmentación social de la ciudad postcrisis. In *Nuevos Escenarios Urbanos: Nuevos Conflictos y Nuevas Políticas: XIII Coloquio de Geografía Urbana*; Castañer, M., Vicente, J., Feliu, J., Martín-Uceda, J., Eds.; Universitat de Girona: Girona, Spain, 2017; pp. 75–90.
97. Temes, R. Valoración de la vulnerabilidad integral en las áreas residenciales de Madrid. *EURE* **2014**, *40*, 119–149. Available online: https://www.eure.cl/index.php/eure/article/view/344/606 (accessed on 8 June 2021). [CrossRef]
98. Temes, R. Visor de Espacios Urbanos Sensibles (VEUS). Una nueva herramienta para intervenir en la ciudad. In Proceedings of the III Internacional Conference ISUF-H. Ciudad Compacta vs. Ciudad Difusa, Guadajalara, Mexico, 18–20 September 2019; Universitat Politècnica de València: Valencia, Spain, 2020; pp. 454–461. [CrossRef]
99. Caravantes, G.M. El Derecho a la Ciudad desde la exclusión residencial: La evolución de los barrios vulnerables de la Comunitat Valenciana. In Proceedings of the XIII International Conference on Virtual City and Territory: "Challenges and Paradigms of the Contemporary City": UPC, Barcelona, Spain, 2–4 October 2019; CPSV: Barcelona, Spain, 2019; p. 8661. [CrossRef]

100. García-Bernal, D.; Huedo, P.; Babiloni, S.; Braulio, M.; Carrascosa, C.; Civera, V.; Ruà, M.J.; Agost-Felip, M.R. Estudio y Propuesta de Áreas de Rehabilitación, Regeneración y Renovación Urbana, con Motivo de la Tramitación del Plan General Estructural de Castellón de la Plana. Memoria (Tomo I). 2017. Available online: https://s3-eu-west-1.amazonaws.com/urbanismo/TOMO_I.pdf (accessed on 15 April 2021).
101. Ruá, M.J.; Huedo, P.; Civera, V.; Agost-Felip, M.R. A simplified model to assess vulnerable areas for urban regeneration. *Sustain. Cities Soc.* **2019**, *46*, 101440. [CrossRef]
102. Cremades, R. Validación de un instrumento para el análisis y evaluación de webs de bibliotecas escolares mediante el acuerdo interjueces. *Investig. Bibl.* **2017**, *31*, 127–149. [CrossRef]
103. Dubé, J.É. Evaluación del acuerdo interjueces en investigación clínica. Breve introducción a la confiabilidad interjueces. *Rev. Argentina de Clin. Psicol.* **2008**, *17*, 75–80.

Article

Present and Future Energy Poverty, a Holistic Approach: A Case Study in Seville, Spain

Mª Desirée Alba-Rodríguez [1], Carlos Rubio-Bellido [1,*], Mónica Tristancho-Carvajal [1], Raúl Castaño-Rosa [2,3] and Madelyn Marrero [1]

[1] Department of Building Construction II, School of Building Engineering, University of Seville, 41012 Seville, Spain; malba2@us.es (M.D.A.-R.); mtristancho@us.es (M.T.-C.); madelyn@us.es (M.M.)

[2] School of Architecture, Faculty of Built Environment, Tampere University, 33720 Tampere, Finland; raulcastano90@gmail.com

[3] Department of Electrical Engineering, University Carlos III of Madrid, 28911 Leganés, Spain

* Correspondence: carlosrubio@us.es

Abstract: Energy poverty is a social problem that is accentuated in a climate change future scenario where families become increasingly vulnerable. This problem has been studied in cold weather, but it also takes place in warm climates such as those of Mediterranean countries, and it has not been widely targeted. In these countries, approximately 70% of its building stock was built during 1960–1980, its renovation being an opportunity to reduce its energy demand, improve tenants' quality of life, and make it more resilient to climate change. In the retrofitting process, it is also important to consider tenants' adaptability and regional scenarios. In this sense, the present work proposes an assessment model of retrofitting projects that takes into consideration energy consumption, comfort, tenants' health, and monetary poverty. For this, the Index of Vulnerable Homes was implemented in this research to consider adaptive comfort in the energy calculation as well as the adaptability to climate change. A case study of 40 social housings in Seville, Spain, was analyzed in 2050 and 2080 future scenarios, defining the impact in energy poverty of the building retrofitting projects.

Keywords: energy poverty; climate change; life-cycle analysis; direct and indirect energy; bill of quantities

check for updates

Citation: Alba-Rodríguez, M.D.; Rubio-Bellido, C.; Tristancho -Carvajal, M.; Castaño-Rosa, R.; Marrero, M. Present and Future Energy Poverty, a Holistic Approach: A Case Study in Seville, Spain. *Sustainability* **2021**, *13*, 7866. https:// doi.org/10.3390/su13147866

Academic Editors: Domenico Mazzeo and Antonio Caggiano

Received: 25 May 2021
Accepted: 9 July 2021
Published: 14 July 2021

Publisher's Note: MDPI stays neutral with regard to jurisdictional claims in published maps and institutional affiliations.

Copyright: © 2021 by the authors. Licensee MDPI, Basel, Switzerland. This article is an open access article distributed under the terms and conditions of the Creative Commons Attribution (CC BY) license (https:// creativecommons.org/licenses/by/ 4.0/).

1. Introduction

In 2010, 32% of global primary energy was employed in buildings, which produced 19% of global emissions, as summarized in the Fifth Assessment Report (AR5) of the United Nations Intergovernmental Panel on Climate Change (IPCC) [1]. In order to reduce emissions and, consequently, stop the temperature increase and mitigate the effects of climate change, a series of agreements have been established at a European level, starting with the new European directive on the energy efficiency of buildings, which toughens its objectives in search of eliminating the use of fossil fuels in the real estate stock before 2050 [2], continuing with the agreement of the Climate and Energy Package 2020 to guarantee that the EU achieves the climate objectives of 2020 [3] and then extends them until 2030 [4], and ending with the European Green Deal [5], a continental tool to combat climate change that aims to make Europe the first climate-neutral continent by 2050.

In European countries with a Mediterranean climate, such as Spain, Greece, and Portugal, with a high percentage of aging existing buildings, built between the 1960s and 1980s [6], the renovation of buildings is a key factor in reducing the environmental and social impact of the housing and in the achievement of the global objective of mitigation of climate change. The mildness of winters in Mediterranean climate areas has resulted in existing homes being energy inefficient and excessively cold, making it very expensive to achieve thermal comfort inside homes. In the same way, the harshness of summers requires that the air conditioners be kept connected for a high number of hours a day, and thus in

143

the Mediterranean climate, the use of refrigeration devices becomes a necessity to achieve adequate comfort inside the houses. This problem is exacerbated in the future projection of climate change, where temperatures will presumably increase, and this will affect more seriously the situations of energy poverty (EP) to which families may be exposed. The energy rehabilitation of existing buildings positively affects both environmental aspects, through the reduction of emissions and energy consumption, and socio-economic aspects, with the savings achieved in the National Health Service (NHS), highly related to energy poverty [7–9].

Although the relationship amongst health, EP, cold homes, and overheating risk has been analyzed in many studies [10,11], it remains difficult to identify the direct impact due to the multidimensional aspects of EP. The effectiveness of current EP indicators is limited, and it is therefore necessary to combine various indicators and to analyze their results together [12].

In the case of Spain, in 2002, the Energy Performance of Buildings Directive (EPBD) [13] was applied in the Building Technical Code (CTE) [14] and in the regulations of thermal installations in buildings (RITE) [15]. More recently, the directive 2018/844 [2] established new energy efficiency targets for 2050, encouraging the energy renovation of existing buildings. This led to a modification of the Basic Document "DB HE Energy Saving" [16]. In the current legislation, which establishes a maximum limit of thermal transmittance to the building envelope elements, the values are becoming increasingly restrictive for winter climatic severity; however, this increase in the restriction does not apply to summer. Moreover, thermal comfort is based on Fanger's thermal model, which presents upper and lower limits with a narrow range, and thus these thermal comfort limits show a static behavior. This supposes overlooking external climate conditions and high levels of energy consumption.

In this sense, international adaptive comfort standards EN 16798-1:2019 [17] and ASHRAE 55-2017 [18] as well as research works studying thermal comfort with field methods in actual buildings based on adaptive thermal comfort [19,20] set an alternative to consider in this research. In addition, various studies gather the potential use of mixed-mode natural ventilation of buildings as an effective strategy to reduce energy consumption [21,22] and fuel poverty [23].

In the present work, an implementation of the Index of Vulnerable Homes (IVH) assesses the vulnerability to EP (pre- and post-intervention) in a housing project in Spain. The implemented IVH adds the evaluation of energy consumption considering mixed-mode and adaptive comfort as well as the future climate change scenarios to the robust indicator, which considered monetary poverty, comfort, and health of tenants. A representative social housing project in Andalusia, Spain is analyzed, which is formed by 40 dwellings of a multifamily building. The main objective is to be able to predict the impact of the renovation projects in the present and future vulnerability to EP. The objective also includes the IVH improvements, that is, the quantification of the energy consumption using an adaptive comfort control system, considering also the climate change predictions for 2050 and 2080.

2. Methodology

EP is commonly defined as the inability of a home to satisfy a minimum quantity of energy for its basic needs, such as keeping the home temperature in a range suitable for its health [12]. This problem has generated interest among the countries, standing out are France, Italy, United Kingdom, Austria, Ireland, and Slovakia [24]. In general, the European Commission (EC) uses three basic criteria to assess EP: the inability to keep the houses adequately conditioned, the delay in the payment of utility bills, and inhabiting unhealthy homes. The EP concept has been evolving to include the deprivation of hot water, lighting, and other domestic needs [25].

Castaño-Rosa et al. [12] reviewed the indicators used to analyze EP and grouped them into two categories: based on household income and expenses and on household perception by surveys. In addition, they identified indicators that analyze, in a broader sense, the most vulnerable consumers through econometric analyses [26,27], identifying overcrowded homes [28,29], measuring thermal comfort [30–32], and analyses based on the energy efficiency of buildings [33,34].

The objective of the work resides in the implementation of the Vulnerable Household Index (IVH) as a comprehensive model of EP assessment. The need for this implementation is because the indices and social parameters required for the analysis of the problem are scattered, and thus it is very complex to carry out a holistic assessment of the problem. In addition, current indicators do not include in the analysis future climate change scenarios, and there is a need to approach the issue from an integrative point of view. Climate change has been analyzed in many economic sectors, especially in construction, which represents approximately 40% of energy consumption by human activities [35]. For this reason, the holistic analysis of energy management is key to guaranteeing minimum conditions of habitability in homes. This makes climate change, together with adaptive comfort, the focus of attention for governments and researchers, generating numerous studies based on the analysis of the adaptive control of thermal comfort for the prediction of consumption in homes. [21,36,37]. The proposed model, in addition to integrating social factors, such as the occupants' health and economic analysis of households [38,39], is capable of integrating energy efficiency factors and their future forecast. As a novelty in this work, climatic adaptability is also included in the model to better predict consumption and comfort levels with respect to habitability conditions.

2.1. Index of Vulnerable Homes

The Index of Vulnerable Homes (IVH) analyses the vulnerability situation of families in relation to the consequences, as well as the possibility of evaluating the energy retrofitting impact in order to improve their life quality. The IVH identifies different situations of vulnerability to EP [12,38,40], becoming a comprehensive measure to better understand EP at the local scale. In its latest application, six buildings located in Seville are analyzed. The index estimates the cost of the National Health Service associated with EP, as well as the corresponding savings after the building renovation project [38]. Additionally, the IVH has been adapted and applied to the British context [39]. The index is formed by four components: Monetary Poverty Indicator (MPI), Energy Indicator (EnI), Comfort Indicator (CI), and Health-Related Quality-Life Cost (HRQLC) (Figure 1).

2.1.1. Monetary Poverty Indicator (MPI)

The monetary vulnerability of the household is analyzed by combining regionally specific indicators, the Monetary Poverty Threshold (MPT) and the Severe Monetary Poverty Threshold (SMPT). The MPT is obtained by extracting 60% of the average operating income of the area under study. In this work, Seville, Spain, is analyzed using Eurostat [41] statistics. The SMTP defines extreme poverty and corresponds to the lowest extraordinary unemployment benefit granted by the Spanish State, called active insertion income [42]. Then it relates the net income to the poverty threshold. A household is in monetary poverty or severe monetary poverty when MPI < 1.00. The calculation procedure is summarized in Figure 1 (Equations (1), (1.1), (1.2), (1.3) and (1.4) of Figure 1).

2.1.2. Implemented Energy Indicator (EnI)

The required energy consumption (EC) of a household is compared to the energy threshold set for the neighborhood (Equation (2), Figure 1) and is obtained according to EN 16798-1:2019 [17] and the works of Sánchez-García et al. [36]; MEC is the median energy consumption required for the type of building in the area of study [43]. Therefore, the housing energy consumption is admissible if it is below the energy threshold, or EnI < 1.00.

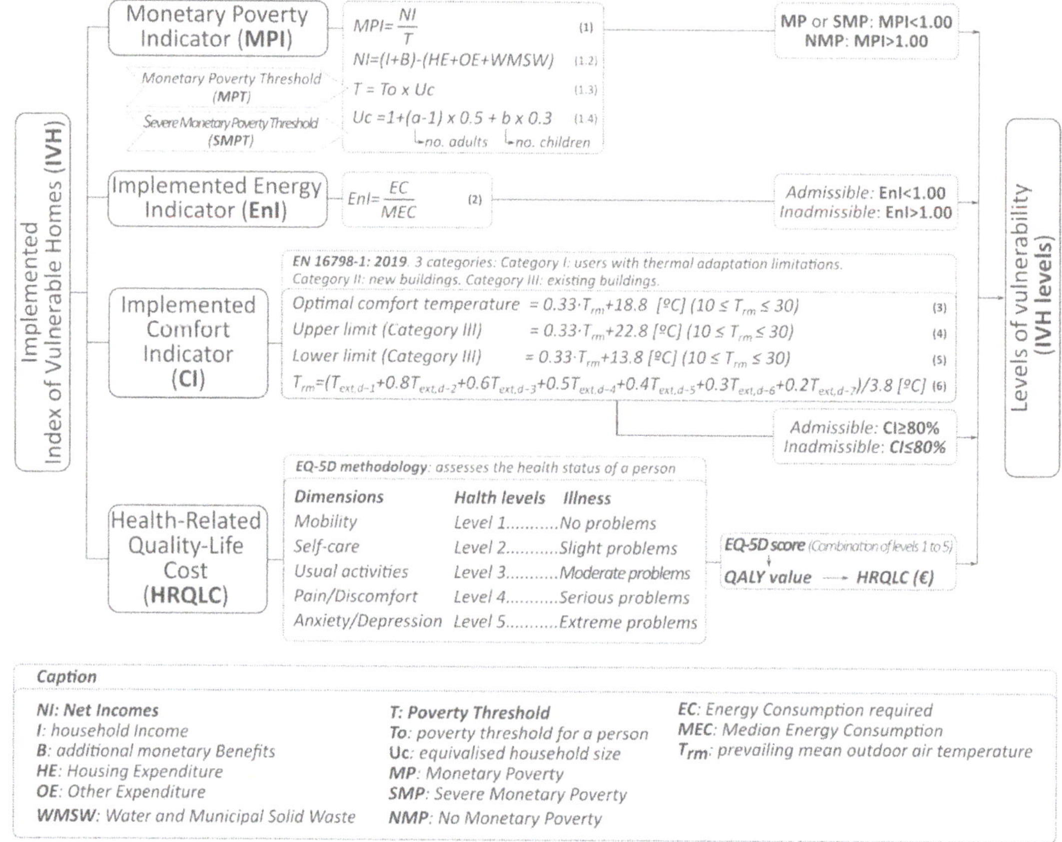

Figure 1. Summary scheme of the methodology for calculating the implemented Index of Vulnerable Homes (IVH).

2.1.3. Implemented Comfort Indicator (CI)

The adaptive thermal comfort model used in the present work considers that if the relationship between the exterior temperature and the interior temperature remains within the established comfort range, the occupants will be in a comfortable situation. IC determines the percentage of hours that the temperature is outside the established comfort range. The comfort threshold is set at 80% because the remaining 20% are considered part of the sleeping hours. This means that the occupants of a home can be thermally uncomfortable for 5 h a day, coinciding with the hours of sleep [44]. To obtain hourly temperatures, advanced dynamic simulations were performed using hourly climate data files in the case study model. Finally, when the hours considered within thermal comfort are in a percentage equal to or greater than 80%, it is established that IC is admissible (IC $\geq$ 80%) (Figure 1).

The EN 16798-1: 2019 standard [17] establishes four categories of comfort temperature range, according to the expectations of the occupants and the age of the building. Due to the type of building under study in this work (existing residential building), category III is considered for the calculation of the limits of the range of thermal comfort.

As can be seen from the explanation developed in the previous paragraphs, the IVH is a model based on the calculation of four indicators. MPI is obtained from the family-specific economic situation, and the comfort and energy indicators are obtained from software modeling. The ones obtained from simulations are not subjected to the tenant's perceptions or actual consumptions. With this new approach, based on the use of adaptive comfort models, it is intended to reduce subjectivity when analyzing EP, using more objective data, which allows opening a new line of research of the EP indicators used so far.

2.1.4. Health-Related Quality-Life Cost (HRQLC)

This health-related cost is defined by the Quality-Adjusted Life Year (QALY), equiv-alized to each level of vulnerability of the IVH (Figure 1). The Spanish National Health Service cost of maintaining a person in good health for a year is EUR 30,000 [45]. The calculation process is explained in detail in Castaño et al. [39].

Table 1 shows the result of the QALYs which depends on the dimensions levels from 1 to 5, 5 being the worst. The example combination 12333, defined in Table 1, is input into the EQ-5 D-5 L Index Value Calculator [46], and its corresponding QALY is obtained. The HRQLC (EUR) is the monetary value assigned to that QALY and is obtained by applying the QALY to the cost of the Spanish NHS to keep a person in good health for one year (EUR 30,000) (Figure 2).

Table 1. Example of QALYs.

Dimensions	Health Levels	Illness	QALY
Mobility	1	No problems	
Self-care	2	Slight problems performing self-care activities	0.642
Usual activities	3	Moderate problems performing usual activities	
Pain/Discomfort	3	Moderate pain/discomfort problems	
Anxiety/Depression	3	Moderate anxiety/depression problems	

MPI	EnI	CI	EQ-5D score	QALY	HRQLC (€)	IHV Level
SMP	Inadmissible	Inadmissible	25555	-0.311	39,330	13
MP	Inadmissible	Inadmissible	24455	-0.096	32,880	12
SMP	Inadmissible	Inadmissible	14455	-0.008	30,240	11
SMP	Inadmissible	Admissible	13344	0.309	20,730	10
SMP	Admissible	Inadmissible	14334	0.358	19,260	9
MP	Inadmissible	Inadmissible	13433	0.484	15,480	8
MP	Inadmissible	Admissible	13333	0.620	11,440	7
SMP	Admissible	Admissible	12333	0.642	10,740	6
MP	Admissible	Inadmissible	11333	0.754	7380	5
MP	Admissible	Admissible	11223	0.786	6420	4
NMP	Inadmissible	Inadmissible	11133	0.825	5250	3
NMP	Inadmissible	Admissible	11122	0.857	4290	2
NMP	Admissible	Inadmissible	11121	0.910	2700	1

MP: Monetary Poverty; SMP: Severe Monetary Poverty; NM: No Monetary Poverty

Figure 2. Levels of vulnerability [39].

In Figure 2, the QALY obtained in Table 1 corresponds to IVH´s level 6 where MPI is severe and EnI and CI are admissible. The subjective information obtained from surveys in Table 1 gave rise to a scale that is defined with objective data measured in terms of Enl, CI, and MPI. The equivalences are summarized in Figure 2.

2.2. Adaptive Comfort and Adaptive Energy Consumption Assessments for Implemented CI and Enl

As stated in the introduction section, energy modeling is usually based on static setpoint temperatures; it overestimates energy consumption because it does not take into consideration the adaptability of building users. The energy-saving prediction is not realistic enough to adequately determine the actions that have high impact on EP and climate change mitigation. The effect called meteorological memory, in which both the expectations of the occupants and their psychological adaptation to different temperatures intervene, is taken into account in the adaptive models [47]. Recently, it has been studied, in relation to PE, how this adaptive approach can influence the use of air-conditioning devices by occupants [37,48,49].

This is supported by the use of so-called daily setpoint temperatures, that is, the temperatures that achieve the highest percentage of acceptability to keep the interior at a set temperature within the daily adaptive comfort limits. If necessary, you can opt for a mixed solution, natural ventilation, when the outside temperature allows it, or use of air conditioning when the outside temperature is not favorable.

The adaptive approach, based on the use of adaptive setpoint temperatures, results in an adaptive energy demand, that is, energy necessary to maintain the interior thermal conditions of the home within the adaptive comfort range. This new definition of energy demand can influence the definition of PE since it allows adaptive comfort to be applied considering the influence of climate change [50,51].

The European standard EN 16798-1:2019 [17] establishes 3 categories according to users' thermal adaptation capacity. More specifically, each category is defined for a type of building or user. Category I is applicable to users with thermal adaptation limitations (e.g., the elderly), category II to new buildings, and category III to existing buildings, the latter having a wider comfortable temperature range. For this research category III is used, in which the optimal thermal comfort temperature (Equation (3) Figure 1) oscillates between the upper and lower limits (Equations (4) and (5), Figure 1). The limits correspond to linear regressions according to the prevailing mean outdoor air temperature T_{rm} (Equation (6), Figure 1). T_{rm} is determined by the weighted average of daily external temperatures; it is useful to determine the values of upper and lower limits and to control whether the adaptive thermal comfort model could be applied. For this purpose, many models establish a range of values among which T_{rm} should oscillate to apply the adaptive model. According to EN 16798-1:2019 [17], T_{rm} should oscillate 10 and 30 °C. These are the thresholds that are applied to the implemented comfort indicator (CI) (Figure 1).

For the quantification of the energy consumption considering adaptive comfort, a combination of setpoint temperatures is considered (Table 2). That is, in the case of T_{rm} below 10 °C or above 30 °C static temperatures are set according to EN 16798-1:2019 [17] for category III. This algorithm is introduced in the dynamic simulations for the implementation of energy indicator (EnI).

Table 2. Setpoint temperatures considering adaptive comfort temperature ranges for category III and the prevailing mean outdoor air temperature (T_{rm}).

Scheme	Prevailing Mean Outdoor Air Temperature T_{rm}—Comfort Temperature		
	$T_{rm}<10°C$	$10°C \leq T_{rm} \leq 30°C$	$T_{rm}>30°C$
Upper setpoint	25.0	$0.33 \times T_{rm} + 18.8 + 4$	27.0
Lower setpoint	19.0	$0.33 \times T_{rm} + 18.8 - 5$	22.0

2.3. Present and Future Scenarios Simulations Considering Global Warming

To assess the vulnerability to EP by means of the implementation of the IVH, the DesignBuilder software is used; this software allows the energy simulation of buildings and is highly reliable as it contains the EnergyPlus calculation engine. Using hourly weather data files, the software develops advanced dynamic simulations, allowing the incorporation of data such as internal loads, construction characteristics, and temperatures adjusted according to the adaptive approach. These are crucial to carry out pre- and post-intervention evaluations in relation to EP. Moreover, to evaluate the degree of households' vulnerability throughout the timespan after the retrofit, future climate scenarios are considered. To this end, the CCWorldWeatherGen tool of the Hadley Centre Coupled Model 3 HadCM3 UK Met Office is used, which, through a morphing process, generates, for any geographical location, meteorological data according to the prediction of climate change. Furthermore, these data are generated in interchange files compatible with a large number of building energy simulation software [52]. The "morphing" of the climatological data used in this work coincides with the A2 scenario of greenhouse gas emissions, as established in the IPCC (Intergovernmental Panel on Climate Change). This has generated the climatic scenarios established for 2050 and 2080 in this work. Energy consumption data can then be extracted from each simulation.

2.4. Case Study

The case study is a residential building formed by 40 social housing apartments, developed on four floors, with 876 m^2 per floor and a total area of 3504 m^2 (Figure 3). The building was constructed in 1950 in Seville, and thus it shares the characteristics of the social housing of the 1950s and 1960s, common to many of the workers' housing developments that were built in the city at that time, in response to demographic and industrial development.

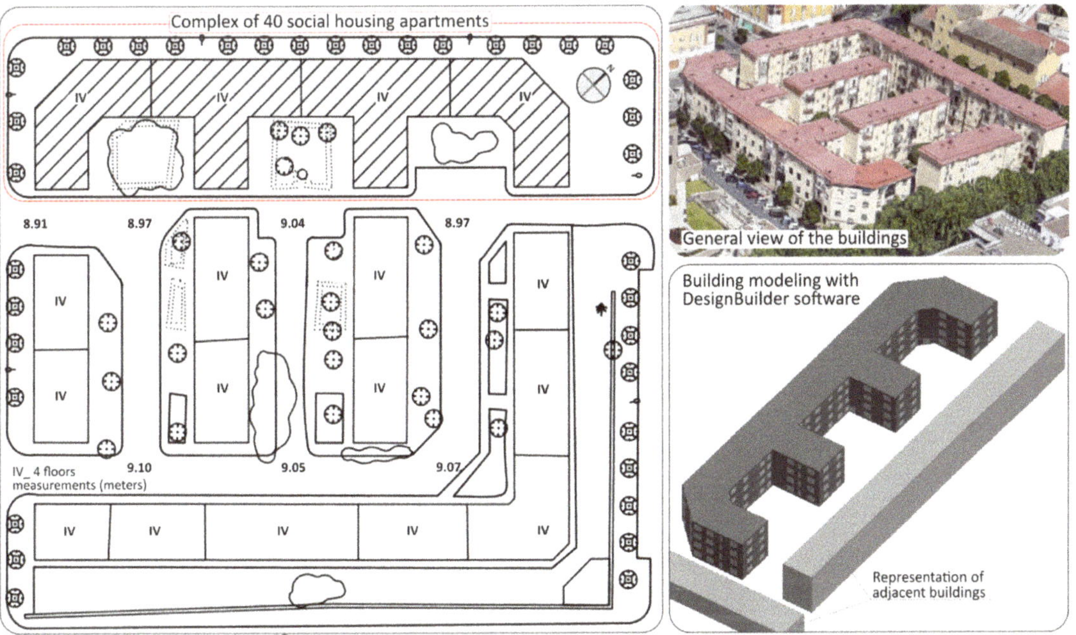

Figure 3. Layout and surroundings of the 40 social apartments building. Image of the energy simulation modeling.

The original foundation consists of a system of concrete pads connected with reinforced concrete beams. The vertical structure is composed of load-bearing walls of solid ceramic brick up to the first floor, and the upper floors are made of alternating solid and hollow bricks. The slabs are made of ceramic lightening pieces and a layer of reinforced concrete. The connection between floors is made of stairs of solid brick vaults. The rooftop is flat, with slope formation and ceramic tile finish. The façade has a final coating of painted cement mortar. The windows are made of sliding aluminum frames and single glass panels.

The interior distribution of the dwellings varies according to the location within the complex; all have a living room, kitchen, bathroom, and two to three bedrooms. This study focuses on the two-bedroom apartments. Domestic hot water (DHW) is independent for each home and is provided by standard combustion gas heaters. The building characteristics and internal loads are shown in Tables 3 and 4, respectively.

Table 3. Building characteristics (baseline case).

	Construction Elements	Thickness (m)	Thermal Conductivity $(w/m^2\ K)$	Transmittance $U\ (W/m^2\ K)$
Envelope	Cement mortar plastering (M5 (1:6))	0.02	0.55	2.1
	Brick wall for facing	0.24	1.04	
	Cement mortar plastering (M5 (1:6))	0.02	0.55	
Rooftop	Ceramic tile floor (14 × 28 cm)	0.02	1.00	1.11
	Bastard mortar	0.02	0.55	
	Lost flooring with ceramic tile	0.02	1.00	
	Protective mortar	0.02	0.55	
	Lightweight slope-forming concrete	0.15	0.41	
	Resistant support with self-supporting beams and ceramic vaults	0.25	0.91	
	Cement mortar plastering (M5 (1:6))	0.02	0.55	
Windows	Sliding aluminum frames, without thermal bridge breakage (4.0 m² K/W)			5.7
	Simple monolithic glass panels (5.7 m² K/W).			
	System			Nominal Performance
Heating	Heat pump			2.10 COP
Cooling	Heat pump			2.00 EER

According to the Spanish building code, the internal load of residential buildings is a low internal load, i.e., electrical equipment, lighting, and occupants generate little heat, with a density of internal sources below 6 W/m².

Table 4. Building characteristics (retrofitting project).

	Energy Retrofitting	Thickness (m)	Thermal Conductivity $(w/m^2\ k)$	Transmittance $U\ (W/m^2\ K)$
Envelope	Mono-layer coating	0.04	0.72	0.38
	Insulation. Rigid EPS panels	0.08	0.04	
	Brick wall for facing	0.24	1.04	
	Cement mortar plastering (M5 (1:6))	0.02	0.55	
	Gypsum plastering	0.01	0.57	

Table 4. *Cont.*

	Energy Retrofitting	Thickness (m)	Thermal Conductivity (w/m^2 k)	Transmittance U (W/m^2 K)
Rooftop	Ceramic tile floor (14 × 28 cm)	0.02	1.00	0.33
	Cement mortar plastering (M5 (1:6))	0.02	0.55	
	Anti-puncture fabric (separator)	0.001	0.05	
	Waterproof layer (EPDM)	0.001	0.25	
	Lightweight slope-forming concrete	0.15	0.41	
	Resistant support with self-supporting beams and ceramic vaults	0.25	0.91	
	Cement mortar plastering (M5 (1:6))	0.02	0.55	
	Gypsum plastering	0.01	0.57	
Windows	Sliding aluminum frames, with thermal bridge breakage (4.0 m^2 K/W)			2.22
	Low-emission glass (6 mm) (1.6 m^2 K/W)			
	System			Nominal Performance
Heating	Heat pump with inverter multi-split system			4.4 COP
Cooling	Heat pump with inverter multi-split system			4.2 EER

According to the Spanish building code, the internal load of residential buildings is low, i.e., electrical equipment, lighting, and occupants generate little heat, with a density of internal sources below 6 W/m^2.

The DHW is replaced by a new system supported by renewable energy. The equipment is centralized for the whole building and is formed by solar thermal panels that contribute to the production of DHW with accumulators and individual auxiliary systems per dwelling that work with electric power. The original exterior windows are replaced by a more efficient one with low-emissive glass and frames with thermal break, with low emissivity and dehydrated air chamber of 12 mm. These reduce the thermal and acoustic transmission.

With respect to the loads schedule, data similar to previous research studies are used [21,23,36] (Table 5). All internal loads vary depending on the day of the week (weekdays and weekends). The airflow is set constant, 0.7 ac/h, due to windows' infiltration.

Table 5. Loads schedule in the case study.

		Loads Schedule						
		Time Period						
Loads		1:00–7:00	8:00	09:00–15:00	16:00–18:00	19:00	20:00–23:00	00:00
Sensible load (W/m^2)	Weekdays	2.15	0.54	0.54	1.08	1.08	1.08	2.15
	Weekend	2.15	2.15	2.15	2.15	2.15	2.15	2.15
Latent load (W/m^2)	Weekdays	1.36	0.34	0.34	0.68	0.68	0.68	1.36
	Weekend	1.36	1.36	1.36	1.36	1.36	1.36	1.36
Lighting (W/m^2)	Weekdays and weekend	0.44	1.32	1.32	1.32	2.20	4.40	2.20
Equipment (W/m^2)	Weekdays and weekend	0.44	1.32	1.32	1.32	2.20	4.40	2.20

3. Results

Vulnerability Comparison: Present and Future Scenarios of the Baseline and Enhanced Case

The energy consumption simulations were applied to the baseline and enhanced case, in the current scenario, as well as in future scenarios considering the predictions of climate change (2050 and 2080, respectively), prior to any intervention. The energy improvement project was developed for the entire building, but the required energy consumption obtained represents a single dwelling on an intermediate floor.

This work aimed to provide a real analysis at the local level to identify the vulnerability of low-income homes and dwellings of poor quality. These results are underpinned by the standard economic situation of homes located in these areas according to the Spanish Household Budget Survey (HBS) as collected by the Spanish National Statistics Institute (SNSI) [53]. Given that the households studied are in a situation of monetary poverty, the same size and typology were assimilated for all scenarios.

The results of applying the indicator to the case study are presented below:

Monetary Poverty Indicator (MPI): As it has been introduced in previous paragraphs, in this study, it was assumed that households, in all scenarios, are in a situation of monetary poverty. To obtain this result, a household size of two adults and two children was considered, and the net income and expenditure correspond to the ones classified as standard in the Spanish National Statistics Institute (SNSI) [53].

The monetary poverty threshold (MPT) used to calculate the MPI corresponds to 60% of median equivalized disposable income in Spain for a person, being EUR 9009 per person in the case of monetary poverty according to Eurostat [41]. In the case of severe monetary poverty (SMPT), the data used correspond to the lowest benefit granted by the Spanish State, which is active income of insertion which is collected as an extraordinary unemployment benefit, with a value of 451 EUR/month per person [42]. Based on these monetary poverty thresholds for one person (MPT and SMPT) and consumption units by household size (Equations (1) and (4) in Figure 1), the MPI value obtained is 0.76 for MP and 1.27 for SMP (after applying Equations (1), (1.2) and (1.3) in Figure 1). It is considered that a household is in monetary poverty when MPI is less than one, and then the result is "MP: Monetary poverty" (see MPI results in the last results table of this section).

Energy Indicator (EnI): Based on the energy consumption data published by the Institute for Energy Diversification and Saving for Spain (IDAE) [43], the average energy consumption threshold was obtained for the type of home analyzed.

Table 6 compares the EnI results. The total consumptions for a 76.22 m^2 dwelling area can be observed, depending on the construction characteristics of the houses of the baseline and enhanced cases (Tables 3 and 4), and in the three scenarios analyzed (current, 2050, and 2080).

Table 6. Results of the energy indicator for baseline and enhanced cases: current, 2050, and 2080.

EnI_Implemented Energy Indicator				
Annual	**Total Consumption kwh**	**MEC_lim. kWh**	**EnI**	**Result**
BASELINE CASE				
Current	5404.15		0.85	Admissible
2050	7953.02	6386.11	1.25	Inadmissible
2080	9631.71		1.51	Inadmissible
ENHANCED CASE				
Current	4901.13		0.77	Admissible
2050	6152.06	6386.11	0.96	Admissible
2080	6841.51		1.07	Inadmissible

Even though the results of the intermediate indicators are binary, CI, EnI, and MPI were combined into the HRQC in the previous work by [40], giving rise to a scale formed by 13 levels of vulnerability (Figure 2). The levels were calibrated with empirical data from surveys, obtained in a neighborhood in England [39].

The Buildings Performance Institute Europe (BPIE) defines as inefficient those residential buildings with an energy demand greater than 200 kWh/m^2; thus we can consider these data as support for the "inadmissible" results for the case analyzed in this study [54].

Comfort Indicator (CI): To obtain this indicator, the percentage of hours within or outside the established comfort range was counted. Each dwelling was studied independently, considering the local climatology (Mediterranean climate, Seville) and the characteristics of the dwellings: in the baseline case, none of the dwellings have been retrofitted, and thus the technical characteristics were maintained in three scenarios (Table 3). In the enhanced case, the improvement measures described in Section 2.4. Case study (Table 4) were implemented in the three scenarios. Table 7 summarizes the comfort hours for each scenario. It may not be possible to replicate this analysis in other countries and/or building typology since thermal comfort situations vary depending on the characteristics of the home. In the results of Table 7, the percentage of hours in comfort (IC) is less than 80% in all scenarios, and thus, according to the limits established by the comfort indicator, the result for both the baseline case and the enhanced case is inadmissible.

Table 7. Results of hours of comfort for baseline and enhanced case: current, 2050, and 2080.

CI_Implemented Comfort Indicator				
Annual	**Total Hours**	**Hours Comfort**	**CI**	**Result**
BASE CASE				
ACTUAL		5289.00	60.38%	Inadmissible
2050	8760.00	4696.00	53.61%	Inadmissible
2080		4078.00	46.55%	Inadmissible
ENHANCED CASE				
ACTUAL		4530.00	51.71%	Inadmissible
2050	8760.00	2739.00	31.27%	Inadmissible
2080		2351.00	26.84%	Inadmissible

Health-Related Quality-Life Cost (HRQLC):

Table 8 shows the results of the IVH in all studied scenarios located in the city of Seville. The final level of vulnerability was obtained from the combination of the results obtained in each of the indicators developed (according to Figure 2). The vulnerability level of the current baseline and enhanced cases is 5.00, derived from inadequate energy efficiency in the home. In the 2050 scenario, for the baseline case, a vulnerability level 8 was obtained, but this vulnerability level was reduced to 5 in the improved case for 2050 due to the energy efficiency achieved. Finally, in the scenario for 2080, the level of vulnerability is 12. It is situated in one of the most critical levels because the worst possible situation of energy poverty is defined, in which the home cannot afford minimal energy consumption due to its low monetary level, representing the "heating or eating" effect (choosing between eating or consuming minimal energy). The HRQLC provided in Table 8 represents cost per life year to the NHS of those of homes analyzed in each scenario.

Table 8. Results of the implemented indicator of vulnerable homes of the baseline and enhanced cases. States: current, 2050, and 2080.

Annual	Monetary Poverty Indicator (MPI)	Implemented Index of Vulnerable Homes (IVH)					
		Implemented Energy Indicator (EnI)	Implemented Comfort Indicator (CI)	Health-Related Quality-Life Cost (HRQLC)			IVH Levels
				EQ-5 D Score	QALY	HRQLC (EUR)	
BASE CASE							
Current	MP	Admissible	Inadmissible	11333	0.754	7380.00	5.00
2050	MP	Inadmissible	Inadmissible	13433	0.484	15480.00	8.00
2080	MP	Inadmissible	Inadmissible	24455	−0.096	32880.00	12.00
ENHANCED CASE							
Current	MP	Admissible	Inadmissible	11333	0.754	7380.00	5.00
2050	MP	Admissible	Inadmissible	11333	0.754	7380.00	5.00
2080	MP	Inadmissible	Inadmissible	24455	−0.096	32880.00	12.00

4. Discussion

From the analysis of the results in Table 8, it is possible to estimate the vulnerability of households by applying the implemented IVH. In addition, these results show that, assuming the same monetary situation for all scenarios, improving the energy efficiency of homes is key to reducing the level of vulnerability of households and, consequently, reducing the cost for the NHS (HRQL).

From the results applied to the case study, it can be noticed that:

— The situation of monetary poverty in which households are immersed is the main cause of the situation of vulnerability.

— The improvement retrofitting carried out in the 2050 scenario contributed to an improvement in the quality of life of the household, reducing the IVH level from 8 to 5; however, it is necessary for the household to overcome the situation of monetary poverty, by means of reducing expenses or increasing their income, in order to get out of the vulnerability situation.

— The implementation of adaptive comfort in the calculation of the energy consumption identified situations of discomfort in a more realistic way because tenants' discomfort is relative to the average outside weather.

— The results show that the improvements implemented in the case studies worsened comfort in the Mediterranean climate as the solutions implemented are too watertight for the local climate.

— From the results, the passive retrofitting proposed by itself does not improve the comfort of the home in the climate under study and makes ventilation necessary to achieve it.

5. Conclusions

The aim of this work was to provide a new approach to energy poverty by identifying vulnerable households, considering economic and social aspects and climate change adaptability of families in a global warming context. The present research can have a big impact technically because it generated a new tool to define priorities in renovation works, and this can be extrapolated to new buildings assessment and to the rehabilitation projects of obsolete ones. The public funding can be allocated in a more efficient way to tackle vulnerability in a climate change scenario.

One of the main contributions of this work lies in the location of the case studies analyzed. Energy poverty has been studied in cold climates since it has traditionally been related to areas where winters are very harsh, but in climates where summers are long and extremely hot, it is not as well studied, although high energy consumption during the summer can cause a situation of energy poverty. The adaptive criteria applied in energy simulation of the building, in future climate change scenarios (2050 and 2080), and the severity of summer in the Mediterranean climate make annual cooling energy consumption much higher than heating consumption in both scenarios.

From the analysis of the results obtained, it can be concluded that, in the Mediterranean climate, energy improvement solutions based mainly on passive building design criteria result in homes that are too tight, making ventilation necessary to reach comfort.

Returning to the objective of this work, the implementation in the IVH of the adaptability of households in the context of climate change provides an evolution of the indicator that allows an assessment of the households' situation in a more complete and complex way by identifying not only which factors have the greatest impact on the situation of vulnerability but also assessing the household's adaptive capacity based on climate variability and how it influences the occupants' quality of life.

The implementation carried out confirms that the IVH can combine information about the monetary situation of the household according to the monetary poverty threshold of the study area and the home energy consumption under adaptive comfort criteria and subjected to the climatic zone where the home is located. New lines of research will be to identify how the climatology of the area defines the comfort levels of homes in relation to the monetary situation, energy costs, and quality of life.

Author Contributions: Conceptualization, R.C.-R.; Data curation, M.T.-C.; Formal analysis, M.T.-C.; Funding acquisition, C.R.-B.; Methodology, R.C.-R.; Project administration, C.R.-B.; Software, M.T.-C.; Supervision, C.R.-B. and M.M.; Writing—original draft, M.D.A.-R.; Writing—review and editing, C.R.-B. and M.M. All authors have read and agreed to the published version of the manuscript.

Funding: This paper and the costs for its publication in open access have been funded by the research project called "Nuevo Análisis Integral de la Pobreza Energética en Andalucía (NAIPE). Predicción, evaluación y adaptación al cambio climático de hogares vulnerables desde una perspectiva económica, ambiental y social (US- 125546)", financed by "Consejería de Economía y Conocimiento de la Junta de Andalucía (Spain)" with the European Regional Development Fund (ERDF).

Institutional Review Board Statement: Not applicable.

Informed Consent Statement: Not applicable.

Data Availability Statement: Not applicable.

Acknowledgments: The authors express their gratitude to the research project called "Nuevo Análisis Integral de la Pobreza Energética en Andalucía (NAIPE). Predicción, evaluación y adaptación al cambio climático de hogares vulnerables desde una perspectiva económica, ambiental y social (US-125546)", financed by "Consejería de Economía y Conocimiento de la Junta de Andalucía (Spain)" with the European Regional Development Fund (ERDF).

Conflicts of Interest: The authors declare no conflict of interest.

References

1. Intergovernmental Panel on Climate Change (Ed.) Intergovernmental Panel on Climate Change Climate Change 2014: Synthesis report. In *Contribution of Working Groups I, II and III to the Fifth Assessment Report of the Intergovernmental Panel on Climate Change*; Cambridge University Press: Cambridge, UK, 2014.
2. European Union. Directive (EU) 2018/844 of the European Parliament and of the Council of 30 May 2018 Amending Directive 2010/31/EU on the Energy Performance of Buildings and Directive 2012/27/EU on Energy Efficiency. *Off. J. Eur. Union.* **2018**, *L156*, 75. Available online: https://eur-lex.europa.eu/legal-content/EN/TXT/?uri=celex:32018L0844 (accessed on 7 April 2021).
3. European Commission. *2020—Europe's Climate Change Opportunity*; Climate Action: Brussels, Belgium, 2008.
4. European Commision. *A Policy Framework for Climate and Energy in the Period from 2020 to 2030*; Climate Action: Brussels, Belgium, 2014.
5. European Commission. A European Green Deal. Available online: https://ec.europa.eu/info/strategy/priorities-2019-2024/european-green-deal_en (accessed on 8 June 2021).
6. Semprini, G.; Gulli, R.; Ferrante, A. Deep regeneration vs shallow renovation to achieve nearly Zero Energy in existing buildings: Energy saving and economic impact of design solutions in the housing stock of Bologna. *Energy Build.* **2017**, *156*, 327–342. [CrossRef]

7.   Anthony, T. *Understanding the Costs and Benefits of Fuel Poverty Interventions: A Pragmatic Economic Evaluation from Greater Manchester*; Greater Manchester Public Health Practice Unit: Manchester, UK, 2011.

8.   Barker, A.V.; Dip, D.M.S.P.G. *Assessment of the Impact on Health and Health Costs due to Fuel Poverty in Bolton*; National Health Service: Bolton, UK, 2011.

9.   Department of Energy & Climate Change (DECC). *Annual Fuel Poverty Statistics Report 2014*; National Statistics: London, UK, 2014.

10. ASSIST 2GETHER. *Report on National and European Measures Addressing Vulnerable Consumers and Energy Poverty*; European Commission: Brussels, Belgium, 2018.

11. Baker, E.; Lester, L.H.; Bentley, R.; Beer, A. Poor housing quality: Prevalence and health effects. *J. Prev. Interv. Community* **2016**, *44*, 219–232. [CrossRef]

12. Castaño-Rosa, R.; Solís-Guzmán, J.; Rubio-Bellido, C.; Marrero, M. Towards a multiple-indicator approach to energy poverty in the European Union: A review. *Energy Build.* **2019**, *193*, 36–48. [CrossRef]

13. European Commision. Directive 2002/91/EC of the European Parliament and of the Council of 16 December 2002 on the energy performance of buildings. *Off. J. Eur. Union* **2002**, *2010*, 65–71.

14. Gobierno de España (Government of Spain). *Royal Decree 314/2006 Approving the Technical Building Code (Real Decreto 314/2006, de 17 de Marzo, por el que se Aprueba el Código Técnico de la Edificación*; Imprenta Nacional del Boletín Oficial del Estado: Madrid, España, 2006.

15. Gobierno de España (Government of Spain). *Royal Decree 1027/2007 Approving the Spanish Thermal Building Regulations (Real Decreto 1027/2007, de 20 de Julio, por el que se Aprueba el Reglamento de Instalaciones Termicas en los Edificios)*; Imprenta Nacional del Boletín Oficial del Estado: Madrid, España, 2007.

16. *Technical Building Code (TBC): Royal Decree 314/2006 of 17 March 2006, Approving the Technical Building Code*; Ministerio de Vivienda: Alcobendas, Spain, 2006; ISBN 843401632X.

17. EN 16798-1:2019. *Energy Performance of Buildings—Ventilation for Buildings—Part 1: Indoor Environmental Input Parameters for Design and Assessment of Energy Performance of Buildings Addressing Indoor Air Quality, Thermal Environment, Lighting and Acous*; European Committee for Standardization: Brussels, Belgium, 2019.

18. American Society of Heating, R.; A.C.E. (ASHRAE). *ASHRAE Standard 55-2017 Thermal Environmental Conditions for Human Occupancy*; ASHRAE Inc., Ed.; American Society of Heating, Refrigerating and Air Conditioning Engineers: Atlanta, GA, USA, 2017; ISBN 1041–2336.

19. Brager, G.S.; De Dear, R.J. Thermal adaptation in the built environment: A literature review. *Energy Build.* **1998**, *27*, 83–96. [CrossRef]

20. Carlucci, S.; Bai, L.; De Dear, R.; Yang, L. Review of adaptive thermal comfort models in built environmental regulatory documents. *Build. Environ.* **2018**, *137*, 73–89. [CrossRef]

21. Bienvenido-Huertas, D.; Sánchez-García, D.; Rubio-Bellido, C. Comparison of energy conservation measures considering adaptive thermal comfort and climate change in existing Mediterranean dwellings. *Energy* **2020**, *190*, 116448. [CrossRef]

22. Bienvenido-Huertas, D.; Quiñones, J.A.F.; Moyano, J.; Rodríguez-Jiménez, C.E. Patents Analysis of Thermal Bridges in Slab Fronts and Their Effect on Energy Demand. *Energies* **2018**, *11*, 2222. [CrossRef]

23. Bienvenido-Huertas, D.; Sánchez-García, D.; Rubio-Bellido, C. Analysing natural ventilation to reduce the cooling energy consumption and the fuel poverty of social dwellings in coastal zones. *Appl. Energy* **2020**, *279*, 115845. [CrossRef] [PubMed]

24. Thomson, H.; Snell, C.; Liddell, C. Fuel poverty in the European Union: A concept in need of definition? *People Place Policy Online* **2016**, *10*, 5–24. [CrossRef]

25. Bouzarovski, S.; Petrova, S. A global perspective on domestic energy deprivation: Overcoming the energy poverty-fuel poverty binary. *Energy Res. Soc. Sci.* **2015**, *10*, 31–40. [CrossRef]

26. Heindl, P. Measuring Fuel Poverty: General Considerations and Application to German Household Data. *Finanz. Arch.* **2015**, *71*, 178–215. [CrossRef]

27. Rademaekers, K.; Yearwood, J.; Ferreira, A.; Pye, S.; Ian Hamilton, P.; Agnolucci, D.G.; Karásek, J.; Anisimova, N. *Selecting Indicators to Measure Energy Poverty*; Trinomics: Rotterdam, The Netherlands, 2016.

28. Legendre, B.; Ricci, O. Measuring fuel poverty in France: Which households are the most vulnerable? *Energy Econ.* **2014**, *49*, 620–628. [CrossRef]

29. Miniaci, R.; Scarpa, C.; Valbonesi, P. *Fuel Poverty and the Energy Benefits System: The Italian Case*; Universita' Bocconi: Milano, Italy, 2014.

30. Hatt, T.; Saelzer, G.; Hempel, R.; Gerber, A. High indoor comfort and very low energy consumption through the implementation of the Passive House standard in Chile. *Rev. Constr.* **2012**, *11*, 123–134.

31. Martínez, P.W.; Kelly, M.T. Integration of performance criteria in the energy-environmental improvement of existing social housing in Chile. *Ambient. Construído* **2015**, *15*, 47–63. [CrossRef]

32. Van Hooff, T.; Blocken, B.; Hensen, J.L.M.; Timmermans, H.J.P. Reprint of: On the predicted effectiveness of climate adaptation measures for residential buildings. *Build. Environ.* **2015**, *83*, 142–158. [CrossRef]

33. Fabbri, K. Building and fuel poverty, an index to measure fuel poverty: An Italian case study. *Energy* **2015**, *89*, 244–258. [CrossRef]

34. Braubach, M.; Ferrand, A. Energy efficiency, housing, equity and health. *Int. J. Public Health* **2013**, *58*, 331–332. [CrossRef]

35. UNEP Global Status Report for Buildings and Constructi. Available online: https://wedocs.unep.org/handle/20.500.11822/34572 (accessed on 9 June 2021).
36. Sánchez-García, D.; Bienvenido-Huertas, D.; Tristancho-Carvajal, M.; Rubio-Bellido, C. Adaptive Comfort Control Implemented Model (ACCIM) for Energy Consumption Predictions in Dwellings under Current and Future Climate Conditions: A Case Study Located in Spain. *Energies* **2019**, *12*, 1498. [CrossRef]
37. Sánchez-García, D.; Rubio-Bellido, C.; Tristancho, M.; Marrero, M. A comparative study on energy demand through the adaptive thermal comfort approach considering climate change in office buildings of Spain. *Build. Simul.* **2020**, *13*, 51–63. [CrossRef]
38. Castaño-Rosa, R.; Solís-Guzmán, J.; Marrero, M. Energy poverty goes south? Understanding the costs of energy poverty with the index of vulnerable homes in Spain. *Energy Res. Soc. Sci.* **2020**, *60*, 101325. [CrossRef]
39. Castaño-Rosa, R.; Sherriff, G.; Thomson, H.; Guzmán, J.S.; Marrero, M. Transferring the index of vulnerable homes: Application at the local-scale in England to assess fuel poverty vulnerability. *Energy Build.* **2019**, *203*, 109458. [CrossRef]
40. Castaño-Rosa, R.; Solís-Guzmán, J.; Marrero, M. A novel Index of Vulnerable Homes: Findings from application in Spain. *Indoor Built Environ.* **2018**, *29*, 311–330. [CrossRef]
41. Eurostat. At-Risk-of-Poverty Thresholds—EU-SILC and ECHP Surveys. Available online: http://appsso.eurostat.ec.europa.eu/nui/show.do?dataset=ilc_li01&lang=en (accessed on 7 April 2021).
42. Gobierno de España (Government of Spain). Real Decreto Legislativo 8/2015, de 30 de Octubre, por el que se Aprueba el Texto Refundido de la Ley General de la Seguridad Social. Available online: https://www.boe.es/buscar/act.php?id=BOE-A-2015-11724 (accessed on 7 April 2021).
43. IDAE. Consumption of the Residential Sector in Spain. Summary of Basic Information. SPAHOUSEC I Project. Institute for the Diversification and Saving of Energy. Eurostat. Ministry of Industry, Energy and Tourism. Available online: https://www.idae.es/uploads/documentos/documentos_Documentacion_Basica_Residencial_Unido_c93da537.pdf (accessed on 7 April 2021).
44. Dear, R.J.; De, G.S. Brager Thermal comfort in naturally ventilated buildings: Revision to ASHRAE standards 55. *J. Energy Build.* **2002**, *34*, 549–561. [CrossRef]
45. Vallejo-Torres, L.; García-Lorenzo, B.; Valcárcel Nazco, C.; García-Pérez, L.; Linertová, R.; Serrano-Aguilar, P. Valor Monetario de un Año de Vida Ajustado por Calidad: Estimación Empírica del coste de Oportunidad en el Sistema Nacional de Salud. Available online: https://www3.gobiernodecanarias.org/sanidad/scs/contenidoGenerico.jsp?idDocument=e690e0c1-cbed-11e5-a9c5-a398589805dc&idCarpeta=ce590e62-7af0-11e4-a62a-758e414b4260 (accessed on 7 April 2021).
46. Van Hout, B.; Janssen, M.F.; Feng, Y.-S.; Kohlmann, T.; Busschbach, J.; Golicki, D.; Lloyd, A.; Scalone, L.; Kind, P.; Pickard, A.S. Interim scoring for the EQ-5D-5L: Mapping the EQ-5D-5L to EQ-5D-3L value sets. *Value Health* **2012**, *15*, 708–715. [CrossRef]
47. Luo, M.; Wang, Z.; Brager, G.; Cao, B.; Zhu, Y. Indoor climate experience, migration, and thermal comfort expectation in buildings. *Build. Environ.* **2018**, *141*, 262–272. [CrossRef]
48. Bienvenido-Huertas, D.; Sánchez-García, D.; Rubio-Bellido, C. Adaptive setpoint temperatures to reduce the risk of energy poverty? A local case study in Seville. *Energy Build.* **2020**, *231*, 110571. [CrossRef]
49. Sánchez-García, D.; Bienvenido-Huertas, D.; Pulido-Arcas, J.A.; Rubio-Bellido, C. Analysis of energy consumption in different European cities: The adaptive comfort control implemented model (ACCIM) considering representative concentration pathways (RCP) scenarios. *Appl. Sci.* **2020**, *10*, 1513. [CrossRef]
50. Sánchez, C.S.-G.; González, F.J.N.; Aja, A.H. Energy poverty methodology based on minimal thermal habitability conditions for low income housing in Spain. *Energy Build.* **2018**, *169*, 127–140. [CrossRef]
51. Sánchez-Guevara Sánchez, C.; Mavrogianni, A.; Neila González, F.J. On the minimal thermal habitability conditions in low income dwellings in Spain for a new definition of fuel poverty. *Build. Environ.* **2016**, *114*, 344–356. [CrossRef]
52. Climate Change World Weather File Generator for World-Wide Weather Data—CCWorldWeatherGen—University of Southampton Blogs. Sustainable Energy Research Group. Available online: https://energy.soton.ac.uk/climate-change-world-weather-file-generator-for-world-wide-weather-data-ccworldweathergen/ (accessed on 12 June 2021).
53. SNSI_Spanish_National Statistics_Institute. Income per Household by Autonomous Communities. Available online: https://www.ine.es/jaxiT3/Datos.htm?t=9949 (accessed on 22 April 2021).
54. BPIE. Europe's Buildings Under the Microscope A Country-by-Country Review of the Energy Performance of Buildings. *Build. Perform. Inst. Eur.* Available online: https://bpie.eu/wp-content/uploads/2015/10/HR_EU_B_under_microscope_study.pdf (accessed on 13 July 2021).

sustainability

MDPI

Article

Key Elements for a New Spanish Legal and Architectural Design of Adequate Housing for Seniors in a Pandemic Time

María Luisa Gómez-Jiménez [1],* and Vargas-Yáñez Antonio [2]

[1] Málaga Law School and (IBYDA), Universidad de Málaga, 29071 Málaga, Spain
[2] Málaga Architecture School, Universidad de Málaga, 29071 Málaga, Spain; antoniovy@coamalaga.es
*   Correspondence: mlgomez@uma.es

Abstract: The provision of housing for the elderly in Spain has been approached from a public policy perspective to understand social housing and the allocation of specialized social services. The lockdowns in cities and the need to remain at home with social isolation and social distance has especially affected the most vulnerable groups, creating situations that widen the gap in the provision of adequate housing. Research is being carried out by a team of researchers at the University of Málaga, funded by European FEDER funds awarded by the Andalusian Regional Government (VIDA project), to analyze the main characteristics connected with "ideal" adequate housing for a vulnerable senior person living alone or in social isolation due to the quarantine period. In this study, we draw a line between the need for adequate housing, the chance to remodel, and the opportunity to propose new Spanish legal approaches from an architectural perspective within the scope of alternative typologies of housing. This article deals with the preliminary findings of the research connected to the architectural review, exploring key elements for senior housing design, and highlighting the need to approach the issue by proposing a new regulation.

Keywords: accessibility 2; housing typology 3; COVID-19 4; social isolation 5; regulations 6; architectural barriers

check for
updates

Citation: Gómez-Jiménez, M.L.; Antonio, V.-Y. Key Elements for a New Spanish Legal and Architectural Design of Adequate Housing for Seniors in a Pandemic Time. *Sustainability* **2021**, *13*, 7838. https://doi.org/10.3390/su13147838

Academic Editor: Pilar Mercader-Moyano

Received: 31 May 2021
Accepted: 1 July 2021
Published: 13 July 2021

Publisher's Note: MDPI stays neutral with regard to jurisdictional claims in published maps and institutional affiliations.

Copyright: © 2021 by the authors. Licensee MDPI, Basel, Switzerland. This article is an open access article distributed under the terms and conditions of the Creative Commons Attribution (CC BY) license (https://creativecommons.org/licenses/by/4.0/).

1. Introduction

The United Nations (UN Habitat) Strategic Plan 2020–2023 [1], published some months before the COVID-19 health emergency declaration by the WHO, included a link to the Sustainable Developments Goals and the effects of sustainability parameters in cities. "Smartness" as a tool for the integration of technology in cities [2], centric smart-cities, and communities, contrasted with the effects of domiciliary lockdowns and isolation measures to prevent the spread of the disease, which ended up affecting the most vulnerable groups of the population.

The fact of the progressive aging of the Spanish population, which in 2019 reached 9.055.580 seniors (that is, 19.3% of the total population) according to INE [3], has overall affected the definition of social intervention in the Spanish socio-democratic state. In addition, this has been projected into a law on the elderly and a law on assistance to the elderly within the Spanish legal framework. This legal framework was heavily impacted by the pandemic. This paper does not explore the legal approach to the housing sector after COVID-19, nor the social change connected to it. It focuses on the presentation of preliminary findings which connect the idea of sustainability and adequacy to the need for housing for seniors. Therefore, it addresses the first part of our study, by checking the suitability of actual regulations and their disconnection from the technical requirements.

Objectives of the Research

The aim of this research is to:

(1) Explore the stock of housing for the elderly in Spain in accordance with the regulations that establish the building standards for this specific segment of the population.

159

(2) Examine housing needs for older people and related architectural requirements from a socio-legal perspective.
(3) Propose a new typological approach that policymakers can promote.
(4) Explore the changes needed to propose new regulations, derived from the above analysis, with particular attention to the impact of the pandemic on the provision of adequate housing for this segment of the population.

This research is divided into three sections. Firstly, the main aspects from the literature review are described, with particular attention to the regulations. Secondly, we explain the main methods used in the study to address the research objectives, present the main hypothesis, and discuss the former with the data provided. Finally, we propose some adjustments of typologies in the housing offered for seniors. Our conclusion shows the evidence of the above data on the need for regulatory changes that impact architectural design.

2. Literature Review

This is not the first time that the rights of the elderly have been addressed in Spanish academic literature [4]. From the legal point of view, the analysis of public policies for the elderly has been integrated into two main areas: civil law and the right to care for the elderly at family level [5] and public administrative law, with care for the elderly as a vulnerable population [6], which involve the protection of the rights of the elderly by the public authorities and their projection in the public sphere. Both elements of public and private law coexist when studying accommodation for the elderly [7].

The Spanish constitution does not attribute a significant number of provisions relating to the rights of the elderly. The opposite is true. Article 50 of the Spanish Constitution [8] refers to the recognition of the rights of older people in a very general way. It does not even mention the need for adequate housing. On the contrary, the Spanish text involves the public authorities in the provision of the public system of social services, which will also include aspects related to housing. Needless to say, this is insufficient to meet their housing needs, and therefore specific regulations and private sector initiatives have had to address this issue.

Since 1978, the Spanish offer of accommodation for the elderly has based its contributions on the definition of public social service residences or homes for the elderly, which are closer to health centers than to autonomous units [9]. Moreover, the number of public places for seniors in homes for the elderly was also insufficient [10]. This has had an impact on the supply of units on the private market and on the need for an agreement to provide space for seniors in private facilities [11]. These units were not affordable, nor culturally accepted by those who were willing to stay at home and age in place [12]. As a result, family members became the primary caregivers of the elderly [13] in this first stage, and the family home became the home of the elderly. Chronologically, this moment correlated with the enactment of the Dependencies Act, 2006, and the recognition of vulnerability to people with any type of disability or impairment. The Dependencies Act did not devote any specific attention to housing, but rather to home services for dependent people ([14] pp. 70–81).

In addition, many of the family units, where seniors were aging, lacked renovations and refurbishments to meet the needs of the elderly [15]. On the other hand, the improvement of the quality of life and life expectancy in Spain in the 1980s and the aging of the general population expanded the need for having a gerontologist and designing a specific policy for seniors.

Thus, from the mid-1990s to 2007, all Spanish regions and authorities established regulations on the care of the elderly, and some began to propose specific housing models that focused more on the removal of architectural barriers [16] than on any other aspect. Housing projects and planning did not include specific typologies for seniors, but economic support aimed at adapting housing to remove architectural barriers. Public subsidies were then allocated to improve affordability and promote housing for private or public entrepreneurs with little interest in refurbishment.

A review of the existing literature shows that the housing crisis that affected Spain from 2007 onwards was followed by a burgeoning interest in active aging at the European level [17], so gerontologists, sociologists, architects, lawyers, and policy makers addressed the need for housing for seniors. Since in Spain older people mostly owned their units free and clear [18], without mortgage charges, their units were less likely to be foreclosed or subject to eviction, demonstrating their important role in supporting families that received more attention to their needs. This led to the first Spanish promotion of assisted housing for seniors. However, new types of families emerged and the composition of the family changed. Thus, many old people ended up living alone in their "inappropriate" units. In this context, the COVID-19 pandemic forced the population to stay at home and clearly demonstrated the challenges and the inadequacy of housing to cope with a healthy social, work, or family life [19].

Within this context, the first part of the VIDA project, carried out during 2020, examined the architectural characteristics of units for seniors and the regulation connected to it, seeking new proposals for policymakers. This paper explores a mere fragment of that research and shows the preliminary findings of studying the existing Andalusian typology of housing for seniors and its disconnection from architectural requirements and seniors' needs.

3. Materials and Methods

To address the goals described above, it is necessary to understand the research context for this piece, and, briefly, the main steps undertaken within the VIDA research project. As a preliminary step towards defining the specific characteristics of households adapted to the needs of this sector of the population, the work carried out in the framework of the VIDA research project during the year 2020 and the first quarter of 2021 established an initial relationship between disability problems and the population aged 65 and over.

This step profited from a Delphi analysis carried out by VIDA sociologists between December 2020 and May 2021. This Delphi analysis also included a diagnosis of the housing needs of the elderly population and focused on the need for a new type of housing for the elderly, drawing conclusions from a sociological perspective. These Delphi results are in press, and will be publicly released at the end of 2021.

The sociological analysis was followed by an extensive review of the existing literature, focusing on legal and architectural sources. The literature reviewed addressed the case of housing adapted to vulnerable groups. This review included an analysis of the legal and, therefore, technical requirements for promoting this type of housing. Secondly, a table was drawn up showing the relationship between the characteristics required by specialized users (companies and associations working with this sector of the population) and the requirements of the regulations in force.

Finally, a correlation of variables was developed. This correlation paid attention to the level of dependency and how disability criteria can be applied to older persons in relation to their age. In order to complete the table and to identify the main points to be addressed for pre-positive ad hoc regulation on the subject, a mixed approach was adopted. This mixed approach was carried out in a systemic chronology combining quantitative and qualitative analysis. This mixed approach was enriched by the results proposed from the Delphi model developed specifically to study the correlation between housing technical arrangements and social needs.

At the same time, a non-exhaustive search was conducted on the web with keywords such as "dwellings," "older people," and "adapted," to locate references of companies or entities that carry out work for the elderly and that have reflected on the characteristics that their homes should have. As a result of this search, a set of 9 electronic sources were selected related to home help companies [20], home services [21], companies with mechanisms for removing architectural barriers [22], companies specialized in renovations, accessible spaces [23], press [24], dedicated websites [25], real estate portals [26], and professional search companies [27].

In order to limit the scope of work of this research, the analysis of technical regulations has focused on CTE and Andalusian legislation: Decree 72/1992, of 5 May, approving the Technical Standards for Accessibility and Removal of Architectural, Urbanistic and Transport Barriers in Andalusia; Decree 293/2009, of 7 July, approving the Regulations on Accessibility, Infrastructure and Accessibility and Transport Standards in Andalusia; the Order of 5 November 2007 regulating the Procedure and Requirements for the Accreditation of Centers for the Care of Older People with Dependency in Andalusia; and the Regulatory Ordinance on Accessibility of the City of Malaga, as an example of regulation in a medium-large municipality.

The development of the methodology described above allowed us to trace two hypotheses:

(a) "The main differentiating factor that a home for the elderly should have is the availability of accessibility conditions appropriate to the problems of dependency that arise in this age group."

(b) "The pandemic has accelerated the need for a new typology of housing for seniors which is more connected and integrated into the health and social services system."

4. Results

The first hypothesis was tested with the following results, which can be summarized as follows:

- The Spanish legal requirements for a housing policy for the elderly are insufficient to meet the needs of the elderly.
- Two key concepts and their projection should be addressed in the new housing typology for older people: accessibility and vulnerability.

4.1. Spanish Legal Requirements for a Housing Policy for Seniors Are Insufficient to Address Their Needs

Spanish legislation is structured on three levels according to the country's administrative structure (state, regional and local regulations) and the 1978 Spanish constitution. In the housing sector, the responsibility for specific designs lies with the regional authorities, while the coordination of technical standards is laid down at national level. The 17 Spanish autonomous communities can regulate the characteristics of buildings in their area of competence, including welfare and accessibility policies. In the exercise of these powers, the communities have enacted bylaws to facilitate access and remove architectural barriers, as well as to regulate residential centers for the elderly. The rules contained in the regional regulations focus on common housing spaces and regulate a minimum reserve of suitable housing for the promotion of sheltered or subsidized housing by the public authorities. It is only in these cases that there are specific characteristics that the dwelling must meet and that make it possible to assess the extent to which a dwelling can be considered capable of meeting the needs of a person with a particular disability.

In this context, the Technical Building Code, CTE, designed as a performance regulation, sets out the minimum requirements for buildings throughout the country. Section 9 of the Basic Document for Safety of Use and Accessibility, DB-SUA, structured into different basic documents according to their subject matter, sets out the accessibility conditions that buildings must meet. In this section, however, the CTE does not regulate the interior conditions of houses, which is usually left to regional regulation.

At the municipal level, municipalities are empowered to supplement these standards with specific ordinances and, in many cases, they have issued municipal ordinances on accessibility. Although each autonomous community and municipality has been able to draw up its own specific regulations, the reality is that there are no essential differences between them in terms of the performance and technical conditions they require of the buildings they regulate.

In fact, it is not easy to find references in the Spanish technical literature to specific studies on the characteristics that homes for the elderly should have. Existing references usually come from the reflections of companies dedicated to the care of this sector of the population that, at some point, express their opinion on the characteristics that these accommodations should have. Regarding the regulation of the subject, different levels must be considered. Thus, in addition to the coexistence of the constitutional right to housing and the general provisions enacted for the allocation of resources for housing projects, a systemic approach to regional housing provisions and addressing the affordability of social housing can be added. In addition, due to the pandemic situation, a package of legal provisions was enacted to prevent foreclosures and protect vulnerable people who have difficulty paying rent or who have been economically affected by the pandemic [28]. One case of particular relevance is the regulation of energy poverty (which is one of the elements found in many old units where the elderly live alone). Thus, the general provisions of the CTE had to be supplemented by public policies on energy efficiency, derived from European directives, with a significant impact on housing promotion, but without adjusting the case of housing for the elderly either in regional or national legislation. With respect to accommodation for the elderly, the closest legislation is a regional regulation on the conditions to be met by residential centers for dependent elderly people, day centers or day-stay units (DUs), and night centers or night-stay units (UENs) [11].

This way, a first correlation is established between the elderly and their situation (level of vulnerability). This circumstance allows a first approach to the problem as an accessibility problem based on the analysis of this legislation. In fact, over and above the provision of services that these centers must have to function properly, these regulations tend to end up referring to the general regulations on accessibility. The reference to the status of dependency included in the above-mentioned regulations focuses the problem or differentiating factor of this sector of the population on its highest level of dependency. Given the fact that the national regulation on dependency was promulgated in 2006 and was not fully implemented until 2010, the lack of assessment of many older people regarding their level of dependency did not affect or improve their quality of life at home.

4.2. Two Key Concepts and their Projection Need to Be Addressed in the New Typology of Housing for Seniors: Accessibility and Vulnerability

4.2.1. Accessibility as a Key Concept in a Model of Housing for Seniors

Accessibility [29], a concept that we will find associated with the reflections and regulation of spaces intended for the elderly, is defined as the ease with which one place can be reached from another; consequently, it is a spatial quality. As a concept it is not an absolute, but a relative one that varies depending on the characteristics of the user, the type of displacement, and other conditions such as climatic factors and the quality of roads. It must therefore be assessed on the basis of a diversity of circumstances that may disrupt mobility in different ways. Similarly, it must be understood holistically and in such a way as to ensure not only access, but also movement, use, orientation, security, and functionality.

The Spanish set of laws has addressed the case of accessibility as an issue to be included in national legislation on disability—the Royal Legislative Decree 1/2013, of 29 November, on the general rights of people with disabilities and the prevention of social exclusion—and in legislation relating to the elimination of architectural barriers, as provided for in Law 15/95 of 30 May. This 1995 provision includes a reference to people over 70 years of age and reminds us of the trichotomy of regulations that involve health provisions affecting older people, as we have already stated [9]. From a legal point of view, lack of accessibility is linked to possible discriminatory behavior by private or public entrepreneurs and owners. This article adds a new element to that trichotomy that we can explore.

The lack of autonomous mobility translates into a situation of captivity of the person concerned, sedentarism, lack of relationships, and segregation, which led the Andalusian Ombudsman to highlight in his 2003 report the need for the integration of "people imprisoned in their homes" [30,31]. This implies the development of processes of social exclusion, understood as multidimensional phenomena that weaken the economic, political, socio-cultural, and spatial links between individuals and society—links that, to the extent that they are weaker, increase the degree of vulnerability of the individual [17]. On the other hand, improving accessibility also means improving the habitability of spaces, a quality which does not only benefit older people. The development of accessible spaces makes them sustainable, healthy, and inclusive—sustainable insofar as the characteristics of the homes and the buildings that house them do not force them to abandon them or the neighborhoods where they are located because they are the oldest population. At the same time, they offer a comfortable environment to all citizens because they are true "livable landscapes." These spaces are healthy because the elimination of the aforementioned barriers prevents accidents and mitigates suffering. Finally, these spaces are inclusive, since they do not segregate or exclude population groups based on their disabilities. This idea of inclusion is reinforced by the Spanish constitution, which states in Article 9.2 that "It is the responsibility of the public authorities to promote conditions in which the freedom and equality of the individual and of the groups to which they belong are real and effective; to raise any obstacles that prevent or hinder their realization…"

Consequently, it can be said that policies to improve accessibility are inseparable from those to keep the elderly in their habitual residence and from the objective of reducing the length of institutionalization and hospitalization, not only of the elderly population, but also of the injured and convalescent [6]. Given this situation, there is an urgent need to reflect on the specific characteristics of housing intended for this sector of the population in Spain, given the lack of general legislation on the typology of housing for the elderly.

Besides this, the promotion of autonomy and self-reliance is linked to active aging policies. As has already been explained in a comparative study on the subject [12], active aging policies focus on prolonging the independence of users by adapting traditional residential types with the idea of postponing the need for relocation to specialized centers or homes for the elderly, avoiding the high economic impact of the increase in the number of people demanding specialized centers. Providing a home tailored to the specific needs of older people therefore means seeking greater personal autonomy and enabling independent living, but also reducing the need for institutionalization and family support.

Furthermore, the progressive aging of the population because of the increase in life expectancy, coupled with the change in family structures, that until 2019 [3,32] favored coexistence in the same family nucleus of several generations, force us to reflect on the specific characteristics of housing intended for the elderly. In addition, it is relevant to note that according to the INE in 2019, 43.6% of people living alone were over 65 years old, which means that when the COVID-19 isolation policies were implemented (in Spain in March 2020) a high percentage of older people were isolated at home.

Given the fact that technology is present in the whole analysis, the role of IT in smoothing communication among generations (living in different units) and the role of devices in improving the connectivity of the house (by the integration of elements of e-health, comfort areas, and safe spaces), proved to be especially important in a pandemic time.

Finally, an assessment has been made of the bodily functions to which legislators and experts pay the most attention when calling for specific household conditions for this segment of the population.

4.2.2. Dependency and the Level of Seniors' Vulnerability as a Key for a New Type of Housing

Dependency is the result of events that cause a physical, mental, or intellectual disability that requires significant assistance in carrying out daily activities. The International Classification of Functioning, Disability, and Health (ICF) goes a little further. For the ICF, it is the result of the interaction between a person with a disability (listening, moving, taking care of oneself, etc.) and the environmental or attitudinal barriers that he or she may encounter. In this way, the concept of functioning is considered as a global term that refers to all bodily functions, activities, and participation, and that of disability, which includes deficiencies, limitations in activity, and restrictions in participation.

The ICF is part of the "family" of international classifications developed by the World Health Organization (WHO) and provides the conceptual framework for codifying a wide range of health-related information in a standardized and unified language. As a classification, it has been accepted as one of the United Nations social classifications and incorporates the Standard Rules on the Equalization of Opportunities for People with Disabilities. The classification defines the health components and some health-related components of well-being, such as education and work. Therefore, the domains included in the ICF (understood as the relevant and practical sets of physiological functions, anatomical structures, actions, tasks, or related spheres of life) can be considered health domains and "health-related" domains, and are described from the bodily and individual point of view in two basic lists: (1) functions and structures of the body, and (2) activities and participation.

As a classification, it groups the different domains of a person in each health status according to what the person affected by a disorder or disease does or can do. Thus, the concept of functioning is understood as an umbrella term that refers to all functions, activities, and participation of the body, while disability encompasses deficiencies, activity limitations, or participation restrictions.

In describing situations related to human functioning and its limitations, the CCF provides a framework organized into two parts:

Part 1. Functioning and disability

(a) Body functions and structures
(b) Activities and participation

Part 2. Contextual functions

(c) Environmental factors
(d) Personal factors

The functional and disability components are divided into two groups. On the one hand, the "body" component differentiates between the functions of the body's systems and the structures of the body. At the same time, in the first case, it distinguishes between bodily functions (physiological functions of bodily systems, including psychological functions) and bodily structures (anatomical parts of the body such as organs, limbs, and their components). The "activities and participation" component covers all areas that indicate operational aspects from an individual and social perspective. Deficiencies are defined as problems in bodily functions or structures associated with a significant deviation or loss. We consider bodily functions as the object of a greater interrelationship between the individual and the home. The ICF classifies them into eight categories:

- Sensory functions and pain: functions of the senses (sight, hearing, taste, etc.) as well as the feeling of pain.
- Voice and voice functions: functions involved in the production of sound and the voice.
- Functions of the cardiovascular, hematological, immunological, and respiratory systems: functions involved in the cardiovascular, hematological, and immune systems, as well as in the respiratory system.
- Functions of the digestive, metabolic, and endocrine systems: functions related to ingestion, digestion, and elimination, as well as those related to metabolism and the endocrine glands.
- Genitourinary and reproductive functions: urinary and reproductive functions, including sexual and reproductive functions.
- Neuromuscular and motion-related functions: related to motion and mobility, including the functions of bones, muscles, joints, and reflexes.
- Functions of the skin and related structures: functions related to the skin, nails, and hair.

The normative requirements arising from the above-mentioned standards have been compared with the assessment of two types of sources, legal and statistical. In addition, on 17 and 18 December 2020, an expert panel was held within the framework of the Fifth International Congress on the State of Smart Cities: Socio-Legal Conditions and Meeting the Needs of Adequate Housing for Older People.

This hypothesis has been verified by analyzing data from the National Institute of Statistics (INE). INE has conducted three macro-surveys on disability in Spain (1986, 1999, and 2008) on disability and deficiencies of the Spanish population (EDDM), the Survey on Disability, Impairments, and Health Status (EDDS 1999), and the Survey on Disability, Personal Autonomy, and Dependency Situations (EDAD 2008). Both EDDS 1999 [18] and EDAD 2008 information is available on the INE website. Consultation of the data for the 2008 EDAD (Tables 1 and 2) confirms that more than 4 million Spaniards suffer from some type of disability, of whom some 2.5 million (2.4494) are over 65 years of age. This represents just over 61% of disability cases and supports the thesis that disability cases are concentrated in this age group.

Table 1. Spanish population in households with some type of disability, according to age and sex.

	Unit: Thousands of People		
	Both Genders	**Males**	**Females**
Total	3.8479	1.5477	2.3002
From 0 to 5 years	60.4	36.4	24.0
From 6 to 64 years	1.5604	754.5	805.9
From 65 to 79 years	1.2013	454.8	746.5
From 80 to older	1.0258	301.9	723.9

Source: Survey on Disabilities, Personal Autonomy, and Dependency Situations. 2008. INE.

Table 2. Spanish population in residences and hospitals with some type of disability, according to age and sex.

	Unit: People over 6 Years Old Residing in some Type of Center		
	Both Genders	**Males**	**Females**
Total	269.139	93.546	175.593
From 6 to 64 years	46.879	29.005	17.874
From 65 to 79 years	59.366	26.439	32.927
From 80 to older	162.894	38.102	124.793

Source: Survey on Disabilities, Personal Autonomy, and Dependency Situations. 2008. INE.

In fact, with a life expectancy in 2008 of 77.77 and 84.11 years for men and women, life expectancy without disability is reduced to 71.27 and 73.75, respectively (Table 3), and 269,139 people have been able to find solutions to these disability problems in specialized centers and residences (Table 2).

Table 3. Health and life expectancy of the population by age and sex.

Unit: Years	LE: Life Expectancy	LEFD: Life Expectancy Free of Disabilities
Both genders	80.94	72.49
Males	77.77	71.27
Females	84.11	73.75

Source: Survey on Disabilities, Personal Autonomy, and Dependency Situations. 2008. INE.

According to data collected in the INE newsletter [15], the main disability groups affecting residents of Spanish households are mobility (6.0% of the population), domestic life (4.9%), and self-care (4.3%). This means that, of the total population with disabilities, 67.2% have mobility or movement limitations, 55.3 % have problems related to housework, and 48.4% to housework and their care and hygiene. The most frequent bone and joint deficiency is bone and joint problems.

This presents a picture in which more than 1.6 million people find it difficult to move outside the home (Table 4).

Table 4. Distribution of the different types of disability among people over 6 years of age in Spain.

Unit: Thousands of People		
	Both Genders	**Males**	**Females**
Total	3.7874	1.5109	2.2765
Vision	979.0	371.3	607.7
Hearing	1.0641	455.7	608.5
Communication	734.2	336.6	397.5
Learning and applying knowledge and developing tasks	630.0	264.5	365.5
Mobility	2.5354	881.5	1.6539
Self-care	1.8245	645.0	1.1795
Domestic life	2.0792	605.8	1.4734
Interaction and personal relationships	621.2	291.7	329.5

Source: Survey on Disabilities, Personal Autonomy, and Dependency Situations. 2008. INE.

However, it should be remarked that mobility is not only limited by problems in the locomotor system, but also by the loss of visual acuity, coordination of movements, or orientation (Figure 1), and that all these deficiencies are accentuated with the increasing age of a person (Figure 2).

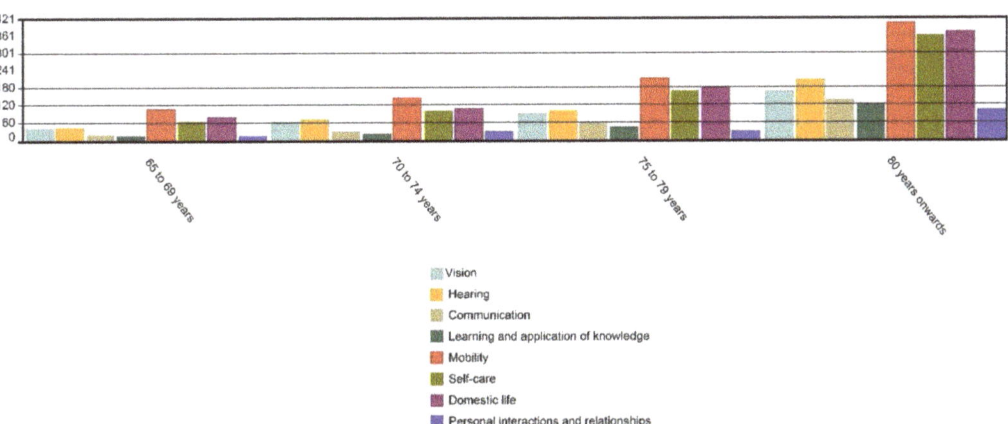

Figure 1. Population rate over 65 years of age with disabilities, distributed according to type of disability. Source: Survey on Disability, Personal Autonomy, and Dependency Situations; ESD 2008.

The panorama presented by the real estate world has been backed up by data obtained from the 2011 Population and Housing Census, the last official census in existence [18]. Although recently many municipalities have carried out numerous actions aimed at improving the accessibility of the built landscape, in 2011 only 23.09% of buildings were accessible and those with accessibility were reduced to 6.27% with elevators (Table 5). As a result, 1.2 million households where people with reduced mobility live have barriers in buildings such as stairs without ramps or mobile platforms.

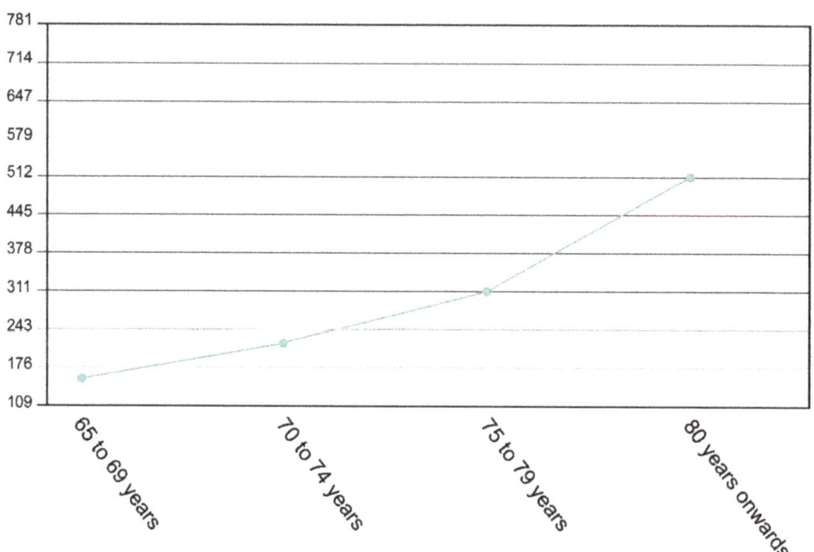

Figure 2. Total rate per 1000 inhabitants of the population over 65 with disabilities in Spain. Scheme 2008.

Table 5. Percentage of accessible buildings and properties with a lift in Spain.

	Unit: Percentage	
	Total	
	Buildings	**Estates** [1]
Accessible (%)	23.09	33.42
With elevator (%)	6.27	41.15

[1] An estate must be understood as each module that a building encompasses. Source: Population and Housing Census 2011. I.N.E.

In the survey "Accessibility and Urban Space" carried out by the ACCEPLAN Project of the Institute of European Studies of the Autonomous University of Barcelona, which served as the basis for the preparatory studies for the I National Accessibility Plan 2004–2012 of the Department of Labor and Social Affairs (I Accessibility Plan 2004–2012), 83% of the residential buildings analyzed had access problems at the threshold, while 96% had inaccessible spaces within their own homes for people with disabilities.

Furthermore, according to the data collected in the INE newsletter [15], 51.5% of disabled people say they have difficulties in getting on with ease in their home or building, especially on stairs (43.3%) and bathrooms (29.8%). Figures increase up to two-thirds in the age group 80 years or older.

Irrespective of whether or not these data are updated, the reality is that they show a picture in which the country's housing stock and buildings suffer from serious accessibility deficiencies. These deficiencies are more visible in the case of older people, although they also affect other age groups. This situation requires an effective response using universal design, which refers to the design of products and environments to be used by all, without the need for adaptation or specific design.

Having confirmed the hypothesis that the distinguishing feature of housing intended for the elderly is greater accessibility, the information from the three reference sources mentioned (expert plans, sources consulted, and reference standards) were cross-referenced in order to structure a table in which the experts' suggestions are related to the different areas and elements of the house—the requirements envisaged in the standards and the eight categories of functional disabilities of the International Classification of Functioning, Disability, and Health.

5. Discussion

The actions that can be carried out on a home can be classified into 10 different areas, which, at the same time, act on a total of 34 different elements (Table 6).

Table 6. Areas and elements of a home susceptible to a design adapted to different disabilities.

Spaces of the House	Items or Indications for Use
Entrances	Access
	House doors
	Corridors
	Lobbies
Terraces, balconies, and roofs	Height of the different elements
Living room and dining room	Clearances and distances
	Furniture
Kitchen	Clearances and distances
	Worktop
	Sink faucet
	Characteristics and recommendations for use of kitchen elements
Bedroom	Free space and distances
	Floor
	Furniture
	Environment control
Bathroom and toilet	Obstacle free space
	Sink
	Toilet
	Shower
	Grab bars
	Design features
	Complementary installation
Windows	Manipulable opening and closing systems
	Sill height
Technology and environmental control facilities	Height and distances
	Extension cords
	Telephony
	Communication system
	Environment control
Illumination	Intensity
	Orientation
	Sensors
Stimulation	Objects
	Color of the wall paint

In many cases, action on these 34 elements of the dwelling is not limited to a single decision, but accessibility is the result of various design corrections that the analyzed regulations have not always taken into account. In total, 133 design proposals have been identified, of which 88 are purely architectural, while the remaining 45 correspond to the areas of furniture or decoration. Of these 88 proposals, 61 have already been included in some type of regulation (a conclusion that refers to the regulations analyzed for this work: CTE, the Andalusian Accessibility Decree, and the Malaga City Council Accessibility Ordinance). In other autonomous communities or municipalities these figures may vary slightly, and most of the unregulated proposals concern issues related to the characteristics of the furniture.

Finally, it has been observed that most of the proposals for adapting housing to the needs of the elderly concern disabilities related to neuromuscular and motor functions, sensory functions (sight) and pain; the rest are treated in the background (Table 7).

Table 7. Number and percentage of proposals associated with the different disabilities of bodily functions.

Body Function According to the CIF	Number of Associated Proposals	% of Total Proposals
Mental functions	6	3.70%
Sensory functions and pain	40	24.69%
Voice and speech functions	5	3.09%
Functions of the cardiovascular, hematological, immune, and respiratory systems	0	0.00%
Functions of the digestive, metabolic, and endocrine systems	0	0.00%
Genitourinary and reproductive functions	0	0.00%
Neuromusculoskeletal and movement-related functions	107	66.05%
Functions in the skin and related structures	1	0.62%
—	3	1.85%
Total	162	100%

The provision of adequate services for the regulation of the accommodation of the elderly, in the scenario examined, has focused on the regulation of residences and care spaces, leaving aside the home itself, as shown in the case of the Andalusian set of laws, whether examining Decree 72/1992, of 5 May, approving the Technical Standards for Accessibility and Removal of Architectural, Urbanistic and Transport Barriers in Andalusia; or the subsequent Order of 12 February 2020, modifying that of 21 July 2008, on the technical requirements applied for subsidized housing in Andalusia. At the national level the same is true, when reviewing, for instance, the regulation included in the Spanish Housing Project and the Spanish Technical Building Code.

Therefore, the provision of so-called housing for the elderly in Spain did not consider all the variables and technical requirements, as was shown above, since the housing-for-seniors-adapted subsidies were targeted to address physical impairments more than other disabilities equally significant for livable and adequate housing. This inadequacy was particularly visible in the case of forced quarantine due to COVID-19 [19].

6. Conclusions

The distinguishing feature of housing for the elderly from the general supply is the need to address the main disability problems experienced by this segment of the population. These problems have not been specifically addressed in the current legislation, which has focused, in general, on accessibility and mobility issues. Thus, the removal of architectural barriers or the integration of health services in units are only some of the elements to be considered. The pandemic has strengthened the role of units as spaces for self-development and healthy recovery, while creating social distance and forcing a dilemma of fast connectivity. According to the surveys carried out by the INE, it can be concluded that a differentiating characteristic of the older population is the increase in disability problems, which places life expectancy without such limitations at 72.49 years.

The preliminary findings of this study show that technical requirements need to be adapted to new standards, while regulations need to address the reality of family structures different from those existing at the time the regulations were adopted.

According to this specific characteristic, the differentiating feature of housing designed for this population sector should be an adapted design able to address seniors' needs. As stated in the Materials and Methods section of this article, in Spain there is no specific regulation on the characteristics that dwellings for the elderly must have or on the obligation to develop them. Spanish regulations are limited to dealing in a cross-cutting manner with a building's accessibility and mobility requirements. In the case of residential buildings, these requirements focus on common areas. Within this context, the accessibility and mobility requirements for dwellings are limited to the percentage of "adapted dwellings" that must be built in each development once it reaches a certain number of dwellings.

The minimum number of dwellings is a requirement established by national regulations (Rulemaking, Article 111, Act 38/1982, of 7 April, on the Social Integration of the Disabled, LISMI, article 57.1; modified by article 19, Act 26/2011, of 1 August, on regulatory adaptation to the International Convention on the Rights of Persons with Disabilities) which may be low, as it does not consider the real percentage of the elderly population that may require this adaptation. The greatest disability problems among the older population are associated with problems of mobility, self-care, and domestic life (Figure 1), although this INE classification does not show which bodily functions are affected.

The analysis of the design proposals and obligations for adapted housing for the elderly shows that all the spaces and facilities in a home can be subject to universal design (Tables 6 and 8). These proposals and regulations focus mainly on neuromusculoskeletal and movement-related functions and, to a lesser extent, on sensory and pain-related functions (Table 7). The preliminary findings of this paper show that the correlation between the variables that connect vulnerability, dependency, and accessibility (previously explained in Section 4.2.2: Dependency and the level of senior's vulnerability as a key for a new type of housing), has not been translated into standards or technical requirements within the units of the Spanish set of laws. Thus, the inadequacy of the existing stock of Spanish homes for older people has had a negative impact on the way they faced isolation due to COVID-19. This means that specific consideration should be given to extending the scope of this regulation to better address accessibility issues and to take into account types of disabilities other than lack of or limited mobility.

Table 8. Conclusion of test parameters to be considered in the design of this type of housing; regulations to be complied with and elements that are not included in the regulations.

Materials/Spaces	Position	Element to Check		Condition Searched	
Floor	Interior	Non-slip material	-	Gradient < 6%	Category 1
				Gradient ≥ 6% and stairs	Category 2
	Exterior	Non-slip material	-	-	Category 3
Accessible itineraries	Lobby	Free circumference not swept through doors		-	Ø ≥ 1.50 m
		Free circumference in front of accessible elevator		-	Ø ≥ 1.50 m
					Ø ≥ 1.50 m
	Corridor	Free width	-	-	≥1.20 m
		Isolated narrowing	-	Length of narrowing	≤0.50 m
				Resulting free width	≥1.00 m
				Separation to doors or changes of direction	≥0.65 m
		Free turning space at the bottom of corridors larger than 10 m		-	Ø ≥ 1.50 m
	Crossing gaps	Free passage of entrance doors and openings	-	-	≥0.80 m
		Horizontal free space on both sides of the doors	-	-	Ø ≥ 1.20 m
		Angle of opening of doors (including exterior doors)	-	-	≥90°
		Opening or closing system	Height of the crank	-	0.80–1.20 m
			Separation of the hatch to the door plane	-	0.04 m
			Distance from the mechanism to the corner meeting	-	≥0.30 m
		Transparent or glazed doors (O3)	Horizontal signage along its entire length	-	0.85–1.10 m; 1.50–1.70 m
			Wide signalling strip (fully transparent doors with automatic opening or without actuating mechanism) perimeter	-	0.05 m

Table 8. *Cont.*

Materials/Spaces	Position	Element to Check		Condition Searched
		Double-leaf doors	Without automatic and coordination mechanism, minimum passage width in one of them	-	$\geq$0.90 m
		Automatic doors	Clearance width	-	$\geq$1.20 m
			Speed reduction mechanism	-	$\leq$0.5 m/s
	Windows in corridors	They do not invade the corridor at a height of less than 2.20 m	-	-	(O2) Window mechanisms shall be located at a height between 0.80 and 1.10 m. (O3) Glass doors shall be made of safety glass, or shall have a protective base 0.40 m high
		Height saved by the flight	With elevator as an alternative	-	$\leq$3.20 m
			Without elevator as an alternative	-	$\leq$2.25 m
		Minimum number of steps per flight	-	-	3 Maximum number of steps without intermediate rest $\leq$ 16
Accessible itineraries		Step	-	-	$\geq$0.28 m
	Stairs	Riser		-	0.13–0.185 m
				-	
		Relation Step/Riser	-		Between two consecutive floors of the same staircase, all the steps have the same footprint and all the steps of the straight sections have the same footprint. Between two consecutive stretches of different plants, the back footprint does not vary by more than $\pm$1.00 cm	0.54 m $\leq$ 2C+H $\leq$ 0.70 m

Table 8. *Cont.*

Materials/Spaces	Position	Element to Check		Condition Searched	
		Free width (In curved sections, the area where the footprint < 0.17 m should be excluded)	-	-	≥1.00 m
		Maximum angle of the partition to the vertical plane	-	-	≤15°
		Landing	Intermediate	With access doors to dwellings. Width	≥Width of stairs
					Ø ≥ 1.20 m free
				No doors to dwellings. Width	≥Width of stairs
					Ø ≥ 1.00 m free
				Back	Ø ≥ 1.00 m free
			Of take-up and landing	Width	≥Width of stairs
				Back	Ø ≥ 1.20 m free
		Distance from the edge of steps to doors	-	-	≥0.40 m
		Banister	Position	Ladders with a width ≥ 4.00 m	Central banister
				Maximum separation of banisters	4.00 m
				Stairs having a height ≥ 0.55, of a width exceeding 1.20 m	Banister on both sides of the staircase and continuous, including landing
			Measures	Greater dimension of the capable solid	0.045–0.05 m
				Height	0.90–0.95 m
		Directrix	-	-	Straight or curve, radius = 30.00 m
					Straight
					-
		Width	-	-	≥1.20 m

Table 8. *Cont.*

Materials/Spaces	Position	Element to Check		Condition Searched	
Accessible itineraries	Accessible ramps	Longitudinal slope (horizontal projection)	-	Length sections < 3.00 m	10.00%
				Length sections ≥ 3.00 m and < 6.00 m	8.00%
				Length sections ≥ 6.00 m	6.00%
		Transverse slope	-	-	≤ 2 %
		Maximum stretch length (horizontal projection)	-	-	≤9.00 m
		Landing	-	Width	≥Width of ramp
				Back	≥1.50 m
				Access ramp to the building. Back	≥1.20 m
		Distance from the edge of the ramp to a door or corridor with less than 1.20 m of width	-	-	≥1.50 m
		Handrail	Position	On ramps having a height major than 0.185 m with a gradient ≥ 6%	On both sides and continuous including landing
			Characteristics	Capable solid dimension	0.045–0.05 m
				Height	0.90–0.95 m; 0.65–0.75 m
				Extension at the ends on both sides (sections ≥ 3 m)	≥0.30 m
		Banister	-	Unevenness > 0.55 m	0.90–1.10 m
				Unevenness > 0.15 m	0.90–1.10 m
		Height of socket or lateral protective element on free edges, on ramps with a maximum height difference of 0.55 m	-	-	≥0.10 m

Table 8. *Cont.*

Materials/Spaces	Position	Element to Check		Condition Searched
Vertical communication	Accessible elevators	Consideration of future installation	-	Housing buildings with PB+1 that have 6 dwellings or less
			-	Buildings in which up to two floors must be saved from an accessible main entrance to the building or to a dwelling or community area or which have 12 or fewer flats on floors without an accessible main entrance to the building (DB-SUA 9)
				Structure estimate for elevator shaft
		Obligación de instalción de ascensor accesible	-	Buildings with more than 6 dwellings that develop at most in PB+1 or with any number of dwellings from PB+2
			-	Buildings where more than two floors need to be saved from an accessible main entrance to a dwelling or communal area, or where there are more than 12 dwellings on floors without an accessible main entrance to the building (DB-SUA 9)
				Installation of accessible elevator
		Free space in front of the lift	-	-	$\varnothing \geq 1.50$ m
		Wide passage of doors	-	-	≥ 0.80 m

Table 8. *Cont.*

Materials/Spaces	Position	Element to Check	Condition Searched	
	Internal measures (Minimum dimensions)	Without accessible houses	One or two doors facing each other	1.10 × 1.40 m
			In elevators with double perpendicular door:	1.20 × 1.20 m
			Two doors at an angle	1.40 × 1.40 m
			In elevators with double perpendicular door:	1.20 × 1.20 m
		With accessible houses	One or two doors facing each other	1.10 × 1.40 m
			In elevators with double perpendicular door:	1.20 × 1.20 m
			Two doors at an angle	1.40 × 1.40 m
			In elevators with double perpendicular door:	1.20 × 1.20 m
	Interior characteristics	Telescopic opening doors	-	Telescopic opening door
		Buttons	H interior	≤1.20 m
			H exterior	≤1.10 m
			Self-relief numbers and Braille system	-
		Handrail	Height	0.800–0.90 m

Table 8. *Cont.*

Materials/Spaces	Position	Element to Check		Condition Searched	
Vertical communication	Accessible elevators	Indicators	Access	Luminous and acoustic arrival indicators	-
			Doors	Indicators indicating the direction of movement	-
			Frame	Number of the plane in braille and Arabic numerals in relief at one height	≤1.20 m
				Voice synthesizer	-
Collective use spaces	Parking	Amount	-	Exclusive use of each dwelling	1 × reserved housing
				Collective use	1 × every 40 or fraction
		Transfer area	-	Perpendicular	Lateral free space
					≥1.20 m
				Parallel	Free space at the rear
					≥3.00 m
		Transfer area shared between two parking places	-	It is allowed if it has a width ≥	1.40 m
	Electric mechanisms	Height of switches	-	-	0.80–1.20 m
		Height of the sockets	-	-	0.30 m
	Lobbies, stairs, doors and exits	Doors	General	Easily identifiable, with sufficient force to open the exit doors	≤25 N (≤65 N when the doors are fire resistant)
				Sliding doors	No risings are available on the pavement
				Indicative band in colour	H 0.60–1.20 m
			Automatic opening with sensitive vertical sweeping devices	Speed-decreasing mechanism	0.50 m/s
				Sensitive devices that open doors in case of imprisonment	-

Table 8. *Cont.*

Materials/Spaces	Position	Element to Check		Condition Searched	
				Devices preventing automatic closing while the threshold is occupied	-
				Manual mechanism of automatic stop	-
		Emergency exits	-	By simple pressure and have double flat bar	H 0.20 m y 0.90 m
		Lighting	Permanent	Minimum intensity	300 lux
			Switches	Equipped with light pilot	-

Source: VIDA research team's own design.

Analysis of the characteristics of existing adapted housing compared with those planned outside this legislation should make it possible to assess the economic impact of the further adaptation of housing in general. This will allow a better match between investments for policymakers and the definition of a new typology.

We propose a complete table of characteristics which will make it possible to draw up a checklist for assessing dwellings from the point of view of their suitability for older people. This table (Table 8) which is now being implemented in a web application, will be presented and completed by the end of 2021. Within Table 8, a comprehensive list of specific features that housing adapted to the needs of older people will require has been established. This table is being implemented as a web application that will be set up as a testing tool.

Within this context, a proposal for a new national housing regulation represents an opportunity to reflect the need for a new type of housing for older people since, at the time of writing, this new typology is not defined. Moreover, the closest approach to regulating the characteristics of housing suitable for the needs of older persons is to be found in the reservation of housing suitable for accessibility standards in promotions with the protections established by the DB-SUA document, belonging to the CTE, for the regularization of good accessibility in wheelchairs, as seen above.

Consequently, it is of as great an importance to consider the housing needs of an elderly population which requires specific care and attention, as much as it is to provide them with adequate financing for their housing needs.

Author Contributions: Conceptualization, V.-Y.A. and M.L.G.-J.; methodology, V.-Y.A.; formal analysis, V.-Y.A. and M.L.G.-J.; investigation, M.L.G.-J. and V.-Y.A.; resources, V.-Y.A. and M.L.G.-J.; data curation: V.-Y.A.; writing—original draft preparation, V.-Y.A. and M.L.G.-J.; writing—review and editing, M.L.G.-J.; project administration, M.L.G.-J. All authors have read and agreed to the published version of the manuscript.

Funding: This paper is the result of the research carried out in the project "Proposals and Innovative Parameters of 'Compliance' applied to Home Automation and Accreditation of the Adequacy of Homes for the Elderly in Andalusia (Home Automation Adapted to the Elderly)," UMA18 FEDERJA-261, co-financed by the FEDER Operational Program 2014–2020 and by the Economy Council and Knowledge of the Junta de Andalucía and the REDIAS thematic network on AI for Health Care funded by the Malaga University Research Plan, Gómez Jiménez, ML (Principal Researcher).

Institutional Review Board Statement: Not applicable.

Informed Consent Statement: Not Applicable.

Data Availability Statement: The main data used in this studies besides the own elaboration derived from Spanish official statistical sources National Institute of Statistics (INE) which were accessed openly in: https://www.ine.es/ (last accessed on 31 May 2021).

Acknowledgments: The authors would love to acknowledge the support for editing this document to Mari-Cruz Moreno Gil, researcher from VIDA project and support with the English proofreading. We would also express the gratitude to the Research Group SEJ-650, PAIDI. PASOS at Málaga University and to the University of Málaga for supporting the realization of this Research.

Conflicts of Interest: The authors declare no conflict of interest.

References

1. Habitat, U.N. *Strategic Plan 2020–2023*; United Nations: Geneva, Switzerland, 2020.
2. Gómez-Jiménez, M.L. Vivienda domótica adaptada a la emergencia sanitaria: Ideas preliminares, retos y propuestas normativas para la sociedad post COVID-19. In *Revista de Derecho Urbanístico y Medio Ambiente*; RDU: Madrid, Spain, 2020; Volumes 337–338, pp. 305–350.
3. INE. Available online: https://www.ine.es/dyngs/INEbase/es/operacion.htm?c=Estadistica_C&cid=1254736176782&idp=1254735573175#!tabs-1254736194716 (accessed on 31 May 2021).
4. Donado Vara, A. *La protección de Las Personas Mayores*; Tecnos: Madrid, Spain, 2007.
5. Garnica, G. La protección jurídica de las personas mayores: Un reto para el siglo XXI. In *Revista de Derecho UNED*; Universidad Nacional de Educación a Distancia: Madrid, Spain, 2018.

6. Abades Porcel, M.; Rayón Valpuesta, E. El Envejecimiento en España ¿un reto o problema social? *Gerokomos* **2012**, *23*, 151–155. [CrossRef]
7. Bazo, M.T. Políticas Socio Sanitarias y el debate entre Público y privado. In *Envejecimiento y Sociedad: Una perspectiva Internacional*; Editorial Medica Panamericana: Madrid, Spain, 1999.
8. Española, C. *Constitución Española*; BOE Diciembre: Madrid, Spain, 1978.
9. Gomez Jimenez, M.L. Public Choices and Housing Opportunities for Senior Citizens: Different Scenarios in the United States of America and Spain. *Chic. Kent J. Int. Comp. Law.* **2012**, *56*, 35.
10. Abellán, A.; Ayala, A.; Pujol, R. Estadística sobre residencias. Distribución de centros y plazas residenciales por provincia y plazas residenciales por provincia, Datos de julio 2017. Consejo Superior de Investigaciones Científicas, Centro de Ciencias Humanas y Sociales (CCHS), Envejecimiento en red. In *Un perfil de las personas mayores en España*; CSIC: Madrid, Spain, 2018.
11. Schmid, A.M. Los Requisitos de Acreditación de Residencias Para Personas Mayores. Normativas Autonómicas Sobre Ratios y Formación Mínima del Personal Para Residencias Privadas de Personas Mayores. Lares. Federación de Residencias y Servicios de Atención a los Mayores, Sector Solidario: Madrid, Spain, 2010.
12. Gómez-Jiménez, M.L.G.; Koebel, C.T. Comparison of Spanish and American Housing Policy Frameworks Addresing Housing for the Elderly. *J. Hous. Elder.* **2006**, *20*, 23–37. [CrossRef]
13. Lazaro Ruiz, V.; Gil López, A.Y. The Quality of old people's housing and their preferences before institutionalization. In *Intervención Psicosocial*; Colegio official de Psicólogos de Madrid: Madrid, Spain, 2004; Volume 14, pp. 21–40.
14. Gómez-Jiménez, M.L. *Active Ageing? Perspectives from Europe on a Vaunted Topic*; Whiting and Birch: London, UK, 2012.
15. INE, "Cifras INE", Noviembre 2009. Available online: https://www.ine.es/ss/Satellite?L=es_ES&c=INECifrasINE_C&cid=1259924962561&p=1254735116567&pagename=ProductosYServicios%2FPYSLayout (accessed on 31 May 2021).
16. Gómez-Jiménez, M. Las barreras arquitectónicas y las personas mayores: Nuevos retos desde la intervención administrativa en la promoción de vivienda. In *Scripta Nova: Revista electrónica de Geografía y Ciencias Sociales*; Universidad de Barcelona: Barcelona, Spain, 2003; Volume 7.
17. Bütow, B.; Jiménez, M.L.G. Social Policy and Social Dimensions on Vulnrerability and Resilience; Verlag Barbara Budrich: 2014.
18. INE. 2011. Available online: https://www.ine.es/dyngs/INEbase/es/operacion.htm?c=Estadistica_C&cid=1254736176992&menu=resultados&idp=1254735576757 (accessed on 31 May 2021).
19. Cuerdo-Vilches, T.; Navas-Martín, M.Á.; Oteiza, I. A Mixed Approach on Resilience of Spanish Dwellings and Households during COVID-19 Lockdown. *Sustainability* **2020**, *12*, 10198. [CrossRef]
20. Gómez-Jiménez, M.L. Ciudades Saludables en la era Post COVID-19. Lecciones aprendidas un apunte sobre los efectos de la pandemia en las políticas de Urbanismo y Vivienda. In *Pandemia y Derecho. Una Visión multidisciplinar*; Laborium: Sevilla, Spain, 2020.
21. López, A. *La Accesibilidad en Evolución: La Adaptación Persona-Entorno y su Aplicación al Medio Residencial en España y en Europa*; Universidad Autónoma de Barcelona, Departamento de Derecho Público Tesis Doctoral: Barcelona, Spain, 2016.
22. Caneiro, O. Así Debe Ser Una Vivienda Adpatada Para Personas Mayores O Discapcidad. 30 May 2018. Available online: https://ocaneiro.com/asi-debe-ser-una-vivienda-adaptada-para-personas-mayores-o-con-discapacidad/# (accessed on 31 May 2021).
23. Junta de Andalucía, Decreto 72/1992, 1992, of May 5 (BOJA 23, may 1992). Available online: https://www.juntadeandalucia.es/boja/1992/44/1 (accessed on 31 May 2021).
24. Doméstiko, 30 Marzo 2017. Available online: https://www.domestiko.com/blog/como-adaptar-la-vivienda-para-personas-mayores-o-discapacitadas/ (accessed on 31 May 2021).
25. TROVIMAP. Available online: https://www.trovimap.com/blog/adaptar-una-casa-discapacitados-gente-mayor/ (accessed on 31 May 2021).
26. VALIDA. Available online: https://es.validasinbarreras.com/blog/post/acondicionar-casa-para-ancianos-11-cosas-que-debes-hacer/ (accessed on 31 May 2021).
27. CUIDUM. Available online: https://www.cuidum.com/blog/envejecer-en-casa-como-adaptar-la-vivienda-para-mayores/ (accessed on 31 May 2021).
28. Quotatis. 24 June 2019. Available online: https://www.quotatis.es/consejos-reformas/Inspiracion/casa-del-abuelo/como-adaptar-su-vivienda-para-personas-mayores/ (accessed on 31 May 2021).
29. Entorno Accesible. Available online: https://www.entornoaccesible.es/productos-y-servicios/una-vivienda-accesible/ (accessed on 31 May 2021).
30. Clarín. 20 August 2019. Available online: https://www.clarin.com/arq/hacer-casa-segura-adultos-mayores_0_FbkInqD-4.html (accessed on 31 May 2021).
31. Ancianos.es. 5 December 2018. Available online: https://ancianos.es/cuidado/adaptar-vivienda-necesidades-ancianos/ (accessed on 31 May 2021).
32. Andaluz, D.d.P. *Informe Personas Prisioneras En Sus Viviendas*; Defensor del Pueblo Andaluz: Sevilla, Spain, 2003.

 *sustainability*

Article

Tools for the Implementation of the Sustainable Development Goals in the Design of an Urban Environmental and Healthy Proposal. A Case Study

Rafael Herrera-Limones [1], Maria LopezDeAsiain [1,*], Milagrosa Borrallo-Jiménez [1] and Miguel Torres García [2]

[1] Instituto Universitario de Arquitectura y Ciencias de la Construcción, Escuela Técnica Superior de Arquitectura, Universidad de Sevilla, Avd. Reina Mercedes 2, 41012 Seville, Spain; herrera@us.es (R.H.-L.); borrallo@us.es (M.B.-J.)

[2] Departamento de Ingeniería Energética, Escuela Técnica Superior de Ingeniería, Universidad de Sevilla, Camino de los Descubrimientos s/n, 41092 Seville, Spain; migueltorres@us.es

* Correspondence: mlasiain@us.es

check for **updates**

Citation: Herrera-Limones, R.; LopezDeAsiain, M.; Borrallo-Jiménez, M.; Torres García, M. Tools for the Implementation of the Sustainable Development Goals in the Design of an Urban Environmental and Healthy Proposal. A Case Study. *Sustainability* **2021**, *11*, 6431. https://doi.org/10.3390/su13116431

Academic Editor: Pilar Mercader-Moyano

Received: 21 April 2021
Accepted: 1 June 2021
Published: 5 June 2021

Publisher's Note: MDPI stays neutral with regard to jurisdictional claims in published maps and institutional affiliations.

Copyright: © 2021 by the authors. Licensee MDPI, Basel, Switzerland. This article is an open access article distributed under the terms and conditions of the Creative Commons Attribution (CC BY) license (https://creativecommons.org/licenses/by/4.0/).

Abstract: This article presents a methodological proposal to address the urban issue from the perspective of the Sustainable Development Goals (SDGs). Different tools have been developed for this purpose: the Aura Method and the Aura Matrix. The Aura Matrix of relationships built from the SDGs, along with the conceptual proposals to which the project must respond, allows for the definition of a methodological framework of action, defined as the Aura Method, applicable to any project that aims to respond to the urban scale from a more sustainable and healthy approach and within the framework of the above-mentioned goals. Two proposals for the Solar Decathlon Latin America of the Aura Team from the University of Seville (2015 and 2019) in Cali, Colombia, and their comparison, are presented as case studies. The scope of the 2019 proposal based on the use of these tools is more rigorous and bold with respect to the requirements defined by the SDGs than the 2015 proposal, based on the millennium goals. This reinforces to a great extent the resilience of the urban scope under study and its capacity to face serious situations in terms of citizens' health, such as the pandemic we are currently suffering, and improves life quality. The main findings lay on the defined Aura Matrix and Aura Method tools as pragmatic opportunities to translate conceptual approaches such as G3: 'Ensure healthy lives and promote well-being for all at all ages' into practical decisions and urban design proposals to improve the quality of life and health of citizens.

Keywords: sustainable development goals; urban design; neighborhood regeneration; competition; architectural education; indoor environmental quality; health

1. Introduction

The Sustainable Development Goals (SDGs) enunciated by the New Urban Agenda [1] in Quito, 2015, are a reference framework recognized worldwide [2,3]. However, its incorporation as a strategy for innovative development in architectural and urban design has been hardly investigated in practical terms, except for the political-normative field, according to some authors [3–9], or the management [10,11] or biodiversity conservation [12] fields, according to others. The references that we can find to research in the urban field focus on establishing a generic framework [13,14] approaching the complexity of the problem and its numerous requirements [7,12,15], without defining specific design tools or strategies that may allow their practical application or their specific assessment by means of indicators [16]. There are also some advances in this framework related to the design of very specific elements of construction systems [2] or fixtures [11,17], also in the building environment [3,18–20], but the review carried out in the main research databases (Web of Science and Scopus), shows that specific urban architectural design aspects are scarcely developed in depth. Thus, the use of the theoretical-practical concept of the SDGs [21] as a design approach strategy represents an innovation in the methodological field of architectural and urban design, as well as an alignment with international urban development policies. The perspective incorporates the way to approach the development of the city from an inclusive point of view [22,23], taking into account issues such as climate change, cultural identity, reduction of poverty levels, drinking water, natural resources and decent life conditions, especially health conditions. In addition, approaching the SDGs in the COVID-19 context, thus giving credit to the unavoidable relationship between urban quality of life and health [24], is a necessity and a responsibility that architecture and urban design must assume. The opportunity offered by this approach is greatly relevant given the current situation. Beyond the search for immediate solutions, there is clearly a need to work on prevention from an architectural disciplinary approach that includes—in practical terms—the capacity of architecture and urban space [25] to guarantee decent conditions of life quality and health [25] for citizens with the least resources.

The Aura Strategy used by the Solar Decathlon Team of the University of Seville in the Latin American Solar Decathlon (both in the 2015 and 2019 editions) seeks to put into practice a research-based educational methodology [26], where the concept of sustainability underlies cross-sectionally the understanding that this approach can serve as a catalyst for student engagement [27]. Likewise, the Aura Strategy was initially formulated based on the Millennium Goals, which remained in force until September 2015, and developed further [21] based on the key transformation of these goals through the 17 Sustainable Development Goals (SDGs) [5], which encourages students to work according to the current international premises and be aware of the challenges that the field of architecture has to face nowadays, due to the pandemic, and in the near future.

It is the sustainable aspect of the urban environment that is considered the main focus of this article, which intends to explain how, in this context, the Aura Strategy produces the subsequent development of the Aura 1.0 and Aura 3.1 Projects (corresponding, respectively, to the Solar Decathlon Latin America Competition 2015 (SDLac15) and the Solar Decathlon Latin America Competition 2019 (SDLac19)), intertwined with the Millennium Goals in the first case and with the SDGs in the second. That is: starting from the implementation of these goals to use them as a tool that can eventually become an Urban Sustainable Healthy Proposal. To carry out the integration of the SDGs in the design of the SDLac19 Aura Project, the Aura Matrix tool is constructed and the Aura Method is defined, which, starting from the experience of the Aura Project 1.0, allows us to improve the initial results for the subsequent Aura Project 3.1 experience.

1.1. SDGs—Sustainable Development Goals and Their Importance

In order to confront the world's rapid urbanization processes in the coming thirty years [28], the 2030 Agenda for Sustainable Development and the SDGs were approved in September 2015, with the acknowledgment of culture and heritage as axes for sustainable urban development. This Agenda is a roadmap to fight poverty and inequality [29], with a focus on people, the planet, prosperity, peace and partnership [30].

The 17 Objectives and 169 Goals defined by the Agenda are based on the Millennium Development Goals (which remained in force until 2015), although they also include new spheres such as climate change, economic inequality [5], innovation, sustainable consumption, peace and justice, and health, among other priorities. The Objectives are interlinked so that the success of one directly affects the rest. The main input lays on income, education and life expectancy in terms of health as key factors which determine the Human Development Index (HDI) [31]. The SDGs convey a spirit of collaboration and pragmatism for choosing the best options to improve life, in a sustainable manner, for future generations. The SDGs are the most inclusive development processes that the world has ever seen. They cover the fundamental causes of poverty and foster collaboration in order to achieve a positive change for the benefit of the world and its inhabitants.

Despite the opportunity it presents, incorporating SDGs into architectural design practice is neither easy nor immediate. The complexity of the objectives planned from a single point of view and the diversity of implications [2] make the extrapolation to the field of architecture extremely difficult. The challenge, therefore, is to establish an objective, technical and architectural strategy, which allows the introduction of SDG concepts from a mutually dependent, yet sufficiently pragmatic, approach, in order to be able to extrapolate these concepts into technical and objective actions and decisions [32]. In order to accomplish this, other previous experiences of important authors have served as a reference for the methodological development [2–4,9,20]. In this sense, this research project has found in the Solar Decathlon competitions a potential resource for experimenting with the methodology defined for the incorporation of SDGs into architectural design practice.

1.2. The Field of Educational Experimentation of SD Competitions

University education is currently undergoing a major transformation. One of the trends in educational innovation with the best results is the set of integrated curriculum proposals [33,34]. In the case of architectural education, the Aura Strategy develops the Bauhaus tradition and the incorporation of new teaching methodologies, including a mix between different disciplines. The Strategy incorporates issues such as integrated curriculum [35], transdisciplinarity [36,37], transversality and learning methods through problem solving [38].

The SD competitions are a great opportunity for exploration, since they are based on a line of research whose fundamental purpose is the search for living models [39] that minimize the environmental impact based on the convergence toward H2020, according to UNESCO's SDGs for 2030 [40] and the related Higher Education Sustainability Initiative (HESI) [27]. This is based on the conviction that in the search for more habitable, healthy, resilient and, therefore, sustainable city models, sensitive to bioclimatic strategies of adaptation to the climate and the supporting ecosystem, and aimed at maximizing efficiency in the use of resources, healthy future building archetypes [41] will be found and, particularly, new urban models of approach to the city will be defined [40]. The time has come to address SDGs in higher education [27,40], meaning that their necessary innovative practical implementation [42] is crucial for this research.

1.3. The Aura Conceptual Strategy

Based on Christopher Alexander's theory of systems that generate systems [43] and applying holistic systems at different scales [44], Project Aura is defined as one of these systems, by promoting the idea of not creating or projecting something specific, but something that can continue to create more systems cyclically, breaking with the typical scheme of relations of the tree city. This theory is strongly supported by the concept of Production and Social Management of the Habitat [45], that starts from the hologrammatic principle according to which 'the whole is part of parts, just as the parts are part of the whole', that is, nothing is understandable if the synergies between its parts are extracted from the whole. It therefore reinforces the idea of a society that makes decisions as a whole about everything that happens in its city [40], based on the collaborative relationship between the three groups of agents that define a city: Politicians, Technicians and Citizens [46]. At present, these have a totally different weight in the decision making process, a fact that the team will aim to balance, in order to give the citizens of Cali the weight they should have. This idea is linked to another concept: the commons, as a result of citizen action in an intermediate space of autonomy between the state and the market, and which is here particularized from our share as technicians betting on open knowledge, a clear example of the commons. Following authors such as Juan Freire [47], we use social technologies (defined as assemblies of people + digital technology) for the different phases of the project: mapping of actors, massive surveys of the population, consultation of active associations in the municipality, dumping of information derived from citizen information repositories, ideas laboratories, participatory workshops, etc. The Aura Project relies on the social, political and economic transformative capacity of open knowledge.

Small-scale urban interventions are now considered necessary as a start for new urban techniques [31]. This is where the concept of Cities in Transition appears, as an urban theory that proposes to raise awareness to the need for social union around shared objectives, health, income and education from an architectural perspective, a fact that supports the citizens' participation [48] in the improvement of their neuralgic center, the city. This theory is based on several applicable principles:

- Collective Management of Housing Rehabilitation, support for heritage and its reuse as the most effective measure to support the environment.
- Collective Management of Public Space, such as the generation or rehabilitation of obsolete or disused spaces within the city.
- Accessibility and Sustainable Mobility, facilitating communications and interrelations between individuals.
- Socio-Environmental Sensitization, Training and Education, generating in the individual a citizen consciousness which relies on a balanced local management of ecosystems that ensures health and resilience conditions.
- Food Sovereignty, strengthening community gardens with ecological management and promoting short marketing channels for local farmers.
- Energy Sovereignty, promoting energy consumption and guarantying habitability in terms of comfort and health.

Everything described above is reflected in the strategic lines that the team defines as essential for its performance in Cali. The experiences of SDLac15 and SDLac19 do not differ regarding this objective, but they do differ regarding the methodology used in their development. In the case of SDLac19, the methodology is systematized by defining the Aura Method and the Aura Matrix, which constitute interesting tools that can help obtain more precise results, clearly involved with the SDGs. This way, the Aura Conceptual Strategy transcends the residential and invokes the urban from a perspective that incorporates the premises that are defined generically from the SDGs. The main strategy is to translate these generic requirements into concrete architectural decisions, not only based in urban forms as usual [14], but also consistent with the reality, the place and the needs of its inhabitants, in order to improve their quality of life in terms of health and well-being. Thus, the research goal is to develop two different tools, which are useful for the strategy presented above. These tools consist of the Aura Matrix tool and the definition of the Aura Method as a methodological process.

2. Methodology

The methodology used is qualitative, based on a constructivist approach that raises the need to understand reality from the subjective field and according to the contributions and approaches of all of those involved in the research [12]. Therefore, the present research adopts a qualitative research approach for data collection and analysis, following other studied authors [12,14]. The methodology is based on the following phases:

- The study of the state of the art in terms of the integration of the SDGs in the architectural design debate is carried out.
- The opportunity presented by the Solar Decathlon competitions is defined as a potential field of experimentation, given its approach linked to the improvement of the habitat in terms of sustainability and its educational capacity, guaranteeing health and increasing resilience of communities.
- The previous experiences developed for the competition are analyzed as a starting context: Aura conceptual strategy in Latin America 2015 and its initial proposal for Latin America 2019.
- The Aura Method is defined as a methodological proposal. This Method includes the development of a tool, the Aura Matrix, for the interrelation between the SDGs and the requirements defined by the competition itself.

- The application of the Aura Method in the Solar Decathlon 2019 is developed as an experiment by the Team of the University of Seville, resulting in the Aura 3.1 Project submitted to the competition.
- The study analyzes the progress that the Aura Method and the Aura Matrix have brought about in the development of the Project for Latin America 2019 compared to the results of the Project for Latin America 2015 (based on the Millennium Goals).
- Given the obtained results, conclusions are drawn and potential extrapolations are made for all types of future and mainly urban architectural design projects.
- These stages of analysis are represented in Figure 1.

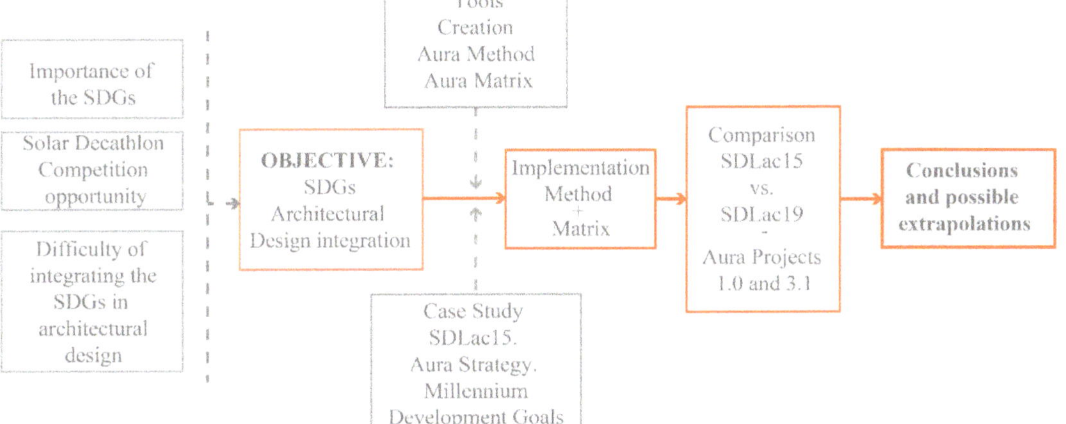

Figure 1. Diagram of the process. Sustainable Development Goals (SDGs). Source: the authors.

2.1. Definition of the Aura Method

Starting from the incorporation of SDGs into the design process, the results obtained in SDLac15 (First Prize Communication, Marketing and Social Conscience, First Prize Comfort Conditions, Third Prize Engineering and Construction, Third Prize Innovation and Runner-up Prize Architecture), based on the previous Millennium Goals, demonstrate that the previous approach is based on a project design strategy [49] distinguished with numerous awards and validated by the design results of the project itself, both urban and building-related. However, in urban matters, many of the proposals are not sufficiently defined or appropriate to these new SDGs. Therefore, it is necessary to develop an 'Aura Method' that allows us to consolidate, with the necessary tools, the process that ensures that these SDGs are explicitly referenced and take shape in the final decisions of project design. Health suitability has recently become a priority in terms of approaching the SDGs, and the Aura Method reinforces the importance of this transversal but crucial aspect as the basis for improving resilience in our building environment and cities.

2.1.1. Generation of the Aura Matrix

The Aura Matrix is a practical, graphically expressed tool for relating concepts that allows the SDGs to be approached in a systematic and direct way, as well as very quickly, with the requirements defined by the SD competition for the completion of its 10 tests. This matrix is a key tool of the Aura Method and is generated for each specific case during its development. Its generation process for the specific case of SDLac19 is explained below.

2.1.2. Phases of the Aura Method

The Aura Method, designed to systematize a way of dealing with the design of an architectural and urban project that allows the SDGs to be rigorously included, follows the scheme shown in Figure 2 and consists of the following phases:

- Framework definition: The Solar Decathlon Competition, Analysis and Requirements
- Case Study: Latin America Cali, Colombia
- Training and Awareness of Agenda 2030 and SDGs
- Development of the SDG Workshop
- Generation of the SD-SDG Interrelationship Aura Matrix
- Development of the Aura Proposal for the SD 2019 Latin America competition

Figure 2. Aura Method diagram. Source: the authors.

The complete development of these phases has made it possible to generate a project for SDLac19, whose characteristics and design decisions explicitly respond to the SDGs, as can be seen below.

2.2. *The Aura Method Application*

2.2.1. Framework Definition: The Solar Decathlon Competition, Analysis and Requirements

The SD competition is centered around the 10 tests mentioned above: Architecture; Energy Efficiency; Engineering and Construction; Comfort, Marketing and Communication; Electrical Energy Balance; House Functioning; Innovation; Urban Design; Affordability; and Sustainability.

The urban issue is part of the initial design phase of the project, and the prototype must be framed within it. However, the prototype construction phase in the Villa Solar requires the adaptation of the urban building proposal to a single-family housing prototype [50] as an experimental case. The definition of the place where the specific urban proposal is made is left to the teams that enter the competition and is always justified based on the needs detected in the city.

As can be seen, the premises and design requirements of the competition itself are perfectly in line with many of the issues defined by the SDGs and are integrated as dynamics or project objectives by the Aura Strategy of the Solar Decathlon Team of the University of Seville. The complexity [51] lies in the multitude of premises that address different scales of action, involve numerous and diverse agents and must respond to the regional, territorial and urban components of the site. To this end, the SDLat19 proposal begins by systematizing the scales of action into: individuals, dwellings, neighborhoods, communities and city. The agents involved are systematized into: citizens, social partners, technicians and politicians. Finally, the regional determining factors are systematized by defining four relevant strategies: the construction and/or improvement of social housing in terms of habitability and health, the increase in density, the use of a rational use of environmental resources and the promotion of regional relevance or progress, the latter being understood as sustainable development and not just economic growth.

2.2.2. Training and Awareness of Agenda 2030 and SDGs

The Aura strategy defined by the present research project is grounded in a global and transdisciplinary university framework that involves the whole University of Seville. Based on this approach, various actions are designed to raise awareness to the SDGs, with the participation of the coordinating members of the Aura Team. These actions, proposed at the university level as an Awareness and Training Program on the new Agenda 2030, give rise to various training events. All these actions are used for the training of the Decathlete students who participate in the Aura project. In particular, the Awareness Days 'The University of Seville for the Sustainable Development Objectives' are the framework in which the students participate in various forums and transdisciplinary working groups, according to their interests and fields of specialization. The Aura team is made up of students from different disciplines [52] and not only architects and engineers, as was traditionally the case. Therefore, the SDGs and their transdisciplinary scope encourage exchange, global and holistic reflection and cooperation among students of the team itself.

The Conferences are held in the six campuses of the University of Seville, with 62 specialized conferences on the most varied topics, covering the spectrum of the 17 Sustainable Development Goals and complemented by subsequent debates. The entire Aura Team also participates in the round tables held at the School of Architecture, whose objective is to deepen the understanding of objectives 9—'Industry, innovation and infrastructure. Building resilient infrastructures, promoting inclusive and sustainable industrialization and fostering innovation'—and 11—'Sustainable cities and communities. To make cities and human settlements inclusive, safe, resilient and sustainable', more specifically with regard to architectural-engineering solutions with a direct impact on the framework of the experimental competition under development—and 'Goal 3: Ensure healthy lives and promote well-being for all at all ages', focused on the great challenges that are being addressed today. These round tables are approached from four thematic areas: production, territory, city and community; and during them, the debate is coordinated by an expert in the field, bringing together both decathlete students as representatives of the local and regional administration and teachers and research experts, enriching the speeches and reflections and outlining practical approaches to the various issues. The conclusions of these debates constitute the basis on which the strategic lines of integration of the SDG of the Aura Project are organized, which allow the concepts transmitted from the Objectives of Sustainable Development to be approached in a broad and inclusive manner.

2.2.3. SDGs Workshop Development and Generation of the SD-SDGs Aura Matrix Interrelationship

After the awareness development process, a workshop is held with the aim of defining the strategic lines of integration of the SDGs in the design of the architectural project of the Aura Team for the SDLat19. This workshop is developed in several phases: Brainstorming session; Definition of priority strategic lines; Categorization and organization of the strategic lines in thematic areas. Decathlete students as well as numerous expert teachers in different specializations and disciplines and the teaching staff who are members or collaborators of the Aura Team attend the workshop. The workshop lasts for six hours, the first two being dedicated to brainstorming and debating ideas, based on the previous approach defined by the Aura Conceptual Strategy, that includes the local premises of the context in Cali, Colombia. The following three hours are dedicated to the construction of the Aura

Matrix, that relates the 17 SDGs to the 10 specific tests of the Solar Decathlon competition, by defining the strategic lines. Finally, the last hour is dedicated to the categorization of these lines in the seven thematic areas defined by the Aura Project. It is relevant to indicate that it is the decathlete students who make the final decisions regarding the strategies and their categorization, with the help and tutoring of teachers and experts, and not the other way around. This dynamic allows the students themselves to develop their own priorities [52] when defining the design of the Architectural Project they are developing, so that they constitute unrenounceable requirements that nourish and condition the Project and not mere declarations of intent without further development.

After the development of the workshop, specialization groups are established within the Aura Team to work on the most relevant issues and their application to the proposal.

2.2.4. Generation of the SD-SDGs Aura Matrix Interrelationship

The matrix tool that enables the SDGs to be related to the project design proposal for the SDLat19 competition is built by defining the 28 strategic lines through the mentioned methodological process of awareness-raising, interiorization, conceptual debate and proposal of the priority strategies that will define the project. This matrix (Figure 3) allows us to verify during the debate and decision making about the strategic lines that all aspects are present and have been taken into account.

These 28 strategic lines, grouped into seven topics (Figure 4), constitute the design basis of the SDLat19 project. Design decisions based on them define the formulation of potential situations of habitability, in terms of comfort and health, and a more sustainable management of the urban space. The architects cannot decide for the user-citizen, but they can design the appropriate conditions of habitability and spatial quality, as well as a more sustainable management, which will incite an improvement in the user-citizen's quality of life [41].

2.2.5. Development of the Aura Proposal for the SD 2019 Latin America Competition

The design process of the Aura Proposal SDLat19 starts from the 28 strategic lines developed thanks to the Aura Method, and is part of the first stage of the process defined by the competition itself and adopted by the Aura Team. Thanks to the involvement of experts and a cross-discipline approach, to the integrated curriculum approach [52] and to the holistic approach provided by the Aura Method, the project design process incorporates educationally the explicit translation of the SDGs into architectural design decisions.

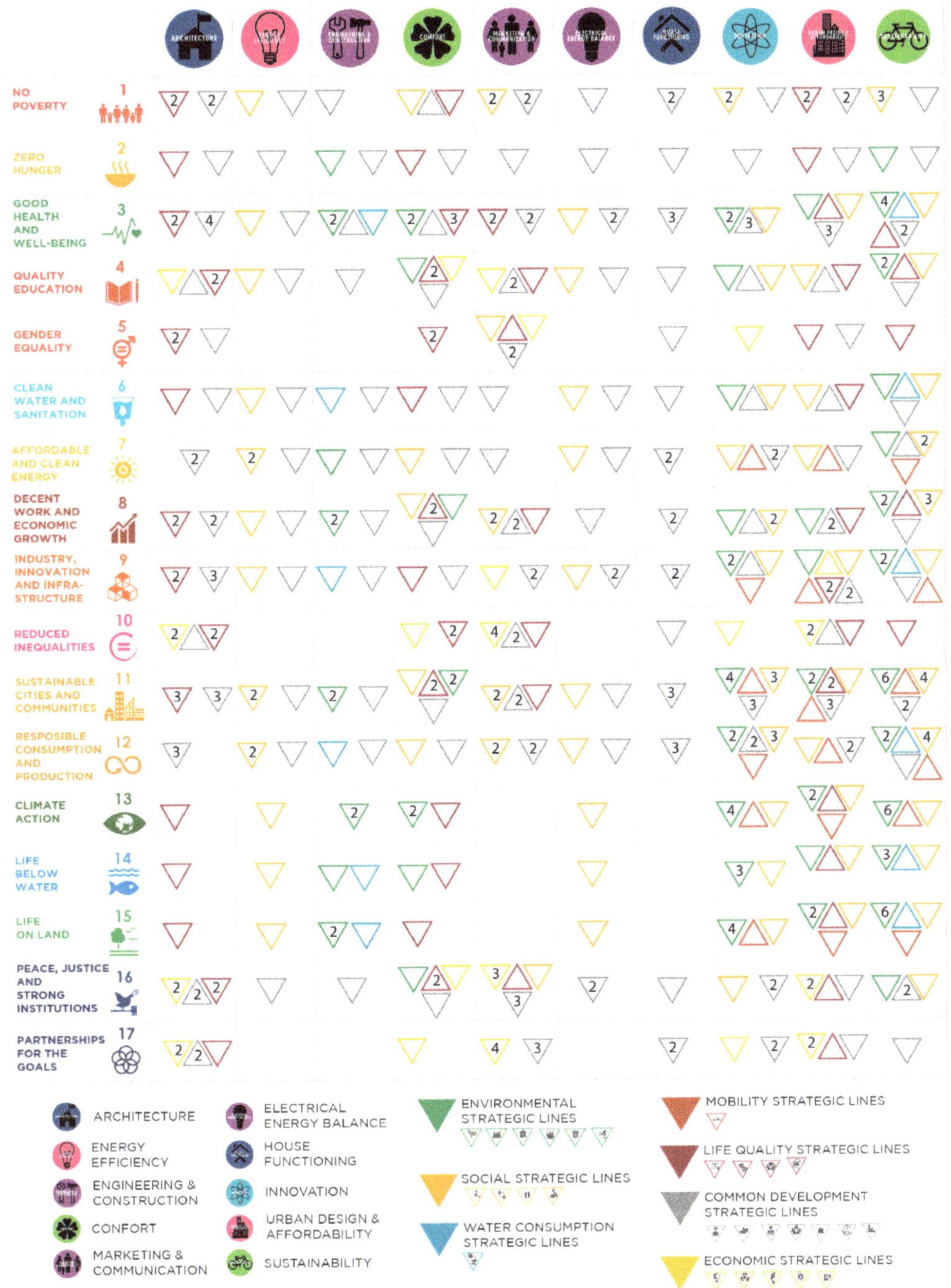

Figure 3. Aura Matrix, 28 strategic lines grouped into seven topics (each with a different color). The numbers inside the triangles indicate the number of specific strategies within the same topic. Source: the authors.

Figure 4. Example of a scheme for checking requirements and relations defined from the Aura Matrix for every strategic line, in this case four life quality strategic lines. Source: the authors.

3. Results

3.1. Aura SD Latin America Project 2015

The Solar Decathlon Latin America and Caribbean competition took place in Santiago de Cali (Colombia) in December 2015. For the first time in the competition, the characteristics of social housing, high-density construction, rational use of natural resources and regional relevance were required, guiding the proposals toward a sustainable social housing solution in Latin America.

The city of Santiago de Cali has a large housing deficit, and the macroproject area of the National Social Interest for the Eco-city of Navarro, a possible area of expansion of public property, has been selected for the development of the proposal [53]. The model of urban growth of the city proposed for Eco-Navarro consists in an organic development where an organized urban fabric will be developed and diluted in a natural space. Thus, the starting point emerges from the edge condition of this location, a meeting point between the city consolidated through the commune 15 and the natural environment of the wetland of the Corregimiento Navarro. The aim is to turn it into a support for an efficient and differentiated habitability (eco-neighborhood) for all the inhabitants of Cali, responding to its favorable and unfavorable conditions. The team aims to tackle the need to provide a solution to the transition space (Eco-Navarro) generated between the city and the eco-park.

Different modes of occupation and relationship with the territory are established through different densities of occupation, which respond to the diversification of neighborhood units (communities) and their location in specific enclaves of the area, attending to different relative and strategic positions in relation to that edge of the city. In each case they have their own articulation elements: a connection with existing or new infrastructures (jetties), treatments of the intermediate spaces between the room, the community and the urban sector to which they belong (social gardens, cooperative workshops, proximity commercial areas, leisure and health points).

The concept of the Equipped Park emerges as a cord, the backbone of the city. Different individual green areas will appear and give rise to communities. These communities will alter the dimensions of the Equipped Park.

The proposal sets in motion a process in which the occupation of this territory, a unique place in the city of Cali, and the traditional and emerging ways of Cali lifestyles go together. The aim is to achieve an efficient housing development based on an agreement between the social interests of various communities and the place itself, preserving its nature, solving its dysfunctions and enhancing its activity in relation to the city and the citizens of Cali. To this end, certain social programs for the promotion and acquisition of housing in Corregimiento Navarro and a mixed program for the rental of social housing and workshops on public promotion premises are proposed as complementary measures. A sustainable city is not only based on the creation of communities in which buildings are sustainable in terms of energy consumption, but also on needs and an economy that allow for the personal development of individuals, linked to the generation of activity and social programs to avoid social exclusion.

3.2. Aura SD Latin America Project 2019

On this occasion, the project is proposed for the district of Siloé, in Cali, Colombia, due to the problems and deficiencies found there, which make it an object of study. It is a self-built district, which allows for the subsequent extrapolation of the Aura Strategy to other situations. The proposed solutions aim to make Siloé a sustainable social and healthy habitat for everyone.

The main innovation of the proposal is that, in this case, we start from a reality that has already been built. Instead of expanding the city beyond its limits or demolishing districts to build them from scratch, the urban fabric is improved with participatory decision-making strategies. The concepts of reuse and recycling of what has already been built are the foundations of the Aura 3.1 proposal. The project starts from the idea of transforming an existing base through fragmented objects and unfinished actions. Thus, the convenience of thinking about interventions made up of fragments as a more realistic and viable option is claimed, as opposed to the unfeasibility, in the current socioeconomic context, of a total renovation of what has been built. This way of intervening, through fragments, will allow for the transformation of a deficient, unhealthy and obsolete starting situation, which no longer responds to current requirements, through microactions that effectively resolve concrete questions linked to contemporary ways of living.

The focus of the approach at the general level arises from the strategy of recovering and valuing the socio-ecosystem support underlying the same neighborhood. To do this, we must work with the 'neighborhood' ecosystem in terms of controlling energy cycles and material resources, and complement it with an inclusive social management strategy that allows its development to become more sustainable over time.

We value the underlying ecosystem, starting from the local water system as a network that allows us to close fundamental material cycles and link them to local energy management by improving both urban and building habitability in terms of comfort and health, linked to climate adaptation and the use of natural energies.

In terms of energy management, the fundamental regenerative strategy is urban naturalization with native species. This strategy allows us to improve air quality, contributing positively to health, to control solar radiation and to reorganize the water cycle, by improving the general habitability of the urban space and the buildings.

In terms of material cycle management, the work is done mainly on the water and soil. Soil management is carried out respecting the building and urban pre-existence, but specifically regenerating those representative points that allow for an autonomous dynamic leading to a subsequent regeneration of the neighborhood.

For an inclusive social management strategy, the services generated from the two previous managements are ultimately defined by the users through participatory strategies, whose nerve centers are the 'Digital Creation Civic Centers', based on the proposals previously defined in terms of health, training, social development and cultural identity.

The 'Aura 3.1' building rehabilitation proposal is configured from a series of specialized devices (fragments/gadgets) capable of coupling to obsolete buildings, which need to be transformed to meet the new living needs of contemporary society. These gadgets are designed, projected and built in the Digital Creation Civic Centers, with citizen participation and with the resources of the nearby industrial fabric. In this way, products and sub-products from the environment are used, promoting recycling and generating, in addition to jobs, the appropriation of the actions by the people of the neighborhood.

These fragments could be included in the following groups:

- Housing units
- Structural elements
- Improvements in water management
- Energy improvements: ventilated plant enclosures and daylighting improvement systems
- Incorporation of Recycling Cycles
- Treatment of public spaces: recovery of roofs as common spaces, accessibility elements, plant elements, natural retaining walls with recycled material from demolitions (gabions).

The city is reactivated through small interventions, so the economic cost compared to a newly built project is significantly lower.

4. Discussion

Comparison of Results between the Aura Solar Decathlon Latin America and Caribbean Projects 2015 and 2019

Once the process has been analyzed by comparing the results obtained in terms of the conceptualization of urban architectural design and the design of the specific urban proposal in relation to the SDGs framework, several conclusions can be established.

From a conceptual point of view, the proposal for SDLat2019 is more appropriate, daring and innovative, since it assumes the built heritage as an undeniable base on which to work, immediately stopping the depredation of the territory with the increase in the extension of the urban fabric. On the other hand, this strategy highlights the value of the urban landscape made up of the most disadvantaged neighborhoods, which are abundant in the Latin American context, and makes it possible to channel their opportunities in terms of greater sustainable development [2], despite their initial differential situation [31].

Both proposals, SDLat15 and SDLat19, improve land use conditions regarding more intensive building densities needed to promote a more compact city and, with it, the advantages in terms of urban metabolism and mobility. The difference lies in the level of social commitment, which in the second case is more in line with an improvement in the social fabric of the involved community.

Both proposals significantly include and adequately resolve energy, material and resource management, linking it in both cases to the management of the supporting ecosystem. The improvement in the SDLat19 proposal lies, once again, in the potential social use, both educational and professional, linked to the management of these resources. This potential social use reinforces the resilience of the neighborhood, not only by improving health aspects related to physical comfort but also by improving psychological comfort, empowering citizens and creating a more creative, resilient and healthy environment [48].

The achievements related to the SDGs that can be drawn from the decisions of the urban architectural project design of both proposals are defined below (Table 1).

Table 1. Comparative results of SDLat15 and SDLat19 for each SDG.

SDG	SDLat15	SDLat19
Goal 1: End poverty in all its forms everywhere		Inclusion of the population in professional educational programs
Goal 2: End hunger, achieve food security and improved nutrition and promote sustainable agriculture	Urban gardens for self-consumption, collective management	Urban gardens for self-consumption, collective and/or family management
Goal 3: Ensure healthy lives and promote well-being for all at all ages	Improved health in new development for relocated people	Improvement of health in the existing city, in terms of physical and psychological comfort and habitability
Goal 4: Ensure inclusive and equitable quality education and promote lifelong learning opportunities for all		Social educational centers, universal Wi-Fi access, professional training linked to the regeneration of the city
Goal 5: Achieve gender equality and empower all women and girls		Participatory involvement of ALL inhabitants. Improvement of urban safety conditions. Enhancing the value of domestic work
Goal 6: Ensure availability and sustainable management of water and sanitation for all	In the new urban development	In the existing city, improving health and quality of life conditions
Goal 7: Ensure access to affordable, reliable, sustainable and modern energy for all	Renewable energies in new development, high energy efficiency	Improvement of the energy efficiency of what already exists. Programs for the incorporation of renewables, professional specialization in energy management
Goal 8: Promote sustained, inclusive and sustainable economic growth, full and productive employment and decent work for all	Potential works during the construction of the new urban development	Vocational training for intervention in existing neighborhoods. Construction level, energy level, urban nature, sanitation, etc.
Goal 9: Build resilient infrastructure, promote inclusive and sustainable industrialization and foster innovation		Contribution to the local industry, valuing the artisanal.
Goal 10: Reduce inequality within and among countries	Work for the city's most disadvantaged population	Work for and with the city's most disadvantaged population
Goal 11: Make cities and human settlements inclusive, safe, resilient and sustainable	Replacing situations of exclusion with quality city situations	Transforming situations of exclusion into quality city opportunities
Goal 12: Ensure sustainable consumption and production patterns	Improvement of consumption conditions linked to buildings	Educational improvement linked to consumption and production patterns for application in neighborhood regeneration
Goal 13: Take urgent action to combat climate change and its impacts	Efficient buildings in terms of energy management, water cycle and material resources	Practical education in efficient buildings in terms of energy management, water cycle, land management, urban green management and material resources
Goal 14: Conserve and sustainably use the oceans, seas and marine resources for sustainable development	Control in water management in order to prevent pollution in seas and oceans	Improved control in water management to prevent pollution in seas and oceans. Education in terms of waste management

Table 1. *Cont.*

SDG	SDLat15	SDLat19
Goal 15: Protect, restore and promote sustainable use of terrestrial ecosystems, sustainably manage forests, combat desertification, and halt and reverse land degradation and halt biodiversity loss	Proper integration into the territory and correct use of the ecosystem	Recovery of the support ecosystem and cessation of the depredation of the territory
Goal 16: Promote peaceful and inclusive societies for sustainable development, provide access to justice for all and build effective, accountable and inclusive institutions at all levels	Improvement of urban social conditions that allow the population to be compared to other neighborhoods in the city	Improving governance by empowering citizens
Goal 17: Strengthen the means of implementation and revitalize the Global Partnership for Sustainable Development	Collaboration between agents, citizens, authorities and technicians	Collaboration between agents, citizens, authorities and technicians, prioritizing the needs of the community

It can be seen from the obtained results that the applied methodology, and therefore the defined Aura Method as a process and the Aura Matrix used as a tool, have involved a transformation and an improvement of the design results obtained for the SdLat19 competition. The approach has been consolidated, but the method has allowed, in addition, for the specific definition of a process design that internalizes the SDGs and provides specific responses to them, as potential opportunities provided by the urban city design. The Aura Matrix is an innovative graphic tool that provides a holistic visualization of the whole situation and promotes complex decision making in a simple way, going further than previous partial approaches based on urban form [14] or general management policies [11]. Thus, the main goal of this research has been achieved and constitutes a step forward regarding new and growing practical and specific approaches that introduce the SDGs concept into architectural design practice and education [15,52,54]. The approach has allowed for the introduction of environmental as well as economic and social aspects, such as those linked to citizen participation [55], an achievement that finds its justification and support in the trans-disciplinary field of the urban approach.

With this methodology, the students have also been able to train autonomously, learn how to work in a team, join multidisciplinary teams and face problem-solving situations [2]. On the other hand, the importance given to the processes of participation and citizen empowerment in terms of governance, thanks to the method itself, constitutes a life learning process for students, beyond the architectural and urban planning discipline, and prepares them to be technicians with greater sensitivity and more competent to deal with SDGs in any personal or professional situation in which they may find themselves in the future, thus responding to the requirements defined by the Higher Education Sustainability Initiative (HESI) [27].

The present study has found a field of educational experimentation of great interest within the framework of the Solar Decathlon competitions [50,52]. However, the process defined as Aura Method and the Aura Matrix tool must be adapted, and their educational capacity analyzed in connection with the Architecture Degree Courses and the disciplines involved. It is necessary to continue research on educational issues and analyze and verify the possibilities that this methodological strategy and this tool—or others that may derive from it—can bring to everyday teaching [56], in order to guarantee the correct transversal incorporation of the SDGs in the curriculum of the universities [52] and in everyday professional practice [57].

5. Conclusions and Possible Extrapolations

The validity of the process, which managed to define a method with excellent results, the richness of the innumerable interdisciplinary and transdisciplinary situations presented along the way, and even the two-way learning between teachers and students, make this educational experiment an innovative demonstration of the translation of SDGs into architectural design project decisions, which are coherent with the current challenges we face in terms of health in particular and in terms of sustainable development of our cities in general.

Likewise, the framework of the experiment represented by the Solar Decathlon Competitions is stimulating as an opportunity for the students to get in touch with the most up-to-date concepts in the field developed in other universities and with the possibility of exchanging concerns, ideas, concepts and materializations. Fortunately, interesting international competitive initiatives targeting higher education students from all over the world have emerged in recent years, such as the Blue Award [58], the Lafarge Holcim Awards for Sustainable Construction [59], the Architecture at Zero Competition [60], the Fassa Bortolo International Prize for Sustainable Architecture, on its 13th Edition [61], the 2020 ASHRAE Design Competition [62] or the Reinventing Cities Competition [63]. Some of these competitions, such as the one presented here, provide an exceptional framework for the exchange of information and an effective forum for the transmission of knowledge and idealizations generated in the academic world.

This proposed method and tool constitute a working base for research in educational matters, which can be extrapolated to all types of international competitive initiatives, such as those described above, since its adaptability is immediate and simple. The Aura Matrix tool is a reference in terms of the translation and applicability of complex and interdisciplinary concepts to the architectural design decisions of the discipline. Its application must be accompanied by a solid and rigorous educational process defined by the Aura Method, which encourages researchers not only to work on the visibility and development of technical knowledge but also to continuously put potential design decisions in relation to the perceptions of transversal disciplines, conjunctural problematic situations such as the current pandemic and the changing needs and priorities of citizen collectives.

This initiative should serve as a starting point to gradually bring about a change in the approach to university teaching methodology, which should not only be adapted to the reality of the profession, but also aligned, like other administrations, with the requirements launched by international organizations, such as the 2030 Agenda and the 17 SDGs.

From the design point of view, even outside the academic field, the incorporation of the SDGs in the development of architectural and urban planning projects is a tool for sustainable and healthy design, an alternative to official sustainability certifications (BREEAM, LEED, VERDE, etc.) and differentiated by its more global nature, which presents an opportunity worldwide to reduce development inequalities between countries and contribute to the fulfilment of the 2030 Agenda.

Author Contributions: All authors have contributed equally to the investigation and final manuscript. conceptualization, R.H.-L., M.L., M.B.-J. and M.T.G.; methodology, R.H.-L., M.L., M.B.-J. and M.T.G.; validation, R.H.-L., M.L., M.B.-J. and M.T.G.; writing—original draft, R.H.-L., M.L., M.B.-J. and M.T.G. All authors have read and agreed to the published version of the manuscript.

Funding: This article is the result of a study in item FADIN 4.0, within the Research support program of the Campus of International Excellence Andalucía TECH, and was backed by the Government of Andalusia (Regional Ministry of Economy, Knowledge, Enterprise and University). The authors are grateful for the support.

Acknowledgments: The authors want to acknowledge all members of the Solar Decathlon Team of the University of Seville.

Conflicts of Interest: The authors declare no conflict of interest.

References

1. Secretaría de Habitat III. *Nueva Agenda Urbana*; Naciones Unidas: Quito, Ecuador, 2017.
2. Omer, M.A.B.; Noguchi, T. A conceptual framework for understanding the contribution of building materials in the achievement of Sustainable Development Goals (SDGs). *Sustain. Cities Soc.* **2020**, *52*, 101869. [CrossRef]

3. Alawneh, R.; Ghazali, F.; Ali, H.; Sadullah, A.F. A Novel framework for integrating United Nations Sustainable Development Goals into sustainable non-residential building assessment and management in Jordan. *Sustain. Cities Soc.* **2019**, *49*, 101612. [CrossRef]

4. Galli, A.; Đurović, G.; Hanscom, L.; Knežević, J. Think globally, act locally: Implementing the sustainable development goals in Montenegro. *Environ. Sci. Policy* **2018**, *84*, 159–169. [CrossRef]

5. Fukuda-Parr, S.; Muchhala, B. The Southern origins of sustainable development goals: Ideas, actors, aspirations. *World Dev.* **2020**, *126*, 104706. [CrossRef]

6. Chirambo, D. Towards the achievement of SDG 7 in sub-Saharan Africa: Creating synergies between Power Africa, Sustainable Energy for All and climate finance in-order to achieve universal energy access before 2030. *Renew. Sustain. Energy Rev.* **2018**, *94*, 600–608. [CrossRef]

7. Botchwey, N.D.; Trowbridge, M.; Fisher, T. Green Health: Urban Planning and the Development of Healthy and Sustainable Neighborhoods and Schools. *J. Plan. Educ. Res.* **2014**, *34*, 113–122. [CrossRef]

8. Campagnolo, L.; Davide, M. Can the Paris deal boost SDGs achievement? An assessment of climate mitigation co-benefits or side-effects on poverty and inequality. *World Dev.* **2019**, *122*, 96–109. [CrossRef]

9. McArthur, J.W.; Rasmussen, K. Classifying Sustainable Development Goal trajectories: A country-level methodology for identifying which issues and people are getting left behind. *World Dev.* **2019**, *123*, 104608. [CrossRef] [PubMed]

10. Etinay, N.; Egbu, C.; Murray, V. Building Urban Resilience for Disaster Risk Management and Disaster Risk Reduction. *Procedia Eng.* **2018**, *212*, 575–582. [CrossRef]

11. França, A.S.L.; Amato Neto, J.; Gonçalves, R.F.; Almeida, C.M.V.B. Proposing the use of blockchain to improve the solid waste management in small municipalities. *J. Clean. Prod.* **2020**, *244*, 118529. [CrossRef]

12. Opoku, A. Biodiversity and the built environment: Implications for the Sustainable Development Goals (SDGs). *Resour. Conserv. Recycl.* **2019**, *141*, 1–7. [CrossRef]

13. Diaz-Sarachaga, J.M.; Jato-Espino, D.; Castro-Fresno, D. Is the Sustainable Development Goals (SDG) index an adequate framework to measure the progress of the 2030 Agenda? *Sustain. Dev.* **2018**, *26*, 663–671. [CrossRef]

14. Jabareen, Y.R. Sustainable urban forms: Their typologies, models, and concepts. *J. Plan. Educ. Res.* **2006**, *26*, 38–52. [CrossRef]

15. Sen, S.; Umemoto, K.; Koh, A.; Zambonelli, V. Diversity and Social Justice in Planning Education: A Synthesis of Topics, Pedagogical Approaches, and Educational Goals in Planning Syllabi. *J. Plan. Educ. Res.* **2017**, *37*, 347–358. [CrossRef]

16. Saha, D.; Paterson, R.G. Local government efforts to promote the "Three Es" of sustainable development: Survey in medium to large cities in the United States. *J. Plan. Educ. Res.* **2008**, *28*, 21–37. [CrossRef]

17. León-Rodríguez, A.L.; Suárez, R.; Bustamante, P.; Campano, M.A.; Moreno-Rangel, D. Design and Performance of Test Cells as an Energy Evaluation Model of Facades in a Mediterranean Building Area. *Energies* **2017**, *10*, 1816. [CrossRef]

18. Domínguez-Amarillo, S.; Fernández-Agüera, J.; Peacock, A.; Acosta, I. Energy related practices in Mediterranean low-income housing. *Build. Res. Inf.* **2020**, *48*, 34–52. [CrossRef]

19. Camporeale, P.E.; Mercader-Moyano, P. Towards nearly Zero Energy Buildings: Shape optimization of typical housing typologies in Ibero-American temperate climate cities from a holistic perspective. *Sol. Energy* **2019**, *193*, 738–765. [CrossRef]

20. Alawneh, R.; Mohamed Ghazali, F.E.; Ali, H.; Asif, M. Assessing the contribution of water and energy efficiency in green buildings to achieve United Nations Sustainable Development Goals in Jordan. *Build. Environ.* **2018**, *146*, 119–132. [CrossRef]

21. United Nations General Assembly. *Seventy-First Session Agenda Item 20 Implementation of the Outcomes of the United Nations Conferences on Human Settlements and on Housing and Sustainable Urban Development and Strengthening of the United Nations Human Settlements Programme (UN-Habitat) Dr*; Naciones Unidas: Quito, Ecuador, 2016. Available online: http://habitat3.org/wp-content/uploads/N1639668-English.pdf (accessed on 14 April 2020).

22. Kooy, M.; Walter, C.T.; Prabaharyaka, I. Inclusive development of urban water services in Jakarta: The role of groundwater. *Habitat Int.* **2018**, *73*, 109–118. [CrossRef]

23. Schwartz, K.; Gupta, J.; Tutusaus, M. Editorial-Inclusive development and urban water services. *Habitat Int.* **2018**, *73*, 96–100. [CrossRef]

24. Serag El Din, H.; Shalaby, A.; Farouh, H.E.; Elariane, S.A. Principles of urban quality of life for a neighborhood. *HBRC J.* **2013**, *9*, 86–92. [CrossRef]

25. Gharib, M.A.; Golembiewski, J.A.; Moustafa, A.A. Mental health and urban design–Zoning in on PTSD. *Curr. Psychol.* **2020**, *39*, 167–173. [CrossRef]

26. Forero La Rotta, A.; Ospina Arroyave, D. Estrategia didáctica para el aprendizaje de la historia y la teoría de la arquitectura. *Rev. De Arquit.* **2013**, *15*, 78–83. [CrossRef]

27. Boarin, P.; Martinez-Molina, A.; Juan-Ferruses, I. Understanding students' perception of sustainability in architecture education: A comparison among universities in three different continents. *J. Clean. Prod.* **2019**, 119237. [CrossRef]

28. ONU-Habitat. *Estado de las Ciudades de América Latina y el Caribe 2012. Rumbo a una Nueva Transición Urbana*; Programa de las Naciones Unidas para los Asentamientos Humanos: Rio de Janeiro, Brazil, 2012; ISBN 978-92-1-133397-8/978-92-1-132469-3.

29. United Nations Development Group for Latin America & Caribbean (UNSDG). *Desarrollo sostenible en América Latina y el Caribe: Desafíos y ejes de Política Pública*; Naciones Unidas: Panama City, Panama, 2018.

30. United Nations. *Transforming our World: The 2030 Agenda for Sustainable Development*; United Nations: Quito, Ecuador, 2015.

31. Sheth, S.; Bettencourt, L.M.A. The Community Human Development Index (CHDI): Localizing Sustainable Development Goals across Scales. *2020 IEEE Conf. Technol. Sustain.* **2020**. [CrossRef]

32. Steinemann, A. Implementing sustainable development through problem-based learning: Pedagogy and practice. *J. Prof. Issues Eng. Educ. Pract.* **2003**, *129*, 216–224. [CrossRef]

33. Iborra, C. Lo que NO hacen los mejores profesores universitarios. *Vivat Acad.* **2007**, 1. [CrossRef]

34. Neuhauser, L.; Pohl, C. Integrating Transdisciplinarity and Translational Concepts and Methods into Graduate Education. In *Transdisciplinary Professional Learning and Practice*; Gibbs, P., Ed.; Springer: Cham, Switzerland, 2015.

35. Barrette, C.M.; Paesani, K.; Vinall, K. Toward an integrated curriculum: Maximizing the use of target language literature. *Foreign Lang. Ann.* **2010**, *43*, 216–230. [CrossRef]

36. Jæger, K. New-Style Higher Education: Disciplinarity, Interdisciplinarity and Transdisciplinarity in the EHEA Qualifications Framework. *High. Educ. Policy* **2018**, 1–20. [CrossRef]

37. Moreno Toledano, L.A. Abordar lo complejo desde el diseño: Una mirada hacia la transdisciplinariedad. *Educ. Y Humanismo* **2017**, *19*, 369–385. [CrossRef]

38. Chau, K.W. Incorporation of sustainability concepts into a civil engineering curriculum. *J. Prof. Issues Eng. Educ. Pract.* **2007**, *133*, 188–191. [CrossRef]

39. Herrera-Limones, R.; León-Rodríguez, Á.L.; López-Escamilla, Á. Solar Decathlon Latin America and Caribbean: Comfort and the balance between passive and active design. *Sustain. (Switz.)* **2019**, *11*, 3498. [CrossRef]

40. Korobar, V.P.; Siljanoska, J. Challenges of teaching sustainable urbanism. *Energy Build.* **2016**, *115*, 121–130. [CrossRef]

41. López-Escamilla, Á.; Herrera-Limones, R.; León-Rodríguez, Á.L.; Torres-García, M. Environmental comfort as a sustainable strategy for housing integration: The aura 1.0 prototype for social housing. *Appl. Sci. (Switz.)* **2020**, *10*, 7734. [CrossRef]

42. Wen, B.; Musa, S.N.; Chuen, C.; Ramesh, S.; Liang, L.; Wang, W.; Ma, K. The role and contribution of green buildings on sustainable development goals. *Build. Environ.* **2020**, *185*, 107091. [CrossRef]

43. Alexander, C. *Tres Aspectos de Matemática y Diseño; y La Estructura del Medio Ambiente*, 2nd ed.; Cuadernos Infimos. Serie Arquitectura y Diseño 3; Tusquets: Barcelona, Spain, 1980; ISBN 84-7223-405-3.

44. Herrera-Limones, R. La Urdimbre Sostenible como Táctica para un Hacer Arquitectónico: De la "Arquitectura de Países Cálidos" Hasta los Nuevos Escenarios y Modos de vida Emergentes, a Través de la Dimensión Dialógica. 2013. Available online: https://idus.us.es/handle/11441/36339 (accessed on 4 June 2021).

45. De Manuel Jerez, E. Construyendo triángulos para la gestión social del hábitat Abstract: Constructing triangles for social management of habitat Key words. *Hábitat Y Soc.* **2010**, *1*, 13–37. [CrossRef]

46. Xiong, W.; Chen, B.; Wang, H.; Zhu, D. Public–private partnerships as a governance response to sustainable urbanization: Lessons from China. *Habitat Int.* **2020**, *95*. [CrossRef]

47. Freire, J. Urbanismo emergente: Ciudad, tecnología e innovación social - Emerging urban planning: City, technology and social innovation. In *Paisajes Domésticos/Domestic Landscapes, Vol. 4 Redes de Borde/Edge Networks*; SEPES Entidad Estatal de Suelo, España: Madrid, Spain, 2009; Volume 4, pp. 18–27.

48. Jackson, L.E. The relationship of urban design to human health and condition. *Landsc. Urban Plan.* **2003**, *64*, 191–200. [CrossRef]

49. Herrera-Limones, R.; Pineda, P.; Roa, J.; Cordero, S.; López-Escamilla, A. Project AURA: Sustainable social housing. In *Sustainable Development and Renovation in Architecture, Urbanism and Engineering*; Mercader Moyano, P., Ed.; Springer International Publishing: New York, NY, USA, 2017; pp. 277–288; ISBN 978-3-319-51442-0.

50. Luna-Tintos, J.F.; Cobreros, C.; Herrera-Limones, R.; López-Escamilla, A. Methodology comparative analysis in the solar decathlon competition: A proposed housing model based on a prefabricated structural system. *Sustain. (Switz.)* **2020**, *12*, 1882. [CrossRef]

51. Villate, C.; Tamayo, B. La práctica de la arquitectura como racionalización sistémica. *Dearq* **2010**, 178–199. [CrossRef]

52. Herrera-Limones, R.; Rey-Pérez, J.; Hernández-Valencia, M.; Roa-Fernández, J. Student competitions as a learning method with a sustainable focus in higher education: The University of Seville "Aura Projects" in the "Solar Decathlon 2019". *Sustainability* **2020**, *12*, 1634. [CrossRef]

53. Herrera-Limones, R.; Gómez, I.; Borrallo, M.; De la Iglesia, F.; Domínguez, A.; Gil, M.A.; Roa, J.; López, E.; Granados, M. Solar Decathlon Latino América y Caribe. Cali 2015 (Colombia) Proyecto Aura. In Proceedings of the II International and IV National Congress on Sustainable Construction and Eco-Efficient Solutions/Congreso Internacional de Construcción Sostenible y Soluciones Ecoeficientes, Sevilla, Spain, 25–27 May 2017; pp. 769–794.

54. Kochan, D. The Prospects and Challenges of Socially Engaged Urban Planning and Architecture in Contemporary China. *J. Plan. Educ. Res.* **2018**. [CrossRef]

55. LopezDeAsiain, M.; Díaz-García, V. The Importance of the Participatory Dimension in Urban Resilience Improvement Processes. *Sustainability* **2020**, *12*, 7305. [CrossRef]

56. LopezDeAsiain, M.; Díaz-García, V. Estrategias educativas innovadoras para la docencia teórica en Arquitectura. In Proceedings of the JIDA'20. VIII Jornadas sobre Innovación Docente en Arquitectura, Málaga, Spain, 12–13 November 2020; pp. 117–127.

57. Borrallo-Jiménez, M.; LopezDeAsiain, M.; Herrera-Limones, R.; Lumbreras Arcos, M. Towards a circular economy for the city of seville: The method for developing a guide for a more sustainable architecture and urbanism (GAUS). *Sustain. (Switz.)* **2020**, *12*, 7421. [CrossRef]

58. Department for Spatial and Sustainable Design. Vienna University of Technology Blue Award: International Student Competition for Sustainable Architecture. Available online: http://www.blueaward.at/home.html (accessed on 18 April 2020).

59. LafargeHolcim Foundation 6th Awards Cycle: 2020-2021 | Lafarge Holcim Foundation for Sustainable Construction. Available online: https://www.lafargeholcim-foundation.org/awards/6th-cycle (accessed on 18 April 2020).
60. American Institute of Architects California (AIA CA) Architecture at Zero. 2020. Available online: http://www.architectureatzero.com/ (accessed on 18 April 2020).
61. Department of Architecture of University of Ferrara Italy Thirteenth International Prize for Sustainable Architecture Fassa Bortolo-built projects division. Available online: https://www.archdaily.com/906316/thirteenth-international-prize-for-sustainable-architecture-fassa-bortolo-built-projects-division (accessed on 18 April 2020).
62. ASHRAE 2020 Integrated Sustainable Building Design (ISBD) Competition. Available online: https://www.ashrae.org/communities/student-zone/competitions/2020-design-competition (accessed on 18 April 2020).
63. C40 Cities Climate Leadership Group Inc. Reinventing Cities Competition. Available online: https://www.c40reinventingcities.org/en/about/ (accessed on 18 April 2020).

Article

Contribution of Roof Refurbishment to Urban Sustainability

Ángel Pitarch, María José Ruá *, Lucía Reig and Inés Arín

Department of Mechanical Engineering and Construction, Jaume I University, CP12071 Castellón, Spain; angel.pitarch@uji.es (Á.P.); lreig@uji.es (L.R.); al341267@uji.es (I.A.)
* Correspondence: rua@uji.es

Received: 10 September 2020; Accepted: 25 September 2020; Published: 1 October 2020

Abstract: Achieving sustainable urban environments is a challenging goal—especially in existing cities with high percentages of old and obsolete buildings. This work analyzes the contribution of roof refurbishment to sustainability, considering that most roofs are currently underused. Many potential benefits of refurbishment can be achieved, such as the improvement of the energy performance of the buildings and the use of a wasted space for increasing green areas or for social purposes. In order to estimate the degree of the improvement, a vulnerable area in Castellón (east Spain) was selected as a case study. A thorough analysis of the residential building stock was undertaken. Using georeferenced information from the Cadastral Office we classified them according to typology, year of construction and roof type. Some refurbishment solutions were proposed and their applicability to the actual buildings was analyzed under different criteria. The theoretical benefits obtained in the neighborhood such as energy and carbon emissions savings were evaluated, together with the increase of green areas. Moreover, other social uses were suggested for neglected urban spaces in the area. Finally, a more accurate analysis was performed combining different solutions in a specific building, according to its particular characteristics.

Keywords: urban regeneration; roof refurbishment; energy performance

1. Introduction

The 11th Sustainable Development Goals (SDG) proposed by the United Nations, focus on cities and sustainable communities to make cities inclusive, safe, resilient and sustainable. To this end, in recent years, many policies have been designed.

The European Union—through the energy performance directives of buildings (EPBD) and their transpositions into the national regulations of State Members—actively works to reduce both carbon emissions and the energy used in buildings. From the first Directive established in 2002 [1] to date, requirements have become significantly stricter [2–4] to accomplish reduction targets (UE key targets for 2030: at least 40% cuts in greenhouse gas emissions from the levels in 1990, 32% share for renewable energy and improved energy efficiency). Currently, the EPBD establishes that all new buildings constructed from 2021 (public buildings from 2019) must be nearly zero-energy (NZEB). As defined in Article 2 of EPBD2010/31/UE, NZEB means a building with very high-energy performance. Moreover, most of the energy required must be covered by renewable sources (sunlight, wind, rain, etc.), including those produced on-site or nearby, and nearly zero or very little energy must be provided by non renewable energy sources (essentially coal, petroleum and natural gas).

Although this is a very positive initiative to improve the building stock's sustainability, it is only applicable to new buildings, and there is a vast number of old and obsolete buildings in terms of thermal properties. According to the data reported by the Spanish National Statistics Office (INE) [5], approximately 54% of existing buildings were built before 1980, when there was still no regulation for the thermal performance that buildings should present. The first standard that regulated the thermal conditions that buildings should meet came into force in 1979 [6] and established quite relaxed thermal requirements. In 2006, the technical building code (CTE, standing for Código Técnico de la Edificación in Spanish) came into force in Spain [7], with new regulations aligned with the European EPBD. The requirements set out in the CTE have been progressively reviewed and updated to date so that, in line with the EPBD, the basic document on energy saving (DB–HE, through Royal Decree 732/2019) aims to build NZEB. However, as reported by the INE and according to the Spanish Government's updated data [8], almost 92% of existing buildings were built before 2006, prior to the CTE being implemented, which means that the contribution of the whole building stock to the NZEB target is minimum. Hence the energy refurbishment of existing buildings is crucial to improve the general energy performance of the building stock. To do so, different refurbishment options may be selected and combined to optimize their potential.

Besides the thermal requirement regulations for buildings, norms on roofing-load capacity should be checked to analyze their construction solutions and buildings' structural capacity should also be observed to specifically calculate overloads. Roldan [9] analyzed the building standards in Spain from the pioneering MV-101/1962 perspective (Decree 195/1963, 17 January). Since then, the overload for roofs of residential buildings has remained at 1 kN/m^2 and 2 kN/m^2 for non-walkable and walkable roofs, respectively, according to consecutive standards: NBE-AE/88 (Decree 1370/1988, 11 November), repealed by CTE-DB SE-AE-06 (Royal Decree 314/2006, 17 March and the 2008 and 2009 updates).

Logically as buildings are already built, applying passive design measures is quite limited, and energy improvement mainly depends on refurbishing the thermal envelope and using very efficient facilities. As improving the thermal envelope lowers energy demand, this should be the first step when undertaking energy refurbishment interventions. However, this is not what happens under actual conditions due to the high economic cost of refurbishment, usually with a long-term investment return. Therefore, non-economic arguments, such as environmental and social benefits linked with refurbishment, would support interventions.

Nowadays, cities are composed of different typologies of residential buildings, most of which were built in the 20th century and some in the 21st century. Their features are conditioned by many factors that can influence a region's urban development, such as historic and political events, regulatory frameworks (new legislation coming into force, general urban plans, etc.) and socio-economic factors (wars, migratory movement from rural to urban areas or new construction and material techniques emerging). The IEE Project "Typology Approach for Building Stock Energy Assessment" [10], devised with the collaboration of 15 partners from European countries, differentiates four main typologies in the Spanish building stock, which are: single-family house, terraced house, multi-family house, apartment block. The Valencian Institute of Building (IVE), which was responsible for the project in Spain, developed a guide which differentiated three climate regions and six construction periods (which depended on normative evolution). Therefore, homogeneous construction solutions were inferred for each period: <1900, 1901–1936, 1937–1959, 1960–1979, 1980–2006, >2006 [11]. As observed by Martín-Consuegra et al. [12], the analysis of the building regulations in force during the different time periods provided quite an accurate idea of the characteristic constructive features of the year of construction.

Objective and Main Conclusions

The present work focuses on the potential benefits offered by the refurbishment of roofs in residential buildings. To do so, constructive solutions were examined to analyze possible refurbishment solutions and the potential achieved benefits. Technical aspects, such as energy performance improvement and its profit, were estimated together with some externalities, such as social benefits linked with the enjoyment of a usually underrated and underused part of the building. This work aims to shed some light on the potential role of roofs in the sustainability of the urban environment.

As summarized by Braulio [13], three periods can be differentiated in Spain according to the construction solutions employed in roofs. Buildings constructed before 1940 generally used timber beams and wattle suspended ceilings in sloping roofs, with Catalan-style ventilated systems on flat roofs. The 1940–1979 period was characterized by the introduction of new materials, such as reinforced concrete, which provided new structural options (new slabs with concrete joists and ceramic or concrete blocks). Flat roofs became more popular during this period, when lightweight concrete was used to form the slope of roofing systems and waterproof materials were introduced. From 1980 onward, roofing systems have continuously evolved to the much more complex systems that can be currently used today, which are composed of different materials that fulfill diverse functions like drainage, waterproofing, thermal insulation, etc.

Although it is more difficult to establish such progressive evolution in roof use terms, in the beginning roofs were also employed for social life purposes. This provided homes with more protection from both weather agents and invading attacks. As explained by Graus [14], in pre-industrial societies a roof always embraced many uses. Given lack of space, the roof emerged as an outdoor room that allowed multiple purposes, such as a better use of space for a settlement on steep mountain slopes, drying grain, hanging food, sleeping on hot nights or even celebrating weddings. With time, social life moved to streets and roofs began to be left aside and even ceased being considered a useful space in the vast majority of cases. Although only the protective function of building roofs is generally valued today, sustainability development trends seek to optimize already built spaces instead of generating new urban expansions. Roofs must be seen as spaces that offer sustainable opportunities to improve the building's energy performance, to install green roofs, to provide renewable energy production systems or for inhabitants' own leisure.

In order to generate practical and applied results, a vulnerable area in the city of Castellón (east Spain) was selected. The city's recent land-use plan defined this area as vulnerable [15–18], and it is characterized by a mixture of building typologies, most of which are old and obsolete. Thus, it could benefit highly by hypothetical intervention. A previous diagnosis of existent roofs should be made to analyze structural limitations, thermal properties and other aspects like orientation, possible uses, and so on.

Some benefits are observed. On one hand, roofs are an important part of the thermal envelope, which is totally exposed to outer conditions. Their refurbishment could contribute to improve buildings' energy performance and to reduce CO_2 emissions. To this end, some representative buildings were selected, and the results were extrapolated to the neighborhood scale. On the other hand, using the space on some roofs could also contribute to increase green areas in the urban environment. Additionally, as roofs are currently underused [19], appropriate use and exploitation would contribute to sustainable cities and positively influence the social resilience of the city. Accordingly, some roofs could accommodate social functions, such as meeting places for residents.

The proposed solutions are meant to place in order the magnitude of benefits and other refurbishment solutions—even combinations of solutions—can be used. An actual intervention would require more accurate conditions to be observed in order to consider the legal framework, property regime issues and exact conditions of each building. A specific building is finally presented as an example to illustrate other options, where the combination of some proposed solutions could apply.

2. Materials and Methods

This work was undertaken in four main stages.

The first stage of the work involved data collection. First, a review was done on the regulations applied to build roofs and about constructive solutions at different times. Second, data on the selected area were collected. The chosen neighborhood is included as a regeneration and refurbishment area (ARRU, standing for Área de Regeneración y Renovación Urbana in Spanish) according to the recently developed land-use plan [12]. ARRU are vulnerable areas according to urban, residential, social and economic features and are meant to be prioritized to make urban regeneration interventions. Third, the data on the buildings in the area were collected using Cadastral Office information. This allowed us to identify buildings by their cadastral reference, and to address and provide information about plot area, built area, number of floors, number of dwellings and year of construction. Using the cadastral cartography and the Google Earth tool, roof types were identified and measured.

The second stage was meant to suggest refurbishment solutions for roofs. A cross-checking of roof types and refurbishment solutions was undertaken by considering parameters like load capacity, orientation, usable area, etc., to not dismiss any possible solution linked with a specific roof type and to select the optimum solution in every single case. A multicriteria analysis was used to assign applicable solutions to different roofs.

The third stage estimated the benefits for the area. Statistically representative buildings were selected to estimate quantifiable benefits. Improving buildings' energy performance by roof refurbishment was calculated with the CE3X energy certification software. This official software is approved by the Spanish Ecological Transition Energy Ministry [20–22], available at: https://energia.gob.es/desarrollo/EficienciaEnergetica/CertificacionEnergetica/DocumentosReconocidos/Paginas/procedimientos-certificacion-proyecto-terminados.aspx (last accessed 5 June 2020). Those buildings that statistically represented the whole building stock were simulated. To simulate the average buildings, the data obtained from dynamic Tables in Excel (Supplementary Materials) were used, which allowed the selection of real buildings whose characteristics came close to average values. This implied that real orientations, measures for façades and windows, and so on, could be inputs in the software by assuming that outputs gave an approximate order of magnitude. The outcome of the simulation provided information on energy demand, energy consumption and carbon emission savings. Other environmental benefits, such as greener areas in the neighborhood, were also calculated. Some social benefits linked to the enjoyment of roofs by dwellers were also considered.

In the fourth stage, these data were extrapolated to the whole urban area so that the benefits of a potential intervention on the neighborhood scale could be estimated. Besides the neighborhood scale, a specific building was selected for simulations to, by way of example, implement a combination of some refurbishment solutions.

This paper is structured as follows: the background section briefly introduces the general goal of achieving sustainable cities and the role of buildings as relevant elements in complex urban ecosystems. Then the evolution and use of roofs in history, how they have become underused and underrated in buildings and how roofs can contribute to sustainability, are highlighted. Moreover, an analysis of the evolution of regulations on roofs construction was done to understand the different construction solutions for each period. Regulations on structural requirements during different periods were reviewed to analyze the technical applicability of the refurbishment solutions. Regulations on the thermal performance of buildings and their evolution in recent decades were also examined as this is a key factor to consider in sustainability terms. The next section explains the methodology and the steps undertaken in this work. Then the urban area selected as a case study is introduced. We present a brief description of the buildings, roof types according to their year of construction and their current use. Then an analysis of the different refurbishment solutions was carried out and they were linked with the physical possibilities of current roofs. The optimum solution to each roof type was selected by considering technical and environmental criteria. Finally, the achieved benefits were analyzed and estimated on the neighborhood scale. The accomplished improvement and the optimum roof-type solution combinations offered valuable information that could help the municipal administration in its decision-making to allocate funds for regenerating the area. Finally, an analysis of a specific building is presented below to illustrate the combination of different solutions.

3. Results

3.1. Case Study

3.1.1. Urban Area

The urban area selected for this study is located in the city of Castellón de la Plana, a medium-sized Mediterranean coastal city located in east Spain, with about 170,000 inhabitants. According to the recent land-use plan, there are 17 vulnerable areas or neighborhoods in this city, called ARRU (standing for areas of regeneration, refurbishment and renewal). These areas are intended to be prioritized by the public administration to allocate funds in order to undertake regeneration and refurbishment [14,23]. Most of the neighborhoods in the city are located inland, about four kilometers from the sea coast, except for the neighborhood called Grao, which is stands by the sea, which especially distinguishes this area.

The Grao neighborhood was selected as the case study. This area corresponds to District 9 in the city and is formed by census Sections 09001, 9002, 9003, 9004, 9006, 09,007 and 9010. Together they form ARRU15 according to urban, building, socioeconomic and sociodemographic indicators [23]. All the vulnerability categories are present in census Sections 9002 and 9003. The worst vulnerability in buildings is present in 9001, 9002, 9003 and 9004, which coincide with a high-density urban area. The remaining area is formed mainly by single-family houses. In order to diagnose, propose solutions and quantify the benefit of refurbishing roofs, the area formed by census Sections 9001 to 9004 (presented in Figure 1) was selected.

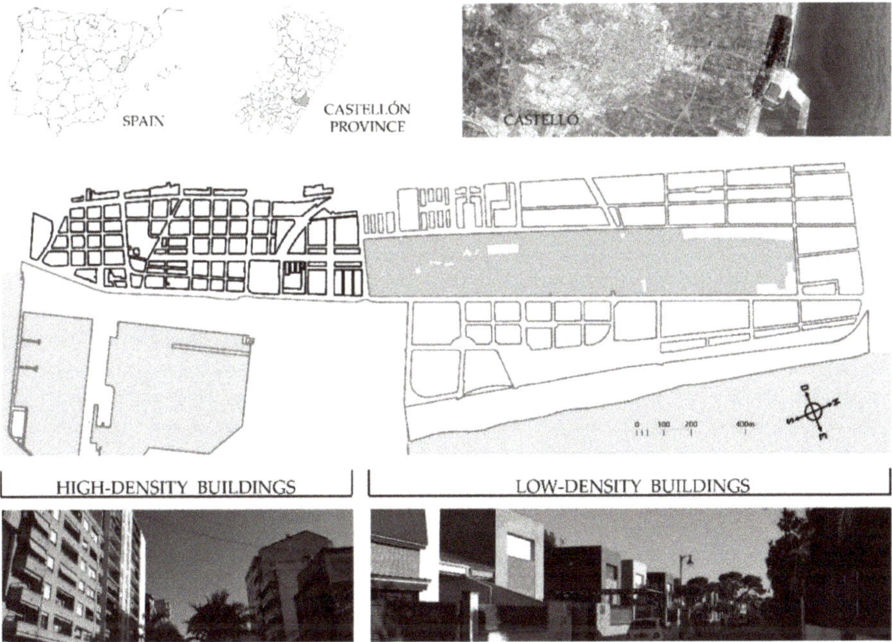

Figure 1. Location of the selected urban area.

3.1.2. Analysis of the Building Stock

In this study, buildings are grouped as typology and building period by considering two typologies, namely single-family house (SfH) and multi-family house (MfH), as suggested by Annex I, Section 5 of the EPBD for residential buildings. At the same time, they were structured according to five different constructive periods having been adapted to both normative periods and cadastral information: 1840–1936, 1937–1959, 1960–1979, 1980–2006, 2007–2012.

The selected area is made up of 794 buildings, of which 775 are residential buildings according to Cadastral Office information, with 335 MfH and 440 SfH.

Data were arranged after contemplating the period when they were built. By considering the total building stock, only about 3% were post-CTE built, which confirms the energy rehabilitation need. As observed in Figure 2, the greatest building activity corresponded to periods 1960–1979 and 1980–2006. About 44% of MfH were built during 1960–1979, followed by 38% during 1980–2006. Almost 70% of SFH were built from 1980 to 2006. During this period, a significant increase took place during 2000–2005, when major developments during the real-estate boom in Spain were built, mostly terraced houses.

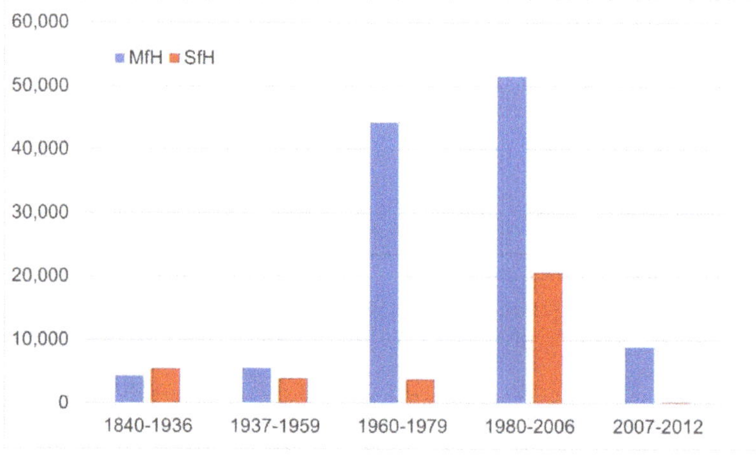

Figure 2. Number of residential building for year of construction and building typology.

3.1.3. Roof Types

To complete the previous quantitative analysis, this section examined roof types. This variable is crucial to assign possible refurbishment solutions, where the original constructive solution needs to be known. Accordingly, two main roof groups were found in the area: sloped and flat. In turn, the latter was subdivided into non-walkable or walkable. Finally, in some cases, roofs are located in inner courtyards. For practical purposes and in relation to potential refurbishment interventions, we also differentiated between roofs for community or private use. Consequently, the roofs in the present study were classified as follows:

- Sloped roofs (SR): roof whose slope is broken by an obtuse angle. A third of the roofs in the area are sloped roofs, finished mainly by ceramic gables
- Flat non-walkable roof (NWR): limited access and sometimes used to contain facilities. It is frequently found in nonresidential uses where facilities are centralized
- Flat walkable roof (WR): a flat roof that is almost level, unlike many of sloped roof types, up to approximately 10° for draining rainwater. Two WR can be distinguished depending on the property regime:

  - Community walkable roof (CWR), with access to all dwellers. They are usually intended for different purposes, such as clothes lines, storage rooms, etc.
  - Private walkable roofs (PWR) for personal use. They were not included in this study because they are terraces for private use.

- Flat roofs in inner courtyard (IC): the inner courtyard is a central space within the building that provides light and ventilation. In some buildings it is located on the ground floor. Two subtypes were considered:

  ○ Private inner courtyards (PIC): often used by first-floor dwellers;
  ○ Community inner courtyard (CIC).

To provide an initial idea of the variability and extension of these four main roof types, their area and location are plotted in Figure 3. Figure 4 quantifies this information.

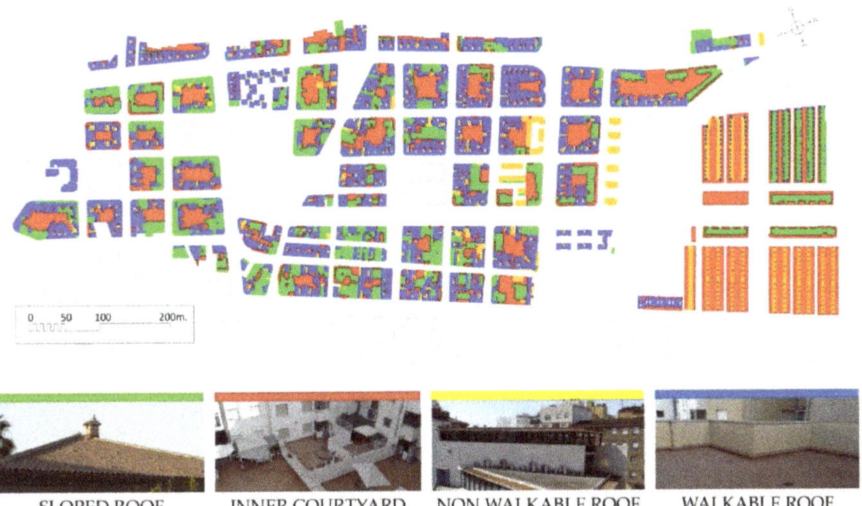

Figure 3. Location of roof types in the area.

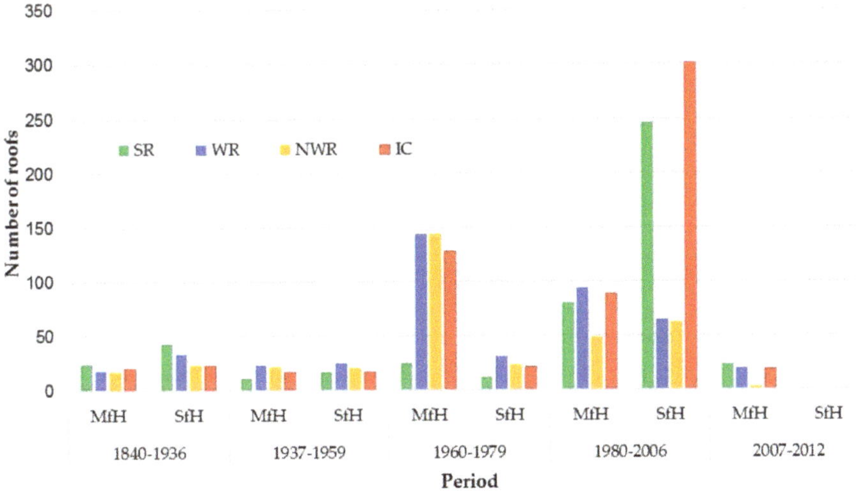

Figure 4. Number of roofs per building typology, roof type and period.

As seen in Figure 4, by considering the number of buildings, inner courtyards represent the highest percentage for period 1980–2006, followed by period 1960–1979 that represents 53% and 23%, respectively, for this type.

Figure 5 presents all the integrated information of the roofs in the neighborhood and considers the five building periods, the two building typologies (in columns) and the six roof types (in lines) that account for 60 groups. All the information is represented proportionally to the area in rectangles. walkable roof presents the highest percentage, 40%, followed by sloped roof, with 27% of the total area, inner courtyard with 23% of the area and non-walkable roof under 10%. Despite the large number of SfH, when considering the roof area almost 77% of this area corresponds to MfH, as we can see in Figure 5.

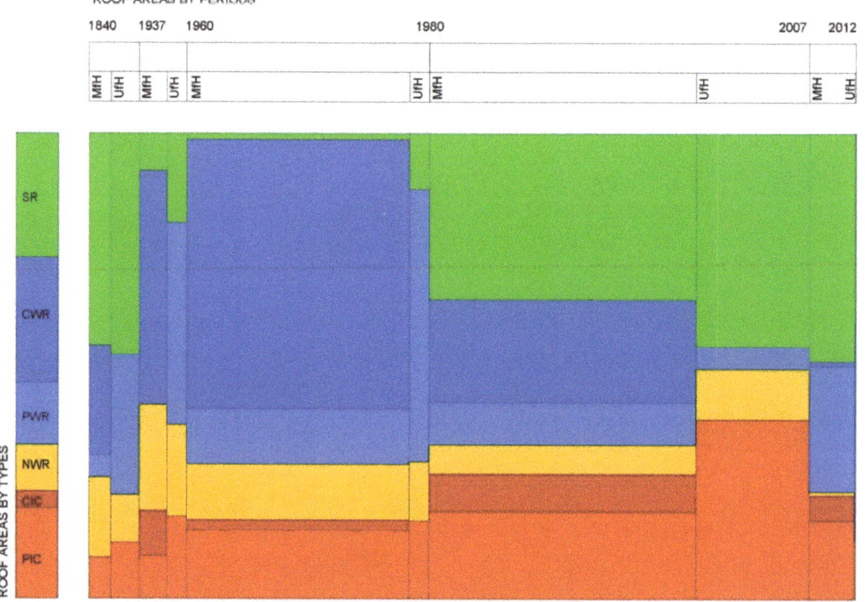

Figure 5. Roof areas per period, building typology and roof type.

The biggest area was found for the community walkable roof in MfH for the period 1960–1979 (1960–1979/MfH/CWR) combination, followed by the sloped roof in MfH for the period 1980–2006 (1980–2006/MfH/SR). When focusing on MfH, we found that they were built mostly during 1960–1979 and 1980–2006, with 45% and 60% of the area of their period, respectively. Regarding SfH, about 60% of the roof area corresponded to period 1980–2006, and the remaining 40% was distributed during the other periods. When considering each roof type, 72% of the area of sloped roof corresponded to period 1980–2006, 54% of the area of walkable roof was built during 1960–1979, most non-walkable roof also corresponded to periods 1960–1979 and 1980–2006, about 39% per period, and 65% of the area of inner courtyard were related to period 1980–2006, which coincides with the periods when building activity was more prominent.

3.2. Refurbishment Solutions

Roofs can benefit from being refurbished in many ways. Maintenance of buildings, energy savings, economic income and ecologic benefits can be obtained depending on the selected solution. However, as the solution must adapt to the specific conditions of the location and existing roof, not every solution would apply to each case.

Hence, depending on the starting conditions, different factors will determine the solution. Some limiting factors are to do with urban and architectonical issues. Sometimes there are legal limitations according to urban plans that determine the interventions to be made to roofs, linked with the allowed buildable area, number of floors, height, etc. Another crucial aspect must do with structural resistance and the loads the system can support, which must be checked. Moreover, the area, orientation and geometry of roofs are decisive for implementing certain solutions; for example, if there is enough room to install some facilities, if the orientation is appropriate for solar radiance gains or if there are solar obstacles to be considered, etc. Other factors are related to the economic cost of the intervention, such as the investment and maintenance cost of the solution.

The benefits obtained from the intervention should be observed in the decision-making process as they may add value to the building. For example, some solutions could improve a building's energy performance and reduce energy bills and carbon emissions from using the building which, therefore, makes the urban environment more sustainable. Other solutions may be beneficial for different reasons, such as their contribution to reinforce the social relations of dwellers, if they consider it convenient to take advantage of this underused space for community purposes.

In nonresidential buildings such as hotels, office buildings or commercial buildings, leisure uses, such as sports, restaurants, cafes, and so on, increasing the usable surface of some businesses is a possibility. Some large roofs can be employed for advertising to provide extra income to owners.

Considering the residential building stock, some refurbishment solutions are described in the following subsections.

3.2.1. Green Roofs

Green roofs (GR) can cover a roof with vegetation either partially or completely. Therefore, it needs a growing medium placed on top of a waterproofing membrane and must also contain a root barrier layer and a drainage and irrigation system. The literature review shows that this entails many benefits, such as providing insulation, which improves the building's thermal performance by making energy savings [24–26] and could be achieved with the addition of thermal insulation. However, this solution entails other added benefits. It also combats noise pollution and can reduce noise annoyance by road traffic [27]. Other environmental benefits are linked with improved urban environment resilience, such as better rainwater absorption, more green areas in the urban environment, carbon sequestration, better air quality, lower urban air temperatures, mitigating the heat island effect, etc. [28–30]. It also contributes to create a habitat for wildlife, reduces people's stress by providing a more esthetically pleasing landscape [30,31] and contributes to solve the shortage of green spaces and land resources [32]. There are many varied technical solutions for installing green roofs. Some can be considered intensive systems and require thicker ground layers to support a wider variety of plants. However, they are heavier and require more maintenance. On the contrary, thickness of extensive roofs can range from 6 to 20 cm, are less maintenance-demanding and lighter than intensive solutions. Finally, some solutions exist in between, which are called semi-intensive [32–34]. Theoretically, this solution could be placed on any roof. However, on sloped roofs, inclination could complicate both installation and maintenance. Moreover, minimum solar radiation should be guaranteed for vegetation. Therefore, flat roofs with solar access seem, a priori, the most appropriated roof type for this solution.

3.2.2. Solar Panels

According to Fuentes et al. [35], domestic hot water (DHW) accounts for about 20% of the total primary energy use by housing. A study by Pomianowski et al. [36] compared the energy use for DHW in some European countries and found that the final energy use percentage falls between 15 and 40%, and around 15% in Spain. The study by Gautama et al. [37] demonstrates the long-term economic feasibility of solar water heating systems, although the payback period depends on many factors like price of fossil fuels, solar insolation, etc. The implementation of solar panels (SP) will depend on the roof area and orientation, as well as on its location in the urban area. In the study area, the proper solar orientation is southern or slightly southeastern [13]. Northern-oriented roofs would not be suitable for this solution or these roofs are surrounded by solar obstacles like taller buildings. The installation of solar panels to contribute to DHW demands can be a suitable solution in the study area, especially for SR for which other options are not considered. They can also be seen as a good solution for some flat roofs, such as non-walkable roofs or small walkable roofs, where other options are inadequate. Installation on sloped roofs is slightly cheaper because they do not need a substructure to support panels. solar panels would increase the renewable energy supply in the neighborhood and reduce energy use and, therefore, energy bills.

Photovoltaic panels are also very interesting—especially in locations where solar radiance is high [38]. Other authors have demonstrated the profitability of this solution in non residential buildings [39,40]. Even the combination of solar panels with green roofs can be found in the study of Berto et al. [41]. They produce energy that is an interesting alternative to achieve economic and environmental benefits. Unfortunately, however, this solution is not completely standardized in Spain for residential building, probably due to the solar energy tax established by Royal Decree 900/2015. Later this law was derogated by Royal Decree 15/2018. As a result, the system has been rarely used—especially in MfH. Nevertheless, it undeniably offers a high potential in the near future and is, in fact, currently being promoted by the government [42].

3.2.3. Residents' Association Meeting Area

The available area misused on walkable roofs can be used for many purposes, such as meetings, social encounters, installation of urban orchards (included in green roofs), solariums or other uses to promote the social interaction of dwellers [31,43,44]. Considering the legal and structural limitations, a residents' association meeting area (RA) could be placed. This could feature various possibilities, from light pergolas to heavier constructions—depending on the starting conditions. Load, complexity and cost will vary depending on the selected solution.

3.2.4. Adding a New Floor on the Existent Building

If this option is allowed by the urban plan—and the structural capacity permits it—adding a new floor (NF) would add a noteworthy economic benefit to ownership.

3.2.5. Selecting the Proposed Solutions

For this work, the new floor solution was rejected because it is a very limited solution that can be applied in a few cases due to urban plan limitations and, moreover, because structural capacity should be enough to support the loads of the added floor. Photovoltaic solar panels were not considered due to lack of experience to date in residential buildings, but they are an interesting and very convenient proposal to be analyzed in the near future as they can imply potential economic profit and energy savings. Therefore, the green roof, solar panels for DHW and residents' area solutions were considered for the scope of this work. Regarding green roofs, an extensive solution was selected over the intensive or semi-intensive options because of the lightest structural requirements and the lowest economic investment and maintenance [31,33,45]. Regarding the solar-panel solution, it is considered to be most appropriate for the neighborhood as it receives high solar radiation most of the year. Castellon is classified solar zone IV in the Spanish territory, with an scale from I to V from the lowest to highest radiation, (according to DA DB–HE/1, document complementing the HE document), which means high mean annual global solar radiation (4.6–5 kWh/m^2). Moreover, this solution supports norm CTE, Section HE4, which requires a minimum contribution of renewable energy to cover the DHW demand, which is 60–70% in the study area. As regards residents' area, we selected a multipurpose light wooden pergola solution to minimize both load and cost. This solution could provide an area where neighbors can gather for their meetings and other agreed uses according to dwellers' interests.

Table 1 summarizes the selection for the undertaken analysis:

Table 1. Selection of refurbishment solutions.

Refurbishment Solution		Pros	Cons	Select	Reason	Selected Solution
NF	New floor	Economic profitability	Scarce applicability	No	Urban and technical limitations	
GR	Green roof	Thermal and acoustic insulation; More green areas Carbon sequestration Reduction of the heat island effect	Maintenance cost Insulation; Noticeable on the top floor of the building	Yes	Entails many environmental benefits	Extensive solution
SP	Solar panels for DHW	Savings in energy and carbon emissions	Savings only for DHW; Maintenance costs	Yes	Frequent solution	Standard model
	Solar photovoltaic panel. Community energy	Huge potential of savings in energy and carbon emissions	Still not standardized in residential uses in MfH; Depends on community agreement	No	Little experience in residential buildings	
RA	Residents association area	Enjoyment of underused space; Multipurpose	Depends on community agreement; Solution depends on final use	Yes	Use of underrated common area in buildings; Social uses.	Light wooden pergola

3.3. Type of Roofs Vs. Refurbishment Solution

The descriptive cadastral data were analyzed using the dynamic Tables in Excel. Some aspects were completed with the available cartography (Spanish Cadastral Virtual Platform) to verify certain aspects, such as orientation or inner courtyards and to identify the private uses to be ruled out for the study.

The applicability of the refurbishment solution to roofs was done by a multicriteria matrix, where all the analyzed information of roofs was crossed with the technical possibilities of the selected solutions (see Supplementary Materials). Of all these premises, some starting hypotheses were put forward:

1. All the combinations representing under 2% of the area were not considered;
2. Roofs that imply private uses (private walkable roofs and private inner courtyards) were not included;
3. Green roofs were not considered a solution for sloped roofs. According to the German norm on green roofs, slopes > 26.8% (15°) require special consideration and slopes > 100% (45°) are not recommended [46,47];
4. Solar panels for DHW are considered in sloped roofs, preferably in solar orientation;
5. For community inner courtyards, only the residents' area solution is considered. As community inner courtyards are located on ground floors, green roof and solar-panel solutions are most likely to be affected by solar obstacles.

After assuming all these hypotheses, five combinations of refurbishment solutions/roof types appeared (see Supplementary Materials):

1. SP-SR: Solar panel/Sloped roof;
2. GR-NWR: Green roof/Non-walkable roof;
3. GR-CWR: Green roof/Community walkable roof;
4. RA-CWR: Resident area/Community walkable roof;
5. RA-CIC: Resident area/Community inner courtyard.

3.4. Estimating the Benefits of Refurbishment

A statistical analysis of typologies provided us with the average values for number of floors, number of dwellings, usable area, roof area and energy requirements for DHW (based on Regulation CTE-DB-HS4) to obtain representative buildings for energy performance simulations. By doing so, the results of energy savings and carbon emission reductions due to the refurbishment could be extrapolated to the neighborhood scale to obtain an order of magnitude of the potential improvement to the whole area. The area is mainly made up of MfH, of 6–7 floor buildings, with 20–30 dwellings ranging from 1440 to 2968 m^2 of usable area and 2016–3427.20 L/day requirements for DHW. The average SfH building has three floors, 168 m^2 and 112 L/day requirement for DHW.

Energy demand will lower in those scenarios that consider the GR solution because roofs form part of the thermal envelope. This must do with the solution's higher thermal resistance (1.969 m^{2}K/W) and, consequently, the lowest thermal transmittance (0.508 W/m^{2}K), which would improve energy performance. Figure 6 shows the composition of the selected extensive solution and the detail of its thermal properties [48].

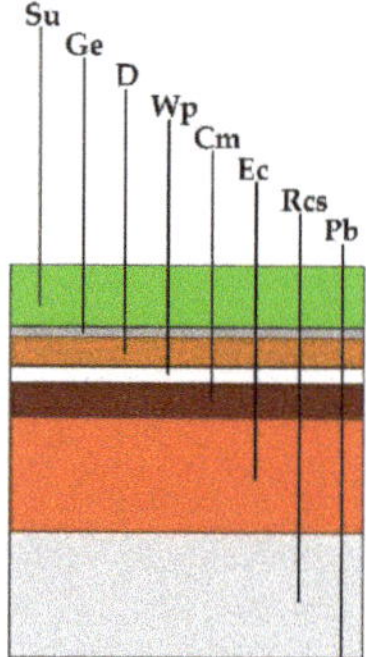

Layer i	Thicknes s e, (mm)	Density (Kgm^{-3})	Thermal conductivity λ_i (Wm^{-1}K^{-1})	Thermal resistance R$_i$ (Wm^{-2}K^{-1})
Substrate (Su)	100	2,000	0.52	0.192
Geotextile (Ge)	1.5	120	0.05	0.300
Non-ventilated air chamber (NvAc)				0.160
Drainage and storage (D)	20			0.090
Root barrier waterproof (Wp)	1.5	1,100	0.23	0.065
Cement mortar (Cm)	20	1,525	1	0.020
Expanded clay (Ec)	100	537.5	0.148	0.676
Slab-reinforced concrete (Src)	250	1,220	0.908	0.275
Plasterboard (Pb)	20	900	0.4	0.050

*Total thermal resistance $R_T = R_{si} + \Sigma R_i + R_{se}$, $R_{si} = 0.1$; $R_i = e_i / \lambda_i$; $R_{se} = 0.04$, $R_T = 0.1 + 1.829 + 0.04 = 1.969$
*Thermal transmittance $U = 1/R_T = 0.508$

Figure 6. Green roofs (GR) extensive solution. Composition and thermal properties (* according to CTE-DB-HE1).

Considering that the residents' area solution does not influence the building's energy performance, and that the thermal resistance value does not vary from community walkable roof to non-walkable roof during the same period, the simulation cases amount to five buildings, represented by Period/Building Typology/Roof Type:

1. 1960–1979/MfH/CWR-NWR: multi-family houses built during period 1960–1979 with community walkable roof and non-walkable roof;
2. 1980–2006/MfH/SR: multi-family houses built during period 1980–2006 with sloped roof;
3. 1980–2006/MfH/CWR-NWR: multi-family houses built during period 1980–2006 with community walkable roof and non-walkable roof;
4. 1980–2006/SfH/SR: single-family houses built during period 1980–2006 with sloped roof;
5. 2007–2012/MfH/SR: multi-family houses built during period 2007–2012 with sloped roof.

Table 2 presents the five combinations, the main features of the representative building (average values for number of floors, dwelling units, usable area, roof area and DHW requirements) for simulation purposes and the total roof area of each combination in the neighborhood.

Depending on the combination, the achieved benefit can be estimated. Some environmental benefits are obtained with the green roof and solar-panel solution as they improve the building's energy performance. The energy demand (kWh/m^2.year), final energy consumption (kWh/m^2.year) and carbon emission ($kg\ CO_2/m^2$.year) reductions can be estimated by the energy performance simulation software CE3X. Moreover, green areas in the neighborhood increase by 8530.70 m^2 (Table 1). Although the Residents Area solution cannot be valued in terms of environmental or economic benefits, it entails social advantages that contribute to the sustainability of the urban area.

Table 2 summarizes the benefits obtained by simulating the refurbishment scenarios using CE3X. Table 2 indicates that energy demand lowers when green roof is implemented as the thermal transmittance of roof decreases (Figure 7); second, energy use savings are made in the green roof and solar-panel solution scenarios, together with lower carbon emissions. All the software outputs are calculated per square meter and then extrapolated to the neighborhood scale by considering the potential improvement accomplished for the total roof area in the neighborhood.

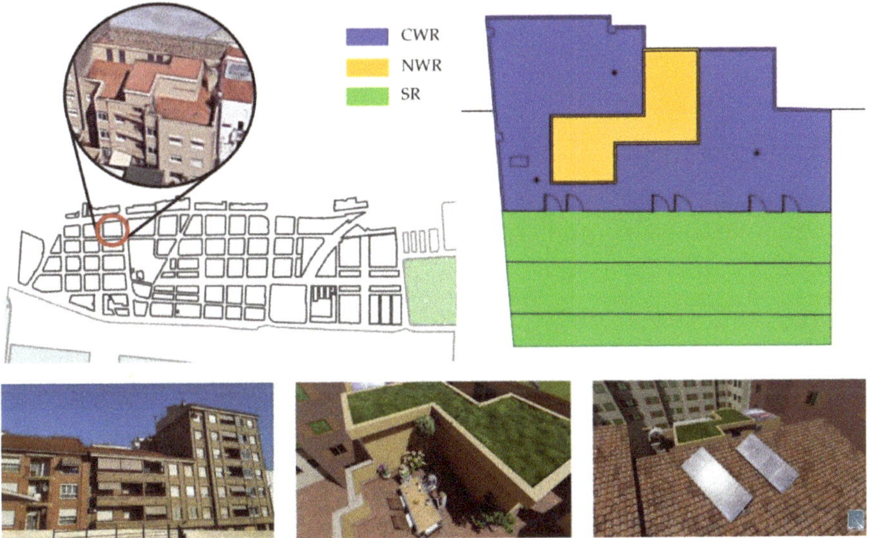

Figure 7. Specific building and refurbishment rendered.

As observed in Table 2, a solar panel for DHW provides greater energy use reductions than green roof. Implementing a solar panel leads to a 10% reduction in SfH, and values rise to around 30% for MfH. Regarding the green roof solution, studies suggest reducing buildings' energy consumption by up to 5% [37]. The results obtained with the case study via simulation ranged from 1.7% to 3.2%, which are similar to those given carbon emissions. Energy demand also reduces with green roof due to lower roof thermal transmittance. All the obtained reductions are worth considering when the results are extrapolated to the neighborhood scale, with savings of 40,782.85 kWh/year in energy demand, 579,595.70 kWh/year in energy consumption and 148,364.94 $kg\ CO_2$/year in carbon emissions.

Table 2. Scenario features of representative buildings and total savings in the area.

Period	Building	Roof	Solution	Representative Building; Cadastral Reference Floors; Number of Dwellings; DHW Requirement (L/day)	Usable Area m^2	Roof Area m^2	Total Roof Area m^2	Energy Demand kWh/year Per sqm % Saving	Energy Demand kWh/year Total	Final Energy Consumption kWh/Year Per sqm % Saving	Final Energy Consumption kWh/Year Total	Carbon Emissions kgCO$_2$/Year Per sqm % Saving	Carbon Emissions kgCO$_2$/Year Total
1960–1979	MfH	NWR	GR	7395404YK5279N 6; 20; 2016	1440	36.70	5285.90	3.51 4.0%	18,553.51	3.87 3.2%	20,456.43	1.30 3.2%	18,553.09
		CWR			1680	254.00	25,596.53		89,843.82		99,058.57		33,275.49
1980–2006	MfH	SR	SP	7496107YK5279N 7; 36; 3427	2912	234.00	18,717.40	–	–	17.99 27.2%	336,726.03	4.5 26.4%	84,228.30
		NWR	GR	7392307YK5279S 6; 30; 2856	2304	67.60	3244.80	0.72 3.6%	2336.26	1.26 1.7%	4088.45	0.3 1.8%	973.44
		CWR			2968	173.90	11,652.45		8389.76		14,682.09		3495.74
	SfH	SR	SP	7697301YK5279N 1; 3; 112	168	41.80	10,292.80	–	–	10.12 10.3%	104,163.14	2.60 10%	26,761.80
2007–2012	MfH	SR	SP	7496404YK5279N 7; 30; 2856	2184	189.33	4354.70	–	–	19.94 36.2%	86,832.72	5.0 34.9%	21,773.50
				Total savings area					119,123.35		666,007.43		189,061.36

The calculations suggest an interesting benefit regarding energy savings and carbon emissions, along with other environmental and social advantages, such as the increase in green areas in the urban environment, 8530.70 m^2 in non-walkable roof and 37,248.98 m^2 in community walkable roofs.

As mentioned earlier, it is an order of magnitude and the particular conditions of each building should be observed to obtain more accurate results. Scenarios may vary to accomplish the optimum solution. A combination of the proposed solutions can even be suggested in buildings where different roof types coexist. By way of example, a building where community walkable roof, non-walkable roof and sloped roof are present and all the analyzed solutions can be implemented, was selected. Knowing the exact building features can offer a more accurate scenario of the refurbishment.

Table 3 presents the main building features and summarizes the benefits obtained from the refurbishment proposal, which is completed by the estimation of the intervention's economic cost. It is an MfH building formed by six dwellings, two commercial premises and 10 parking lots. It was built in 1994 in accordance with both Regulation NBE-AE-88 on structural requirements and Regulation NBE-CT-79 on thermal requirements. The building combines three roof types: non-walkable roof over the staircase, community walkable roof and sloped roof facing southeast, with 25, 108.50 and 134.64 m^2, respectively. After checking the structural capacity with NBE-AE-88, which came into force in 1994, the following solutions are proposed: green roof suggested for non-walkable roof, as shown in Figure 7; the resident's area is suggested for community walkable roof by estimating a wooden pergola of 60 m^2. Finally, solar panels for DHW are proposed for sloped roof with the proper orientation. The load resulting from the solution and the improved transmittance value in green roof are also presented in Table 3. The cost of the intervention was estimated from the construction prices database (CYPE) [49], with a total amount of 18,142.45€ (2267.80 €/owner, 6 dwellings + two commercial premises, not including professional fees, licenses and taxes). The combination of these roof refurbishment solutions provides economic and social benefits that are all relevant in sustainability terms.

Table 3. Description of the specific building.

Building					
Roofs					
MfH 6 dwellings 2 commercial premises 10 parking lots Address: Alcalá Galiano, 1 Cadastral reference: 7191303YK5279S Orientation: East–West Year 1994	**Type**	**Area (m^2)**	**NBE-AE-88 Load/Overload (kg/m^2)**	**NBE-CTE-79 Transmittance (kWh/m^2)**	
	NWR	25.00	618/140	0.93	
	CWR	108.25	618/190	0.93	
	SR	134.64	346/134.64	1.12	
Refurbishment					
Roof	**Solution**	**Added load (kg)**	**Transmittance (kWh/m^2)**	**Budget * (€)**	**Benefits**
NWR	GR	2735.50	0.51	2421.25	E. demand: 18 kWh/year E. consumption: 31.5 kWh/year C. emissions: 7.5 kg CO_2/year 25 m^2 increase in green area Others: sound insulation, heat island effect reduction, etc.
CWR	RA	2043.14	–	3658.80	Social
SR	SP	5900.00	–	12,062.40	E. consumption: 2422,17 kWh/year C. emissions: 605.88 kg CO_2/year

* Prices: RA—60 m^2 wooden pergola, unit price—60.98 €/m^2; GR—25 m^2, unit price 96.95 €/m^2; SP—8 panels, unit price 1507.80 €/panel.

The economic analysis of the viability of the global cost (investment, running and disposal costs and the entailed energy savings), according to the comparative methodology framework for calculating cost-optimal levels of minimum energy performance requirements for buildings and building elements (Commission Delegated Regulation (EU) No 244/2012 of 16 January 2012, supplementing Directive 2010/31/EU of the European Parliament and of the Council on the energy performance of buildings), is not very optimistic. Even when considering the macroeconomic perspective of the methodology that assigns a price for carbon emissions (public cost) that would be included as savings when energy performance improves, the payback is too long, over 30 years. So, the proposed solution is not very appealing from the economic perspective. However, that term could be shortened if some funds were allocated to the intervention (up to 40% of the cost of the investment and up to 75% for low incomes, according to Order 1/2019, 18 February, of the Regional Administration in Charge of Housing, Conselleria de Vivienda, Obras Públicas y Vertebración del Territorio, which approves funds for ARRU for the National Housing Plan 2018–2021) (see Supplementary Materials). Moreover, there are other gains that would be obtained from the activation of the refurbishment economic sector, e.g., from the generated taxes (VAT, licenses, etc.), together with increased property value due to the intervention. Then there other costs that should be included in the balance of payments in sustainability terms, namely public costs, which means that they are difficult externalities to monetize. Those could be the benefits linked with an increase in green areas in the neighborhood or social benefits associated with the use of wasted space.

Other solutions that were ruled out in the proposal of this work could also be considered. For example, the use of photovoltaic panels could be supported by considering the economic profitability of the investment. A basic simulation was performed for this specific building using the European Commission's simulation tool, Photovoltaic geographical information system (available at https://re.jrc.ec.europa.eu/pvg_tools/es/#TMY, last accessed 12 August 2020). According to the obtained results, installing 25 PV panels on the sloped roof would save 13,232.61 kWh/year. This facility would increase the energy and carbon emission savings significantly compared to the 2422.17 kW/year obtained with DHW panels. The global cost economic approach shows that, when DHW solar panels are replaced with PV panels, the payback period shortens substantially, with positive values for the net present value and the investment return is even obtained in about five years in the funded scenario (see Annex A, tabs 10 to 13).

4. Discussion and Conclusions

Current urban areas are complex environments that include most of the population and need varied and combined solutions to enhance sustainability. Buildings' improved energy performance has become a crucial target in recent years to seek low carbon economies. Moreover, today's roofs are underused and have been devalued compared to their use of old. From this viewpoint, roofs can be seen as an opportunity because they form part of the thermal envelope and can contribute to improve buildings' energy performance—especially old and obsolete buildings where energy refurbishment and updates are necessary. Besides the environmental benefits, some roof refurbishment solutions offer many other advantages that can be hard to monetize but are extremely interesting in sustainability terms by contributing to ameliorate citizens' lives as social benefits.

This work analyzed the improvement potential of consolidated urban environments through roof refurbishments. It focuses exclusively on roofs to analyze the possibilities linked with an underused part of the building. One of the achieved benefits is the building's energy performance improvement, which is especially needed in vulnerable dwellings that usually present very poor thermal performance. In this regard, the roof's contribution is partial and integral energy refurbishment of housing would be advisable to achieve optimal results. Some suggested solutions allowed savings to be estimated in terms of energy and carbon emissions and other more social-oriented benefits. To consider the neighborhood scale, an area was selected, and a thorough analysis of the building stock was performed. The existent constructive solutions of roofs were analyzed with the georeferenced information from the Cadastral Office of the actual buildings. The actual constructive solution is crucial for suggesting adequate refurbishment solutions in practical applicability terms (load capacity, orientation, geometry, surrounding conditions, etc.). refurbishment solutions were assigned depending on the roofing system. To estimate some figures, the study focused on a specific vulnerable residential area in the Grao neighborhood in the Spanish city of Castellón, defined as ARRU in the Urban Plan. Three simple solutions were selected as being most likely implemented, although some others could also be included. Based on these assumptions, the study obtained savings of 119,123.35 kWh/year in energy demand, 666,007.43 kWh/year in energy consumption and 189,061.36 kg CO_2/year in carbon emissions. In addition, 35,779.58 m^2 of new green area were added to the area. The suggested solutions include other social benefits linked to the use, enjoyment and advantage of some wasted spaces in the urban area. However, we must bear in mind that the study provides an order of magnitude based on representative buildings, and a more accurate calculation would require an individual analysis of the parameters in every single building. This is illustrated by the example performed in the selected building, which combined three roof refurbishment solutions. Considering the economic analysis of the intervention, and based on the global cost calculation, it shows that the payback period is too long, and the investment is not profitable. However, the term could be shortened if some funds were allocated to the intervention when the area is considered to be ARRU. Moreover, in sustainability terms, other costs should be placed in the balance of payments when considering public costs, such as those linked with the increase in green areas in the neighborhood or those connected to the social benefits associated with the use of wasted space.

This work is based on a simplified and fixed theoretical scenario to gain a potential quantifiable amount of energy savings, an increase in green areas and carbon reduction and other non quantifiable benefits linked with the non economic factors that affect urban sustainability. A myriad of scenarios can occur under today's conditions and this is the main limitation of our analysis. In real applicability terms, the legal framework, property regime, etc., should also be observed as actual conditions because they are too complex, and an accurate estimation is unattainable. Further research is needed to explore different solutions and scenarios. Rooftop architecture could become an interesting solution for the energy retrofit of the entire building and could represent an important economic opportunity in multifamily buildings whenever applicable. Real projects with different solutions can also shed some light on current results. The influence of the photovoltaic energy option should be further studied because it potentially provides renewable energy for a real NZEB scenario. The development and standardization of this technology in residential buildings could be a means to achieve higher energy savings and to improve the economic viability of refurbishment. Hence further research should be conducted to seek new market strategies—especially in MfH—where a community meter could be installed in the building [46]. However, the results found in this study could be useful for decision-making to allocate funds for urban regeneration interventions, especially in vulnerable areas that usually present an old and obsolete building stock.

Although a previous analysis of the characterization of buildings, roof types and boundary conditions in the area is necessary, this estimation method can be replicated in other case studies.

Supplementary Materials: The following are available online at http://www.mdpi.com/2071-1050/12/19/8111/s1, Excel file.

Author Contributions: Conceptualization, M.J.R.; Formal analysis, M.J.R. and L.R.; Investigation, Á.P. and I.A.; Supervision, L.R. All authors have read and agreed to the published version of the manuscript.

Funding: The APC was funded by the Department of Mechanical Engineering and Construction of the University of Jaume I, Castellón, Spain.

Conflicts of Interest: The authors declare no conflict of interest.

References

1. Directive 2002/91/EC of the European Parliament and of the Council of 16 December 2002 on the Energy Performance of Buildings. Available online: https://eur-lex.europa.eu/LexUriServ/LexUriServ.do?uri=OJ:L:2003:001:0065:0071:EN:PDF (accessed on 16 June 2020).
2. Directive 2010/31/EU of the European Parliament and of the Council of 19 May 2010 on the Energy Performance of Buildings (recast). Available online: https://eur-lex.europa.eu/LexUriServ/LexUriServ.do?uri=OJ:L:2010:153:0013:0035:EN:PDF (accessed on 16 June 2020).
3. Directive 2012/27/EU of the European Parliament and of the Council of 25 October 2012 on energy efficiency, amending Directives 2009/125/EC and 2010/30/EU and repealing Directives 2004/8/EC and 2006/32/EC. Available online: https://eur-lex.europa.eu/legal-content/EN/TXT/?uri=OJ:L:2012:315:TOC (accessed on 16 June 2020).
4. Directive (EU) 2018/844 of the European Parliament and of the Council of 30 May 2018 Amending Directive 2010/31/EU on the Energy Performance of Buildings and Directive 2012/27/EU on Energy Efficiency. Available online: https://eur-lex.europa.eu/legal-content/EN/TXT/?uri=OJ:L:2018:156:TOC (accessed on 16 June 2020).
5. Spanish Statistic Institute INE. 2011. Available online: https://www.ine.es/jaxi/Datos.htm?path=/t20/e244/viviendas/p01/&file=01011a.px#!tabs-tabla, (accessed on 18 May 2020).
6. *Royal Decree 2429/79, 6 July 1979, Approving Basic Housing Regulation on Thermal Conditions, NBE-CT-79*; The Ministry of Public Works and Urbanism of Spain: Madrid, Spain, 1979.
7. Codigotecnico.org. Available online: https://www.codigotecnico.org/index.php/menu-documentoscte.html (accessed on 18 May 2002).
8. Ministry of Transports, Mobility and Urban Agenda. 2020. Available online: https://www.fomento.gob.es/BE2/?nivel=2&orden=33000000 (accessed on 18 May 2020).
9. Roldán, J. Evaluación de Sobrecargas de Uso de Vivienda en Estructuras de Edificación. Ph.D. Thesis, Universitat Politècnica de València, Valencia, Spain, 28 October 2003. [CrossRef]
10. Intelligent Energy Europe, Typology Approach for Building Stock Energy Assessment—TABLE Database (National Building Typology). 2016. Available online: http://episcope.eu/building-typology/ (accessed on 15 May 2020).
11. Generalitat Valenciana; Instituto Valenciano de la Edificación. *Catálogo de Tipología Edificatoria Residencial*; Conselleria de Obras Públicas, Urbanismo y Vertebración del Territorio: Valencia, Spain, 2016.
12. Martín-Consuegra, F.; de Frutos, F.; Oteiza, I.; Hernández-Aja, A. Use of cadastral data to assess urban scale building energy loss. Application to a deprived quarter in Madrid. *Energy Build.* **2018**, *171*, 50–63. [CrossRef]
13. Braulio, M. Propuesta Metodológica Para la Caracterización del Comportamiento Energético Pasivo del Parque Edificatorio Residencial Existente Considerando su Contexto Urbano. Ph.D. Thesis, Universitat Jaume, I., Castellón, España, 9 June 2016. [CrossRef]
14. Graus, R. *La Cubierta Plana, un Paseo por Su Historia*; Texsa y Universitat Politècnica de Catalunya: Barcelona, Spain, 2005.
15. García-Bernal, D.; Huedo, P.; Babiloni, S.; Braulio, M.; Carrascosa, C.; Civera, V.; Ruá, M.J.; Agost-Felip, R. Estudio y Propuesta de Áreas de Rehabilitación, Regeneración y Renovación Urbana, con Motivo de la Tramitación del Plan. General Estructural de Castellón de la Plana. 2017. Tomo I. Available online: https://s3-eu-west-1.amazonaws.com/urbanismo/TOMO_I.pdf (accessed on 18 May 2020).
16. García-Bernal, D.; Huedo, P.; Babiloni, S.; Braulio, M.; Carrascosa, C.; Civera, V.; Ruá, M.J.; Agost-Felip, R. Estudio y Propuesta de Áreas de Rehabilitación, Regeneración y Renovación Urbana, con Motivo de la Tramitación del Plan. General Estructural de Castellón de la Plana. 2017. Tomo II. Available online: https://s3-eu-west-1.amazonaws.com/urbanismo/TOMO_II.pdf (accessed on 18 May 2020).
17. García-Bernal, D.; Huedo, P.; Babiloni, S.; Braulio, M.; Carrascosa, C.; Civera, V.; Ruá, M.J.; Agost-Felip, R. Estudio y Propuesta de Áreas de Rehabilitación, Regeneración y Renovación Urbana, con Motivo de la Tramitación del Plan. General Estructural de Castellón de la Plana. 2017. Tomo III. Available online: https://s3-eu-west-1.amazonaws.com/urbanismo/TOMO_III.pdf (accessed on 18 May 2020).
18. García-Bernal, D.; Huedo, P.; Babiloni, S.; Braulio, M.; Carrascosa, C.; Civera, V.; Ruá, M.J.; Agost-Felip, R. Estudio y Propuesta de Áreas de Rehabilitación, Regeneración y Renovación Urbana, con Motivo de la Tramitación del Plan. General Estructural de Castellón de la Plana. 2017. Tomo IV. Available online: https://s3-eu-west-1.amazonaws.com/urbanismo/TOMO_IV.pdf (accessed on 18 May 2020).

19. Corcelli, F.; Fiorentino, G.; Petit-Boix, A.; Rieradevall, J.; Gabarrell, X. Transforming rooftops into productive urban spaces in the Mediterranean. An LCA comparison of agri-urban production and photovoltaic energy generation. *Resour. Conserv. Recy.* **2019**, *144*, 321–336. [CrossRef]

20. Aguacil, S.; Lufkin, S.; Rey, E.; Cuchí, A. Application of the cost-optimal methodology to urban renewal projects at the territorial scale based on statistical data—A case study in Spain. *Energy Build.* **2017**, *144*, 42–60. [CrossRef]

21. Gangolells, M.; Casals, M.; Ferré-Bigorra, J.; Forcada, N.; Macarulla, M.; Kàtia, G.; Tejedor, B. Office representatives for cost-optimal energy retrofitting analysis: A novel approach using cluster analysis of energy performance certificate databases. *Energy Build.* **2019**, *206*, 109557. [CrossRef]

22. Semple, S.; Jenkins, D. Variation of energy performance certificate assessments in the European Union. *Energy Policy* **2020**, *137*, 111127. [CrossRef]

23. Ruá, M.J.; Huedo, P.; Civera, V.; Agost-Felip, R. A simplified model to assess vulnerable areas for urban regeneration. *Sustain. Cities Soc.* **2019**, *46*, 101440. [CrossRef]

24. Saadatian, O.; Sopian, K.; Salleh, E.; Lim, C.H.; Riffat, S.; Saadatian, E.; Arash, T.; Sulaiman, M.Y. A review of energy aspects of green roofs. *Renew. Sust. Energy Rev.* **2013**, *23*, 155–168. [CrossRef]

25. Maiolo, M.; Pirouz, B.; Bruno, R.; Palermo, S.A.; Arcuri, N.; Piro, P. The role of the extensive green roofs on decreasing building energy consumption in the Mediterranean climate. *Sustainability* **2020**, *12*, 359. [CrossRef]

26. Bevilacqua, P.; Bruno, R.; Arcuri, N. Green roofs in a Mediterranean climate: Energy performances based on in-situ experimental data. *Renew. Energy* **2020**, *152*, 1414–1430. [CrossRef]

27. Weber, M. Healthy urban living: Integration of noise in other local policy domains. In Proceedings of the 10th European Congress and Exposition on Noise Control Engineering, Euronoise 2015, Maastricht, The Netherlands, 1–3 June 2015; pp. 387–392.

28. Hirano, Y.; Ihara, T.; Gomi, K.; Fujita, T. Simulation-based evaluation of the effect of green roofs in office building districts on mitigating the urban heat island effect and reducing CO2 emissions. *Sustainability* **2019**, *11*, 2055. [CrossRef]

29. Shafique, M.; Xue, X.; Luo, X. An overview of carbon sequestration of green roofs in urban areas. *Urban For. Urban Gree.* **2020**, *47*, 126515. [CrossRef]

30. Akther, M.; He, J.; Angus, C.; Huang, J.; Van Duin, B. A Review of green roof applications for managing urban stormwater in different climatic zones. *Sustainability* **2019**, *10*, 2864. [CrossRef]

31. Wilkinson, S.J.; Ghosh, S. Roles of a roof top garden in enhancing social participation and urban regeneration in Sydney CBD. In Proceedings of the RICS COBRA Conference, Sydney, NSW, Australia, 8–10 July 2015. Royal Institution of Chartered Surveyors.

32. Catalano, C.; Laudicina, V.A.; Badalucco, L.; Guarino, R. Some European green roof norms and guidelines through the lens of biodiversity: Do ecoregions and plant traits also matter? *Ecol. Eng.* **2018**, 15–26. [CrossRef]

33. Hong, W.; Guo, R.; Tang, H. Potential assessment and implementation strategy for roof greening in highly urbanized areas: A case study in Shenzhen, China. *Cities* **2019**, *95*, 102468. [CrossRef]

34. Cascone, S. Green roof design: State of the art on technology and materials. *Sustainability* **2019**, *11*, 3020. [CrossRef]

35. Fuentes, E.; Arce, L.; Salom, J. A review of domestic hot water consumption profiles for application in systems and buildings energy performance analysis. *Renew. Sust. Energy Rev.* **2018**, *81*, 1530–1547. [CrossRef]

36. Pomianowski, M.Z.; Johra, H.; Marszal-Pomianowska, A.; Zhang, C. Sustainable and energy-efficient domestic hot water systems: A review. *Renew. Sust. Energy Rev.* **2002**, *128*, 109900. [CrossRef]

37. Gautama, A.; Chamolib, S.; Kumara, A.; Singhc, S. A review on technical improvements, economic feasibility and world scenario of solar water heating system. *Renew. Sust. Energy Rev.* **2017**, *68*, 541–562. [CrossRef]

38. Alghamdi, A.S. Potential for rooftop-mounted PV power generation to meet domestic electrical demand in Saudi Arabia: Case study of a villa in Jeddah. *Energies* **2019**, *12*, 4411. [CrossRef]

39. Jorge, C.M.; Ester, H.G. Energy earnings and carbon dioxin reduction achieved by the application of photovoltaic cells, case study of shopping light mall in Sao Paulo. *J. Eng. Appl. Sci.* **2018**, *13*, 7051–7056.

40. Aste, N.; Adhikari, R.S.; Tagliabue, L.C. Solar integrated roof: Electrical and thermal production for a building renovation. *Energy Procedia* **2012**, *30*, 1042–1105. [CrossRef]

41. Berto, R.; Stival, C.A.; Rosato, P. The valuation of public and private benefits of green roof retrofit in different climate conditions. In *Values and Functions for Future Cities*; Springer: Cham, Switzerland, 2020; pp. 145–166.

42. Spanish Ministry for Economic Transition; Instituto de Diversificación y Ahorro Energético IDAE. *Guía para el Desarrollo de Instrumentos de Fomento de Comunidades Energética Locales*; Instituto de Diversificación y Ahorro Energético, IDAE: Madrid, Spain, 2019.

43. Yuliani, S.; Hardiman, G.; Setyowati, E. Green-roof: The role of community in the substitution of green-space toward sustainable development. *Sustainability* **2020**, *12*, 1429. [CrossRef]

44. Wilkinson, S.J.; Dixon, T. Innovation in the built environment. In *Green Roof Retrofit: Building Urban Resilience*; John Wiley and Sons: Chichester, UK, 2016.

45. Teotónio, I.; Cabral, M.; Oliveira-Cruz, C.; Matos-Silva, C. Decision support system for green roofs investments in residential buildings. *J. Clean. Prod.* **2020**, *249*, 119365. [CrossRef]

46. Dvorak, B. Comparative analysis of green roof guidelines and standards in Europe and North America. *J. Green Build.* **2011**, *6*, 170–191. [CrossRef]

47. FLL. *Guidelines for the Planning, Construction and Maintenance of Green Roofing*; Forschungsgesellschaft Landschaftsentwicklung Landschaftsbau e.V.: Bonn, Germany, 2008.

48. Ministry of Transports, Mobility and Urban Agenda. Documento de Apoyo al Documento Básico DB-HE Ahorro de Energía Código Técnico de la Edificación. Available online: https://www.codigotecnico.org/images/stories/pdf/ahorroEnergia/DA_DB-HE-1_Calculo_de_parametros_caracteristicos_de_la_envolvente.pdf (accessed on 17 June 2020).

49. Generador de Precios de la Construcción. Available online: http://www.generadordeprecios.info/ (accessed on 1 May 2020).

 © 2020 by the authors. Licensee MDPI, Basel, Switzerland. This article is an open access article distributed under the terms and conditions of the Creative Commons Attribution (CC BY) license (http://creativecommons.org/licenses/by/4.0/).

 *sustainability*

Article

An Approach to Environmental Criteria in Public Procurement for the Renovation of Buildings in Spain

Teresa Soto [1], Teresa Escrig [1], Begoña Serrano-Lanzarote [2,*] and Núria Matarredona Desantes [2,3]

[1]   Instituto Valenciano de la Edificación, 46018 Valencia, Spain; tsotov@five.es (T.S.); tescrigm@five.es (T.E.)
[2]   School of Architecture, Universitat Politècnica de València, 46022 Valencia, Spain
[3]   Conselleria de Vivienda y Arquitectura Bioclimática, 46018 Valencia, Spain; matarredona_nur@gva.es
*   Correspondence: apserlan@mes.upv.es

Received: 31 July 2020; Accepted: 10 September 2020; Published: 15 September 2020

Abstract: In order to contribute to climate neutrality within the EU in 2050, it is necessary for administrations to play a driving role through green public procurement for building renovation (GPPBR). Among the main barriers that slow down the GPPBR, a lack of knowledge of the parties involved can be highlighted. Faced with this scenario, the aim of this article is to provide a compilation, as a preliminary state of the art, of the most important environmental measures to bring to the GPPBR specifications. The methodology used for this compilation and critical analysis consisted of a systematic search for laws, regulations and guides prepared by Spanish public administrations, as well as looking into other international information sources, mainly collected from the EU. Despite the fact that related technical information is abundant, it is scattered and at times impractical. This study can be useful as a basis for both drafting specifications and highlighting the need to develop other specific and advanced technical procedures to assist GPPBR professionals.

Keywords: environmental; green public procurement; renovation; construction; energy; circular economy; innovation; green; sustainable

---

1. Introduction

There is a clear opportunity to promote building renovation through Green Public Procurement (GPP). The implementation of environmental measures in buildings for energy efficiency and transition to a circular model depends on the active involvement of all the agents around. Although the Public Administration, by function and regulatory work should first take these measures and play a leading role in the private sector.

1.1. Green Public Procurement for Building Renovation (GPPBR) in the EU

The main policies in the European Union are framed in the current climate and environmental emergency [1]. It is essential to achieve climate neutrality for the EU by 2050, and meet environmental goals committed by governments through the Climate Agreement: likewise, constructing play a relevant role, among other sectors such as transport.

On the one hand, it is imperative to reduce energy consumption caused by buildings themselves [2], since this sector represents around 40% of total consumption in the world [3,4]. Furthermore, 75% of all the buildings in the EU are energy inefficient, whereas their annual renovation rate is only 0.4–1.2% [4,5]. According to the European Green Deal [5], in order to achieve the EU goals, the intervention rate on buildings should at least double. Directive 2012/27/EU [6] requires that new buildings have nearly zero energy consumption and sets an annual renovation of 3% of the total area of buildings when it comes to central administrations, so that they meet current energy efficiency requirements. According to Directive 2010/31/EU [7], every Member State should set a long-term strategy to support the renovation of the building stock.

On the other hand, it is urgent to implement decisively the principles of a circular economy in buildings [3]. Awareness and knowledge on the characteristics of circular public procurement, based on the implementation of circular policies and strategies, are essential [8]. According to the New Action Plan for the circular economy in the EU [3], the building sector demands huge amounts of resources, uses around 50% of all extracted materials and generates more than 35% of the total waste. Directive 2008/98/EC [9] provides that Member States shall take action to prevent waste generation, by promoting reuse, repair and recycling of construction and demolition materials and products, among others.

Within this context, public procurement is a key action for implementing the EU environmental policies, since it makes up around 19% of the gross domestic product [10]. Thus, an effective green public procurement is needed, as a force for change. Directive 2014/24/EU [11] already introduces environmental considerations regarding award criteria, technical specifications and construction conditions. The EU also offers Member States an optional tool, the GPP criteria, as a reference for developing environmental clauses [12]. Despite the fact that these criteria offer great potential to carry out environmental policies, based on studies from an international background, it is found that the level of implementation of the GPP is varied, and it is generally limited [8,13–15]. The lack of knowledge and definition when it comes to introducing environmental criteria in tender papers, and the absence of practical information for designing, are some potential causes. Moreover, further development of the GPP criteria in relation to innovative solutions is required [15,16].

1.2. Green Public Procurement in Building Renovation (GPPBR) in Spain

In Spain, the regulation of the GPPBR starts from the environmental policies of the EU [17], as well as energy renovation needs of the existing building stock.

Around 21% of existing buildings in Spain are older than 50 years, and 55% were built before 1980, when there was no obligation to be thermally insulated [18]. Residential buildings are the highest consumers of final energy consumption caused by construction. However, buildings in tertiary sector, which occupy only 13% of total square meters of building space, consume 42% of final energy [19]. That is why these buildings and particularly those for Public Administration (such as health care, education, assistance, administrative centers...) show a very significant potential to reduce carbon footprint and other environmental impacts. In this respect, and given its exemplary role, Public Administrations establish specific obligations for building renovation in energy matters and concerning a transition to a circular economy model. The renovation of buildings themselves is part of this model, since one of the principles is to keep products and materials in use [20]. According to the National Action Plan for Energy Efficiency [21], the Central Administration should annually renew 3% of the total area of buildings—a goal nearly achieved in Spain during the period 2014–2019 [19].

It is estimated that public procurement in Spain represents between 10 and 20% of GDP, depending on the year [22]. As it is a key economic sector, public procurement in terms of building should include environmental criteria, which are collected in various regulatory tools. Among the provisions of the General State Administration affecting the GPPBR, it is worth mentioning the following:

- The Law on Public Sector Contracts (LPSC) [23], which refers to the possible consideration of environmental measures in technical prescriptions and award criteria. It states that the evaluation of the best value for money can include environmental factors. It also refers to life cycle cost (LCC) as a means of assessing the best cost-effectiveness. It also considers environmental labels as a means for testing.

- The Draft Law on climate change and energy transition [24], which promotes and allows an efficient use of energy from renewable sources in the area of energy efficiency and building renovation.

- The Ecological Public Procurement Plan (2018–2025) [17]. It refers to the GPP criteria set by the European Commission, highlighting the carbon footprint reduction.

- The Draft for National Integrated Energy and Climate Plan 2021–2030 [25], with measures aimed at reducing primary energy consumption and promoting renewable energies. It provides for energy renovation over 3% of the built-up area in public buildings of the General State Administration, and at least 3% of the buildings of regional and local administrations.

- The Draft for the Long-term Decarbonization Strategy 2050 [26], which estimates that, for that time frame, 80% of the built stock will be made up of buildings already built. For this reason, the actions on the existing building are prioritized. On the other hand, most of new buildings (representing the remaining 20%) will have nearly zero energy consumption.
- The Circular Economy Strategy [27], which considers building sector and building renovation with own relevance in the materials cycle, climate change, resilience and local economy, among others. It also has a high level of improvement concerning treatment of construction and demolition waste (CDW) and optimization of resources used.
- The Long-term Strategy recently updated for energy renovation in the building sector in Spain. It includes the need to provide policies and actions in all public buildings [19].
- The Royal Decree, which modifies the Basic Document on Energy Saving from the Technical Building Code (TBC) [28], which includes basic requirements related to: restriction of energy consumption, conditions for the control of energy demand, conditions of thermal and lighting installations, minimum contribution of renewable energy to cover the demand for sanitary hot water and minimize electrical power. Currently, it requires revision to introduce the requirement of charging points for electric vehicles.
- The PAREER program [29], to encourage energy renovation of buildings so as to improve the energy rating.

This state framework is as a reference for the different regional Public Administrations.

Regarding public procurement in Spain, the percentage of contracts in which the best quality-price ratio is valued as an award criterion is high if related to the EU average [30]. However, the implementation of environmental policies in the public procurement of buildings in Spain is low if compared to other countries. The group of countries known as "Green 7" (Austria, Denmark, Finland, Germany, the United Kingdom, the Netherlands and Sweden) [31] shows a higher number of tenders in terms of ecological criteria, rather than the rest of the EU countries. It also reaches the highest qualifications in questionnaires about environmental measures. One of these measures is the introduction of environmental management systems as well as the use of innovative tools for public procurement, such as the one related to life cycle [13,31]. Likewise, the Nordic countries, through their 2010 Action Plan, are pioneers in promoting sustainable production and eco-innovation [14].

However, in all EU countries, there is still a long way to go. An example is the lesser importance of environmental measures in the GPP, approximately 10% in the European context [14], if compared to other criteria. These can be costs, construction process report and delivery time. In some cases, the weighing of environmental criteria is even negligible. Especially in lower budget tenders, it can be seen that the financial approach tends to a higher valuation with regard to other criteria, such as the environment-related ones [31]. Likewise, the lack of adoption of the LCC criterion is generally observed [32].

There are several reasons why innovative or environmentally responsible public procurement are not yet fully implemented. The main barriers found are the lack of resources, knowledge, staff training, motivation and funding [8,31,33,34]. This has been confirmed in various studies about the implementation of GPP policies [30,31], and reflected in the Report on public procurement in Spain - 2017, prepared by the Ministry of Finance and Public Function [35]. Thus, the mentioned barriers have increased in small towns with fewer resources to set up specialized contracting departments. Without them, it is not possible to gain knowledge and obtain developed strategies for GPP [31]. As a consequence, the implementation of circular economy principles is often the result of random chance or unsystematic experimental models [8].

For this reason, it is necessary to develop suitable technical procedures for the award of contracts [35]. The dissemination of these procedures and the training of professionals who have to put them into practice, together with coordination and cross-cutting strategies of organizations are required, since the model design-tender-build in sustainable buildings requires minimal training and management to be adopted [36].

1.3. Objective of the Article

On this basis, the aim of this article is to open up new avenues so as to improve implementation of environmental criteria in the GPPBR in Spain. Among the main barriers, the lack of knowledge of the agents involved can be mentioned.

As a first step to facilitate knowledge, a compilation is made in Section 3, as a preliminary state of the art about environmental measures included in the regulations and guides of all Spanish regions, in line with the European context for implementation of GPP. In Section 4, an analysis is carried out concerning the feasibility of applying such measures, alongside the identification of potential barriers. Section 5 contains the review of measures along with conclusions, limitations and possible future research. Given the abundant but scattered information, there is a need to develop advanced technical procedures for professionals responsible for public procurement, including evaluating indicators.

All of this seeks to move towards an objective: the GPPBR should be truly "ecological". Thus, the Administration should have an exemplary role. Knowledge generated should help increase the motivation among professionals and evaluate proposals, not just according to initial costs but the best value for money. This would allow to gain the benefits of sustainable renovation throughout the life cycle and promote climate neutrality of the EU in 2050.

2. Methodology

As explained above, the purpose of this article is to contribute to the improvement of knowledge so as to carry out the GPPBR in Spain. That is why methodology used is based on an approach on the state of the art of the major environmental measures proposed by the Spanish Public Administrations, regarding the goals set by the EU about building renovation. Due to the particular features of each region and the territorial organization in Spain, these measures were set out by the State and regional administrations through various papers: laws, regulations and guides. Each of them within the scope of its competencies. The present article is an in-depth study of the regional GPPRB.

The methodology used is summarized in Figure 1 and described below.

In order to obtain documentation about the GPPBR at regional level, internet sites of all Spanish regions and Autonomous Cities were checked first. All of them can be accessed through the "General Access Point" web site, created by the General State Administration:

https://administracion.gob.es/pag_Home/atencionCiudadana/SedesElectronicas-y-Webs-Publicas/websPublicas/WP_CCAA.html#.Xzu2Pugza00

Next, by entering each site, a search on subjects about sustainability in construction and GPP was carried out: building, renovation, climate change, energy efficiency, circular economy, waste management, public procurement of innovative solutions (PPI), environment, territory, sustainability, and urban planning. This search led to different web sites of regional governments: departments, councils and organizations for energy, business competitiveness, building, housing, land, etc. General and specific tools were found about the implementation of measures in the GPP and building renovation. Appendix A contains a broad list of laws (L), regulations (R) and guides (G) found (a total amount of 145 papers). Figures 2 and 3 provide a summary of the main topics and the percentage that each type of tool represents over the total analyzed. Figures 4 and 5 show the main topics related to the documents found.

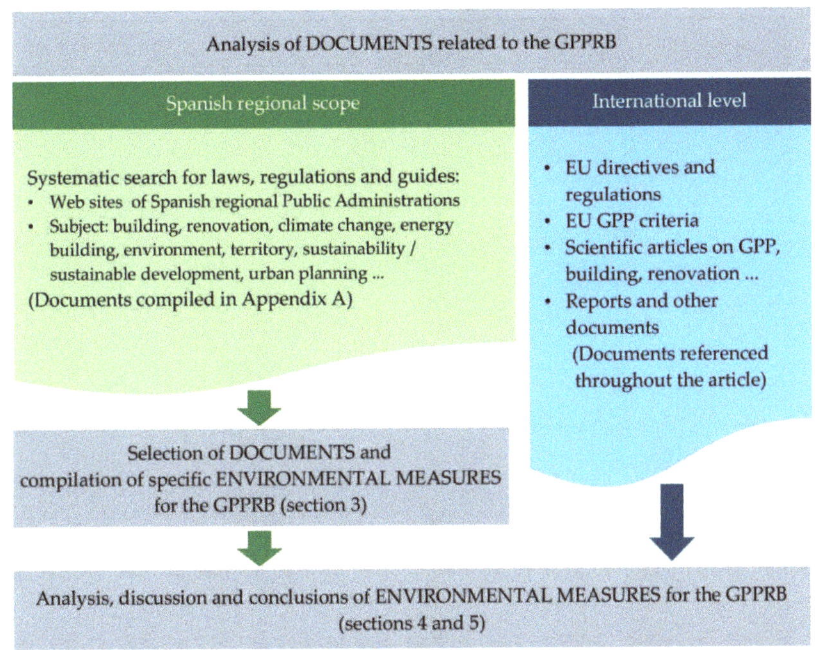

Figure 1. Methodology.

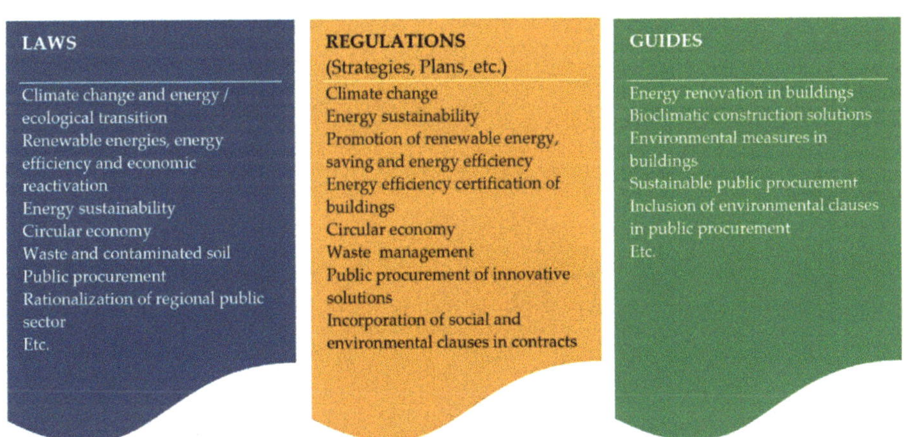

Figure 2. Documents considered for the compilation of environmental measures in the green public procurement for building renovation (GPPBR) of the Spanish regions and Autonomous Cities.

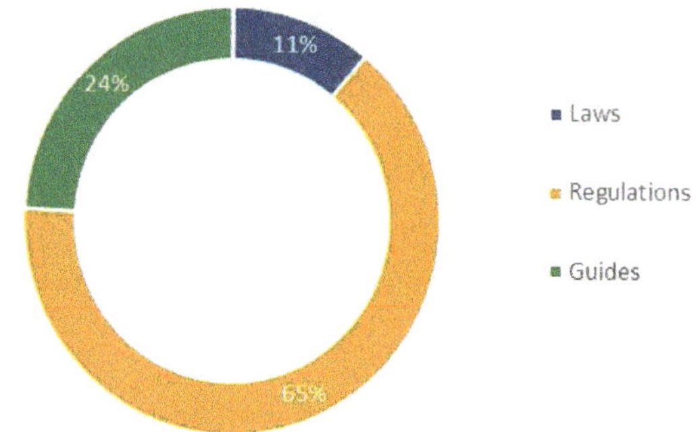

Figure 3. Percentage of papers collected according to the type of document.

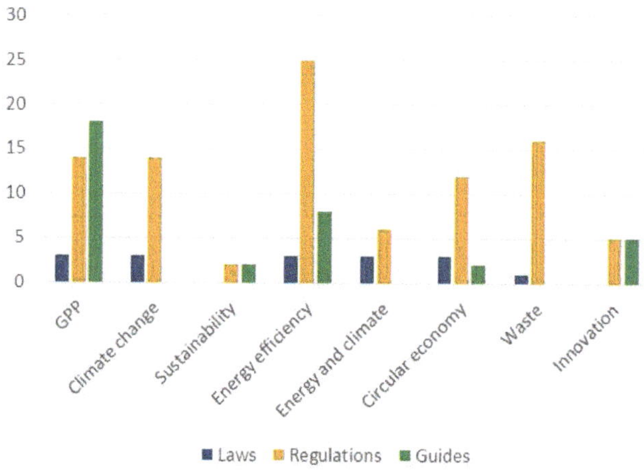

Figure 4. Number of laws, regulations and guides compiled according to topics.

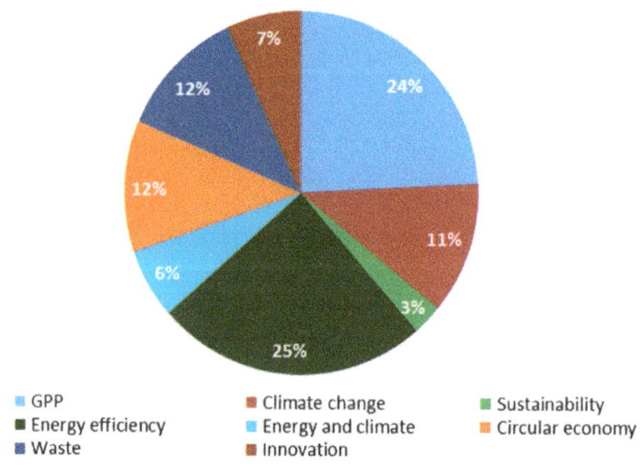

Figure 5. Percentage of documents collected according to topics.

Once papers were collected, they were analyzed to draw environmental measures specifically applicable to the GPPBR.

The environmental measures are set out in Section 3 together with a list of documentation as close as possible to the different subjects, or documentation developing advanced and useful aspects to bring into contracting procedures. Although various laws, regulations and guides were compiled from all regions and autonomous cities, the list of these papers as well as measures included in this article is not exhaustive. Yet, it is enough to obtain consistent results as a reference, and open up avenues for further research. Moreover, it is necessary to mention that a comparative study among regional Administrations is not the purpose of this approach.

At the same time, other international sources on implementing environmental measures in buildings and GPP were analyzed. Particularly, legal framework and criteria for the EU GPP, as well as other articles and papers outlined in this article, were taken as a basis. The study of all documentation allowed a rough idea of the situation and issues of the GPPBR, as explained in Section 1. All of this made it possible to enrich the analysis, discussion and conclusions of the measures collected within a broader context, as Sections 4 and 5 show.

3. Results. How is GPP Implemented in Spain?

3.1. Regional Administration Papers

Based on the state regulatory framework set out in the Introduction, regional Administrations have drawn up a large number of laws, regulations and guides related to the subject of this article. The following are papers concerning climate change, energy transition and circular economy, closely related to the GPPRB.

- Climate change laws: whereas Spanish climate change law is still to be approved [24], three regions have previously approved own laws Illes Balears in 2019 [37], Andalucía in 2018 [38] and Catalunya in 2017 [39]. Likewise, Islas Canarias, the Comunidad de Madrid and the Comunitat Valenciana are still drafting respective laws on climate change and energy/ecological transition.
- Laws, strategies and plans for energy saving and efficiency in public buildings: in general, all regions, as well as the autonomous cities of Ceuta and Melilla, have implemented actions to improve the energy efficiency in public buildings. The National Plan describes the latter [21], although it is worth highlighting regional plans with comprehensive objectives for public buildings and public facilities, as well as specific measures to include in procurement documents; it is worth mentioning the Energy Sustainability Law of the País Vasco, which binds public administrations to improve energy efficiency and use renewables in buildings [40], and additionally, the plan of Catalunya during the period 2018–2022 [41], the plan of the Comunidad de Madrid, approved in 2017 [42], the plan of the Comunitat Valenciana in 2016 [43], and the plan of Navarra [44]. On the other hand, strategies indicate lines of action with an impact on public contractual measures. The most recent ones include the Euskadi 2030 Energy Strategy [45] and that of Extremadura, for public regional administration buildings [46].
- Circular economy and waste regulations: as in energy matters, these documents also mention constructing sector and even public procurement; for its current interest, it is worth mentioning the Extremadura 2030 Green and Circular Economy Strategy [47], as well as strategies for the circular economy of Galicia, País Vasco, Aragón, and Navarra [48–51], all of them with targets set for 2030. Furthermore, the Circular Economy Law of Castilla-La Mancha [52], the first region to legislate circular economy [53], deserves special attention. Additionally, Andalucía and Islas Canarias are currently drawing up own laws. In addition, regional waste regulations also contain actions to manage the waste stream from public buildings; that is the case of the Comunidad de Madrid Waste Strategy [54], which has also developed the MADRID7R Circular Economy platform [55] and the Cantabria Waste Plan [56].

Although the GPP model specifications are very helpful for contracting authorities, there are few that have been prepared. These include the following:

- Model specifications for specific administrative clauses recommended by the Consultative Board for Administrative Contracting of the Comunidad de Madrid [57], with references to the inclusion of mandatory environmental clauses.
- The model for contracting energy services in public administration buildings of the IDAE [58] and the guide for drafting specifications on energy performance contracts with guaranteed savings of the ICAEN [59], which allowed the first contracts to be put out to tender [19].

In recent years, since the approval of Directive 2014/24/EU, of which the LPSC [23] is transposed, many guides related to GPP have been elaborated, even if these papers address the subject broadly, and also include building-related aspects. Among them, it is worth mentioning the following:

- The 2020 "General Guide on Aspects Related to Environmental Protection", developed by the Comunidad de Madrid [60]. In this paper, information is provided on obligations in force in this regard. The latter are included in the specifications for public contracts.
- The guide "Frequently Asked Questions and Quick Answers for Responsible Public Procurement", 2017, from the Comunidad de Madrid [61]. It is a general paper with a practical purpose, which helps the contracting body prepare specifications. Along with interesting strategies to achieve the objectives pursued, it also offers environmental considerations illustrated with success stories.
- The "Guide for Including Social and Environmental Clauses in Contracting", by the Junta de Andalucía, 2016 [62]. This paper contains mandatory clauses and others recommended as an example for any type of contract.
- The 2018 "Practical Guide for Including Social and Environmental Responsibility Clauses in Administrative Contracting", from the Principado de Asturias [63]. It is a general guide that considers water saving, materials, energy and recycled materials as potential environmental clauses.
- The Environmental Clause Sheets of Navarra, 2017 [64], with generic technical content but standing out a clear exposure of contractual procedure phases and how to verify measures.
- Likewise, various regions have committed to continue developing guides. The Government of Cantabria is to draft a guide including environmental criteria for contract valuation: a measure already established in the Strategy for Action against Climate Change in Cantabria 2018–2030 [65]. The government of the Illes Balears is to develop a technical guide to meet requirements set forth in the Law on climate change and energy transition in public works: reduction in greenhouse gas emissions (GHG), energy efficiency, water saving and waste reduction [37].

It should be noted that some regions collaborate with international projects to develop technical measures and evaluation criteria within the GPP. A remarkable example is the Basic Guidelines for applicating environmental criteria in the purchase of products/services by public administration of the GreenS project [66]. The Regional Energy Agency of Cádiz and the Andalucía Federation of Municipalities and Provinces have taken part in it.

Some guidelines were found, containing environmental criteria for buildings. Due to the relevance for adding environmental measures to the GPPRB, the following should be mentioned:

- The "Green Guide", published by the Generalitat Valenciana and the IVE in 2020 [67]. This paper addresses globally environmental measures to be included in public procurement specifications concerning the field of constructing of the Generalitat Valenciana. It also extends the scope of energy efficiency to saving water and circular economy. It is a paper designed for continuous updating, and greatly facilitates the addition of measures in accordance with the LPSC [23].
- The "Public Services" guide from the series "Guides for Energy Efficiency", published by the Government of Cantabria in 2013 [68]. It compiles a series of recommendations for decision-making in building renovation, quantifies energy saving, investments and the adequacy of measurements for financing, through a Company for Energy Services (CES).
- The "Environmental Criteria for Buildings", published by the Public Society for Environmental Management of the Government of the País Vasco (Ihobe) in 2020 [69]. It covers aspects to

enhance public procurement in building, considering environmental impact during the life cycle, with good practice examples. Although the content focuses on office buildings, recommendations are applicable to buildings with different uses.

- The 2015 "Guide for Sustainable Construction and Renovation for Housing in the País Vasco" [70]. It is aimed at professionals in the constructing sector, and offers relevant information to assess environmental impacts. Although it is specific to residential buildings, many actions can be taken for the renovation of tertiary sector buildings.
- The newsletter for "Saving and Energy Efficiency in Public Buildings", published by ICAEN in 2009 [71], informs and trains energy managers. This paper shows technology, facilities, habits and examples to adopt saving and efficiency measures in buildings. Measures to boost energy efficiency, renewable energy systems and water saving can be drawn from this publication.

It should also be noted that the guidelines set out in this section are not mandatory, so the application of environmental criteria proposed for the GPPRB is left to contracting bodies.

However, mandatory application of laws and regulations is bringing about results. Specific actions result from compliance with regional energy efficiency plans for buildings and public facilities. For example, in Catalunya, nearly all the energy acquired by the regional Administration is certified as 100% renewable [72]. In the Comunidad de Madrid, heating renovation and air conditioning installations in public buildings represented 64% of actions of the regional plan during the first year [73]. In the Comunitat Valenciana, it was possible to reduce energy consumption in public buildings by 11 million kilowatt hours in one year [74].

Results after putting into practice regional circular economy strategies are also observed. For example, in Extremadura, advice at the local level to identify municipal green initiatives is being given, as those related to carbon sinks [75]. In Galicia, the first public building built entirely using certified native wood has been put out to tender, within the framework of the Lugo Biodinamic project [76]. In the País Vasco, the evolution of CDW recycling in recent years is upward (42% in 2013, 59% in 2015 and 61% in 2016), but there is still room for improvement until reaching 70% set as a target for 2020 [77].

Finally, it is interesting to mention that several regional administrations carry out annual monitoring of the GPP results. For example, in the País Vasco, 24% of ecological tenders were carried out in 2019, which represented 22% of the total economic amount. In the building sector, the GPP represented 27%, with an economic volume of 28% with respect to the total tendered in that sector [78].

3.2. Environmental Measures

In this section, the most relevant measures set by Spanish regional Administrations are compiled and grouped according to the following subjects: passive energy saving measures, proactive measures for energy efficiency and water saving, life cycle, products and waste management. Those measures required by administrations as an example to be incorporated into the GPPRB, are presented next. Figure 6 shows the number of measures in each subject.

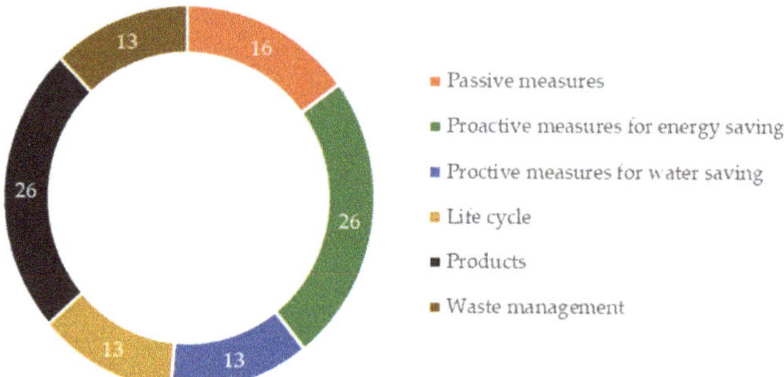

Figure 6. Environmental measures (total: 107).

3.2.1. Passive Measures for Energy Saving and Environment Protection

Passive strategies should be adopted from the beginning of the architectural design phase:

- If volume of buildings can be modified, shape should be designed to maximize the capture of solar energy; in some cases, a high level of compactness is valued [67,69].
- Optimize orientation of different building areas, according to temperature, evaluating performance of sun exposure studies [70].
- Openings design (number, size and distribution), according to orientation to increase heat gains during winter and reduce them in summer [67].

So as to reduce energy demand, it is essential to act on the thermal envelope:

- Reduction in the demand for energy consumption by lowering wall thermal transmittance, floors, roofs, openings and removing thermal bridges [68,69].
- Promote ventilated roofs and facades with greater ability to cool the building envelope in summer or provide thermal stability in winter, depending on the degree of ventilation [67,68,79].
- Evaluate landscaped roofs and facades to minimize energy flows and obtain other benefits; these solutions can remove pollutants through vegetation, mitigate heat island effect and benefit the ecosystem and the urban landscape, etc. This measure is conditioned to the resistant capacity of the renovated building structure [67,79,80].

To obtain heat gains, solar capture solutions can be added: glazed galleries, Trombe walls, thermal inertia walls, etc. [67–69,79,80].

With regard to sun protection, the following recommendations are provided:

- Supply solar protections to prevent indoor overheating, designed according to orientation.
- Use vegetation and its location to improve atmosphere quality: generation of shade, damping of noise, etc. [67,68,79–81].

It is advisable to provide natural ventilation through building design. For this purpose, choosing the best location on the plot should be a factor to consider; the most suitable shape, size and orientation of the building and its openings; the provision, if necessary, of screens (natural or artificial); interior layout; the shape of the cover, etc. Concerning solar protection, the following types of ventilation can be adopted:

- Cross ventilation, by wind action through openings in the façade or other open spaces located on different sides.
- Use thermal draft, either by chimney effect or static aspiration (Venturi effect).
- Induced ventilation, with devices allowing cold air to pass through, such as wind towers [67,79,80].

To provide the building with cooling, natural air cooling is recommended: by introducing treated air to lower temperature or increase humidity content. Depending on site conditions, different types of air treatment are proposed:

- Evaporative cooling by using water or vegetation that absorbs heat and increases humidity content.
- Heat sinks to precool air, such as buried ducts, patios or night cooling [67,69,79,80]. Measures on buried or semiburied construction are also proposed [80].

The incorporation of natural habitats in the vicinity of buildings or in the interior, such as gardens, groves or patios with vegetation, contribute to all kinds of benefits; the design and selection of plants and materials should be consistent with the intended purpose and friendly to the environment [67,69].

Bringing in space for bicycle storage and personal mobility vehicles (such as scooters and others) helps promote use and reduce carbon emissions, with better air quality [37,67].

Some examples obtained from the analysis performed are shown in Figure 7.

Passive measures for energy saving and environment protection	
Architectural design	• Building shape and compactness • Orientation, shape and compactness of building • Building openings design • Optimization of orientation of building areas
Thermal envelope	• Reduction of transmittance and elimination of thermal bridges: walls, floors, roofs, holes • Roofs and ventilated / landscaped facades
Solar heating	• Glazed galleries • Trombe walls /thermal inertia walls
Solar protection	• Fixed / movable shaft protections • Vegetation
Natural ventilation	• Cross / with thermal draught / induced
Natural air cooling	• Evaporative cooling • Temperature reduction: buried ducts, patios, night cooling • Buried / semi-buried construction
Habitats in patios and gardens	• Landscaped areas / patios with vegetation • Efficient and non-invasive plants
Parking spaces	• Parking for sustainable means of transport: bicycles, personal mobility vehicles

Figure 7. Examples of passive measures for energy saving proposed by Spanish regional administrations.

3.2.2. Proactive Measures for Energy Saving

It is advisable to limit energy consumption in public buildings, with specific goals:

- To reduce energy consumption in public buildings by 7.7% in 2022 compared to the 2017 consumption, as proposed by the Generalitat de Catalunya [41] or by 18% in 2025 compared to the 2014 consumption, as proposed by the Comunidad de Madrid [42]. These reductions are equivalent to the annual energy renewal of 3% of public building area, established by Directive 2012/27/EU. In the Comunitat Valenciana, the target set is to reduce energy consumption by 25% in 2025 compared to 2014 consumption [43]. In the "Green Guide" [67], it is proposed that existing buildings meet regulatory conditions established for new construction or, in case of impossibility, to reduce by at least 40%, both the consumption of nonrenewable primary energy and the total.
- To improve the energy rating of public buildings before 2025: at least 25% of buildings, as proposed by the Comunidad de Madrid [42] and 25% of buildings with energy consumption over 200,000 kWh/per year proposed in the Comunitat Valenciana [43].
- To integrate renewable energy systems—both for thermal use and electricity generation—into 25% of buildings before 2025 [42].

- To contract energy audits: to obtain knowledge of energy consumption profile during a building's use phase and quantify the possibilities of cost-effective saving energy [37,41,67,68,81–83].
- To contract energy service companies (ESCs) or an energy manager to monitor consumption and energy efficiency improvements [37,79,81,82,84];

Promoting self-consumption can be achieved by incorporating renewable energies:

- To generate electric power, for own use or network supply, such as photovoltaic/solar thermal energy, mini-wind energy or micro cogeneration or thermal installations, by using renewable-source systems, in situ or in the vicinity of the building: low temperature solar thermal energy, biomass, geothermal, hydrothermal, micro cogeneration or aerothermal (subject to the consideration of renewable energy in electrically driven heat pumpsand) [67].
- The acquisition of photovoltaic systems and solar thermal energy installations should be implemented considering the LCC [66].

So as to set objectives including renewable energies, as the following cases:

- In Catalunya, a forecast has been established to implement until 2022 solar energy installations on building roofs, which generate renewable energy for 2% of energy consumption in buildings and facilities of the Generalitat de Catalunya [41].
- The Junta de Castilla y León establishes the objective of reducing fossil fuel use in administrative buildings, considering biomass as a reference fuel [85].
- In the Illes Balears, the administration is to discard nonrenewable energies, providing for the progressive replacement of installations using fossil fuels by others using renewable energies [37].

Various administrations are acquiring electric power 100% from a renewable origin:

- The electrical power contracted to supply buildings and departments of the Generalitat de Catalunya should be from 100% a renewable origin, and may be reduced to a minimum of 70% established by the Climate Change Law [39], depending on whether certain conditions are present.
- Electricity supply contracts tendered in the Illes Balears should certified energy from 100% recent renewable origins, according to Law 10/2019 [37]; as far as possible, they should be self-supplied using renewable electricity with self-consumption or bilateral contracts. The latter contract with renewable energy producers from the Illes Balears should be promoted.
- The "Green Guide" by the Generalitat Valenciana proposes the addition of electricity generation systems from renewable sources and/or signing a contract with electricity suppliers coming from guaranteed renewable sources, accredited by the National Market Commission and Competency (NMCC) [67].

In order to upgrade the energy efficiency of interior lighting installation, the following recommendations should be adopted: replacing luminaires with LED lighting systems, using control systems, regulating the level of lighting according to natural light, as well as sectorize circuits [67,68,70,81,84].

In terms of energy efficiency, it is also advisable to incorporate energy-efficient sanitary taps [67].

To promote a greater use of electric vehicles, some Administrations are planning a major endowment of charging points than established in the Electrotechnical Regulation for Low Voltage (ERLV) [86]. An example is the Generalitat Valenciana, through the Green Guide [67], and the Generalitat de Catalunya, which contemplates the installation of 200 charging points in Administration buildings [41,87].

It is also advisable to contract energy through a central public procurement service of Administration [30]. The Junta de Castilla y León created the OPTE Platform to facilitate the energy purchase of different Board bodies and energy contracts optimized annually [84].

Other measures related to energy saving are: performing reactive energy correction, applying inmotics with IT tools for monitoring consumption, proposing regular studies of tariff optimization or an analysis and adaptation of contracted power, as well as training users in good energy practice.

Figure 8 shows some examples of proactive energy efficiency policies obtained from the analysis.

Proctive measures for energy saving	
Limitation of energy consumption	• Reduction of a certain percentage of energy consumption in a given percentage of Administration buildings • Improvement of energy rating • Annual energy renovation, around 3 percent of the area occupied by Administration buildings • Nearly zero energy consumption buildings
Energy audits	• In all public buildings or with energy consumption> x kWh / per year
ESCs	• Energy Service Companies or energy manager
Renewable energy	• Promotion of self-consumption: by using renewable energies (photovoltaic, thermosolar, mini-wind, microgeneration, solar thermal, biomass, geothermal, hydrothermal, aerothermal ...) for generating electricity or thermal installations (DHW production and air conditioning), in situ or in the building vicinity • Installation of photovoltaic panels in certain percentage of Administration buildings • Regulation to enable installation of solar / photovoltaic panels on existing building roofs • Procurement of photovoltaic and solar thermal systems considering LCC • Electricity with a guarantee of 100 percent renewable origin
Interior lighting	• LED lighting • Presence control systems and regulation systems • Circuit sectorization
Energy-saving efficient sanitation fitting	• Maximum temperature limiters • Thermostatic adjustment / cold opening • Heat exchangers • Efficient recirculation elements
Charging points for electric vehicles	• Provision of charging points higher than established in the ERLV • Minimum provision in Administration buildings
Energy contracting	• Central recruitment service • Contract optimization
Other measures	• High-efficiency elevators and escalators • Reduction in engine consumption • Reactive energy correction • Inmotics: IT tools for consumption monitoring • Users training in good energy practice

Figure 8. Examples of proactive measures for energy saving proposed by Spanish regional public administrations.

3.2.3. Proactive Measures for Water Saving

Some of the measures proposed are the following:

- Installation of: consumption reduction devices and systems for reuse gray water and/or rainwater [67,69,88], specifying maximum flow rate in taps and maximum flush volume in toilets and urinals, as well as saving devices [67,88,89]; treatment of waste water, efficient irrigation systems even with recycled water [67]; components for saving water such as flow restrictors, the overflow of rainwater collection system with discharge to surrounding surface waters or underground; sustainable urban drainage systems (SUDS) to collect rainwater [70].

- Tap and cistern maintenance to avoid leaks or water loss, and where appropriate, replacement by others that are environmentally advanced [69].
- Systems for network water leak detection [69], specifically in buried pipes [70].
- To conduct a water audit after the first year of building use, in order to continuously monitor performance [69,70].
- Water pressure regulation in collective supply systems to reduce consumption [70,88].
- To landscape green areas of buildings with efficient plants for water saving [67,70].

Some examples of these measures obtained from the analysis are summarized in Figure 9.

Proactive measures water saving	
Water use systems	• Gray water reuse • Rainwater harvesting • Water purification according to supply needs: to flush sanitary fittings, private garden irrigation
Efficient sanitary fittings, toilets and flush urinals	• Faucets with flow limiters, diffusers in sinks and showers, timed taps, mixer taps • Half-load tanks • Toilets and urinals with sensors / urinals without flush
Leak detection	• Detection systems, specifically in buried ducts • Taps and tanks maintenance
Consumption accounting	• Sectorized meters • Audit after one year from the start of using the building
Water pressure regulation	• Pressure regulation in collective systems to reduce consumption
Gardens and vegetation	• Efficient irrigation systems • Native plants with low water requirements

Figure 9. Examples of proactive measures for water saving by Spanish regional public administrations.

3.2.4. Life Cycle

The advisable measures regarding a building's life cycle, which are, therefore, more ambitious, are the following:

- To use LCC for cost calculation, and analyze profitability of sustainable products from a global cost-effectiveness viewpoint. This is already established by regulation in Andalucía [38] and suggested in guides and other provisions [69,90,91].
- To include award criteria and special execution conditions considering the environmental impact generated by products or service during the entire life cycle, as established by regulation in Andalucía [38].
- To have available planning tools against climate change, with binding determinations for different Public Administrations. It is the case of the Andalucía Climate Action Plan (ACAP), being prepared at the time of writing this article [92], or the Strategy for mitigation and adaptation to climate change in the Región de Murcia [93].
- To contemplate measurements of circular economy strategies or plans. For example, the Galicia Circular Economy Strategy [94] sets the following issues for public works: a study of taxes for dumping and waste disposal; training, design and construction with reused materials; the design and use of buildings through passive strategies, clean energy and local materials to reduce consumption of natural resources (energy and water). In addition, it is worth highlighting the building design to increase resilience against extreme meteorological phenomena, such as the frequent floods in the eastern regions [95].

- To design buildings so that future changes of use and distribution can be performed simply [69]: For example, the layout of structural walls, interior partitions with quick and removable mechanical joints, distribution of installations (preferably through technical ceilings and/or floors), etc. [70].
- To design buildings in a flexible way, with spaces allowing new future facilities; in this way, slashes and generation of waste are avoided, and access to facilities for removal is facilitated [70].
- To dimension buildings based on an estimate of uses and needs in the medium and long term, for future reuse [69].
- To study space and equipment needs and decide whether they can be met by internal redistribution of space or a partial/total renovation of existing buildings [69].
- To make resistance demands and durability in structure calculation compatible with sustainability demands, such as optimizing the amount of material [70].
- To calculate the carbon footprint for products, supplies and services available. For example, in the Illes Balears, this requirement has become mandatory for large and medium-sized companies [37]. Additionally, construction materials marketed in Catalunya should include assessment of the carbon footprint on labelling [39].
- To include the need of carbon footprint in products, services and supplies during all phases of the procurement procedure, proving registration in the Carbon Footprint Registry, as established by regulations of the Junta de Andalucía [38].

Figure 10 shows some examples of measurements obtained from the analysis performed.

Life cycle	
Building design and life cycle analysis	• Adaptation to future changes of use and distribution, flexibility • Estimation of medium-long-term uses and needs • Optimization of materials in structural calculation • Calculation of the LCC to obtain total building costs
Product life cycle analysis	• Impact of products throughout entire life cycle • Manufacturing process: reduction of raw materials use and resources, waste reduction in production chains • Distribution: efficient transportation • Promotion of using tools for product LCA
Carbon footprint calculation	• Products and materials with evaluation of visible carbon footprint on labelling • Calculation of carbon footprint of products, services and supplies subject to public tender • Calculation and registration of carbon footprint in large and medium-sized companies
Productive processes	• Production processes reducing GHG emissions • Ecological production process

Figure 10. Examples of life cycle measures proposed by Spanish regional public administrations.

3.2.5. Products (Measures Related to Circular Economy)

In general, many measures were found on specific products following a circular economy model:

- To reuse materials in situ [69].
- To reuse materials from another work, recovery plant or demolition [39,67,69,88,94,96]; to take advantage of materials reusable for building renovation [70].
- To use easy-dismantling products and reusable with minimal alteration [70,88,89,97,98].
- To use recycled products—a measure proposed generically in all regions. In the Green Guide of the Comunitat Valenciana [67], a further step is taken, specifying that using recycled products for building elements reaches a certain percentage of execution cost of the work, justifying the

percentage according to the weight of recycled material in products. This guide also recommends using concrete with recycled aggregate, and includes an annex with the approximate proportion of recycled material in particular products, as well as a list of building elements and recycled materials. In this area, the Sustainable Building and Renovation Guide of the País Vasco [70] suggests that at least 25%, by weight, of raw material has recycled origin.

- To use easily recyclable end-of-life materials [69,88,97,98]. The Sustainable Building and Renovation Guide of the País Vasco [70] offers examples of potential recyclable materials easily separable in the different waste streams after removing from buildings (glass, plastics, wood, metal, stone, etc.).

- To consider components previously repaired [96] or easily repairable [99], as well as remanufactured or renewed ones to extend their useful life [49].

- To use standardized, prefabricated and/or industrialized construction products and elements, due to greater durability, CDW reduction, easy decommissioning and possible reuse [67,69]. The Sustainable Building and Renovation Guide of the País Vasco [70] contains examples of the latter.

- To use quality components and materials with easy replacement, for a maximum durability of buildings [69].

- To avoid using heavy metals due to high pollution [100], particularly in materials and coatings for roofs, facades and installations [70], since rainwater can release heavy metals to rainwater—a vehicle for pollutants.

- To use products that do not contain harmful or dangerous substances [61,69,99], such as chipboards with low formaldehyde emissions and paints that do not contain lead, chromic substances or organic solvents [70]. The Green Guide of the Comunitat Valenciana [67] recommends limiting content and emissions of harmful substances in paints, wood coatings (low VOC and formaldehyde emissions), cork and bamboo flooring, as well as sealants and adhesives.

- To use products from sustainably managed sources [69], such as sustainably managed wood [67,98,100,101].

- To use products that do not require dangerous cleaning products [69], such as ceramic tiles [67]. It is recommended to incorporate ceramic products with heavy-metal free enamels to avoid pollution of water, subsoil or air [70].

- To use products for environmental impact reduction during the different phases of life cycle, including production [69], and certified according to the ISO 14006 standard for Ecodesign [37,70,100].

- To incorporate innovative materials with environmental performance: ceramic tiles with photocatalytic effect, and with bacteriostatic, bactericidal or viricides properties [67]; floors and facade elements that capture CO2, piezoelectric materials, etc. [70].

- To use wood because of its natural, renewable and recyclable properties, with accreditation of legal origin from sustainable forest management [99]. It is advised to consider the limitation of harmful or dangerous substances, treatments against biodegradation, as well as the reaction to fire conditions according to location [68]. Preferably the use of local wood to reduce the impact associated with transportation, as well as reused/recycled whenever possible [70].

- To incorporate recycled aggregate resulting from CDW treatment to minimize the demand for raw material and waste [70,89].

- To use materials extracted and manufactured in the region to reduce environmental impact resulting from transportation [70,88].

- To use eco-products, as they comply with the eco-label criteria [69,102].

- To use products with type I environmental labels, which guarantee environmental impact reduction—a measure suggested in most Spanish regions [70].

- To use products with type III environmental labels, reporting on potential environmental impacts, and promoting transparency in the products for constructing market [67,70].

Some examples of product measurements obtained from the analysis are shown in Figure 11.

Products	
Reused	• Use of reused materials on site • Use of reused materials from another work, recovery plant or demolition • In renovation of building materials themselves
Reusable	• Incorporate with easy dismantling, for reuse with minimal alteration
Recycled	• Elements with recycled materials representing a percentage of total building elements • Recycled material content: percentage by weight (> 25 percent) • Concrete with recycled aggregate • Recycled aggregates from construction and demolition waste
Recyclable	• Use of easily recyclable materials at the end of their useful life
Lifetime extension	• Easily repairable products • Incorporate repaired components • Remanufactured or Refurbished Products • Quality materials and easy replacement to extend the life of buildings
Standardized, prefabricated and / or industrialized	• Standardized, prefabricated and / or industrialized products and elements: greater durability, reduction of waste generation on site, ease of dismantling and possibility of reuse
Sustainable	• Wood from sustainably sources
Limitation of harmful or dangerous substances	• Avoid using heavy metals due high pollution • Boards with low formaldehyde emissions • Paints that do not contain lead, chromic substances or organic solvents • Wood, cork and bamboo coverings
Easy cleaning and hygiene	• Use of products that do not require dangerous cleaning products, such as ceramic tiles (containing heavy metal-free enamels)
With environmental benefits	• Flooring and façade elements that collect CO2 • Ceramic tiles with catalytic effect • Piezoelectric materials
Proximity	• Use of materials extracted and manufactured in local places to avoid environmental impact resulting from transport
Environmental evaluation	• Products friendly to criteria set by eco-labels • Products with environmental labels (type I or III)

Figure 11. Examples of product measurement proposed by Spanish regional public administrations.

3.2.6. Waste Management Measures

A proper management of CDW is essential for a transition to circular models in the building sector. The recommendations found in this area are:

- To include public procurement in sustainable waste management strategies [97], with specific clauses promoting circular economy: reuse and recycling of waste, sustainable products, service purchase instead of product purchase, etc.
- To minimize generation of waste during construction works (with on-site reuse) and at the end of the building's life (facilitating selective deconstruction) [67,69]. To elaborate a waste management plan to identify opportunities for reuse, recycling and recovery [67,103]. Selective collection of all waste and further recovery by authorized managers. This is a requirement suggested in all regions, especially the recovery of mineral waste [70] or used products by the work contractor [88,91].
- To prepare a percentage of CDW for reuse, recycling and other ways of recovery. To do this, an audit prior to demolition is necessary to determine materials to be reused or recycled [67,103].

- Building design to favor selective demolition [69].
- To reuse stone waste and soil remains in the same work as filling material [69,70,104,105]; this measure was already stated in Decree 200/2004 of the Comunitat Valenciana [106]. It should be noted that the use of nonhazardous waste as filling material for regeneration of degraded areas (mining operations, landscaping engineering works) is considered as a recovery operation, and meets the requirements of Royal Decree 105/2008 [107].
- To reduce packaging waste: large capacity packaging [69]; reusable packaging [70,99]; packaging collection by providers for reuse or recycling [67,88]; bulk products [99], such as concrete and mortars prepared in plant; reuse pallets on site [67]; supply products in containers and dispensers [70,101].
- To promote separation of ordinary waste generated in buildings with facilities for this purpose [67].

Figure 12 shows some examples of waste-management measures obtained from the analysis.

Construction and demolition waste (CDW)	
Strategies	• Building design to favour selective deconstruction • Substitution of product purchase by service contract
Sustainable waste management during construction works	• Reduction of waste generated (reuse in situ) • Waste management plan identifying opportunities for reuse, recycling and waste recovery • Selective collection of all waste and management through authorized managers, in particular mineral waste • Reuse and recycling of waste • Reuse of stone waste and remains of land from site work as filling material • Pre-demolition and clearing audit
Reusable, recycled and bulk packaging	• Collecting containers and packaging by suppliers suitable for recovery • Supply Product supply in paper, cardboard, plastic or wood packaging, using recycled material: percentage of total products • Ready-mixed concrete prepared in plant • Mortars in dry premixed form
Waste storage	• Space in buildings for ordinary waste separation

Figure 12. Examples of construction and demolition waste (CDW) management measures proposed by Spanish regional public administrations.

3.2.7. Environmental Assessment and Monitoring Measures

In this area, some recommendations found are the following:

- To request an environmental management plan for work execution. This should consider water accounting and energy consumption, as well as a calculation of CO_2 emissions associated with material and transportation, even a minimization of impacts on the close environment (noise, dust, dumping, damage to vegetation, etc.) [69,70,88,100]. Actions to avoid those impacts can be washing vehicle wheels, land watering or shielding works [70].
- To prepare specific documentation containing all the environmental measures of the project to ensure implementation during the construction phase [70].
- To install sectorized meters for different networks (water, electricity and/or gas supply) and circuits (lighting, plugs, exterior lighting, etc.) to test the effectiveness of the project's environmental measures [69].

3.3. Application of Environmental Measures in the GPPRB: Examples

As mentioned above, the tenders incorporating environmental measures with relevance in award criteria are still insufficient [14]. However, both in the EU and Spain, there are many success cases in which these measures are defined and valued. Below, three examples of good practice in Spanish regions are described:

- Refurbishment of the façade in a warehouse converted into a youth center in Alzira, Comunitat Valenciana [108]: in the tender, the adoption of comprehensive solutions was established as a priority criterion, as well as a demonstration of energy efficiency and LCC, by using common verification tools for all participants. Based on a theoretical starting solution, each bidder was asked to prove the energy performance improved on the solution proposed. Compared with the baseline model, the solution contracted for this refurbishment showed a 15% reduction in heating demand (from 63 to 54.6 kWh/m2 per year) and a 19% reduction in cooling demand (from 34.1 to 27.5 kWh/m2 per year). Furthermore, the determination to sustainably renovate an existing building rather than undertake a new construction also reduced CDW and raw material extraction. The tender emphasized the ease and low-energy intensity of maintenance as well as material guarantees, therefore minimizing future costs.
- Energy self-sufficient hospital in Ourense, Galicia [109]: the main challenge for the contracting body was to optimize energy consumption in the extension of the existing hospital. For this, two complementary tenders were called, focused on both improving the energy facilities and the software monitoring energy management system. The main award criteria used were based on the performance of the energy equipment proposed (electrical and thermal), CO2 emissions produced by facilities, electrical efficiency of the cogeneration engine, thermal efficiency of the biomass boiler, as well as chiller absorption.
- Refurbishment of the town hall building in Amorebieta-Etxano, País Vasco [110]: prior to the tender, the City Council commissioned a preliminary study to analyze three possible scenarios: complete refurbishment, partial reform or new construction. It was concluded that the best option was the first one but maintaining the original two-story structure and adding a third one. The criterion for this choice was the saving of 20% of resources, compared to the new construction. Based on these results, the City Council launched the tender, taking into account "sustainability criteria, thermal and light comfort and a building integration in urban landscape", among others. The successful tenderer designed and built a building with a high environmental sustainability rating, according to the building rating system in the País Vasco. Some of the environmental measures to favor this rating were: maintenance of the main facade; reuse of materials from the building itself according to the concept "cradle to cradle" or "the best product is the one that is already there". For this, a selective demolition of the building was carried out for material recovery, such as stone.

In the three examples, the valuation of environmental measures on other criteria was much higher than 10% of the current EU average, as mentioned in the introduction.

4. Discussions

From the analysis of the source information on energy efficiency and circular economy in the GPPBR within Spanish regions, a series of questions arise about barriers to application and opportunities for improvement.

4.1. Are All the Measures Feasible? Is the Sector Well-Prepared?

In the measures analyzed themselves, barriers to application of different kinds are observed.

In the case of passive energy efficiency measures, there might be difficulty in implementing them in building renovation. Notably, bringing in thermal insulation or solar shadings in facades as well as renewable energy facilities on roofs may be restricted by municipal regulations. Therefore, local regulations should be modified, based on higher ranking guidelines and standards. For example, the Decree-Law 16/2019 [111] of Catalunya provides the placement of facilities for solar energy on roofs of existing buildings. For that purpose, specific terms were imposed, with no limitation of urban plan requirements.

Regarding renewable energies, there is controversy over the technical solvency criterion referring to the fact that electric power trading companies have the A label issued by the CNMC. This requires the electricity marketed to be of 100% renewable origin [112]. On the other hand, in order to make energy 100% renewable by 2050, not only the technology and price reduction are needed, but also political will. Regarding decarbonization in the Spanish building sector, it is essential to increase the level of electrification in energy demand, and prioritize renovation.

It is found that there are not very many measures related to the life cycle analyses, although these are vital to optimize results [3,4]. Furthermore, carrying out the LCC requires experience and competence. However, the sector appears not to have fully matured [69].

In relation to gray water reuse systems, these demand the installation of a treatment tank. Therefore, the practical application of this measure can be conditioned by available space [70].

Additionally, barriers caused by shortcomings in the construction product market itself are detected. In the case of products using recycled material, raw materials can have a lower cost. This is due to the additional costs of dismantling, separating and transforming the waste. On the contrary, there are no additional charges on raw materials extraction or CDW disposal in landfill. This last issue should be tackled, as specified in the Galicia Circular Economy Strategy 2019–2030 [94], although this measure is planned to be implemented in some regions [99].

With reference to concrete using recycled aggregate, there might be a lack of production and supply in the vicinity of the work or the region itself. Besides, the CDW recovery plants might not exist to manufacture recycled concrete aggregate in the concrete plant environment. In view of these conditioning factors, environmental impacts of transportation can exceed the benefits achieved through the measure itself [67]. The drawbacks might come up in relation to sustainable products, such as products using recycled material, sustainably managed wood, industrialized components, etc. For this reason, it is advisable to check availability in the environment around the site work before ordering in the contracting specifications.

There are also other scarce and scattered products on the market, such as eco-products guaranteed by Type I eco-labelling. To make their demand feasible in the GPPBR, it would be necessary to create an updated registry of sustainable products to easily identify companies [67] that market them. In addition, the continuous evolution of the market makes it necessary to permanently review measures related to eco-products, reused or renewed products. There may even be regulatory changes such as the introduction of requirements on recycled content in the Regulations for Construction Products [3].

There are also difficulties in collection by suppliers of products, containers and packaging for reuse or recycling. This measure requires the product supplier to guarantee logistical facilities for reuse or recycling. If not, the supplier should deliver waste to an authorized manager. However, this practice can increase environmental impact due to transportation.

Finally, bringing in reused products with a short useful life can cause negative effects, since this implies an early replacement or repair. Thus, to propose this measure, useful life of products in their new use should be considered.

4.2. How to Apply the GPPRB in the Rehabilitation of Historic Buildings?

To the difficulties mentioned above, some limitations must be added in interventions within historic buildings. In this case, instead of indiscriminately applying criteria established in the GPP, the contracting bodies should draw up a detailed study of each building requiring, if appropriate, advice. In general, feasible measures should be implemented, with flexibility in case they conflict with the essence of the building itself and, if appropriate, compensate with additional measures. Likewise, communication between all those involved in the rehabilitation should be fostered in order to better integrate environmental criteria.

4.3. Does Public Procurement Itself Favour Sustainability and Innovation? How to Increase the Participation of Small and Medium-Sized Enterprises (SME)?

Barriers are also observed in the technical terms proposed in procurements, which prevent environmental criteria from having a greater impact on proposal evaluation. Therefore, it may be appropriate to increase the relevance of these measures in valuation criteria, so that the total cost is not higher than 50%, so as not to be decisive for tenders.

On the other hand, as shown in the results, a greater impulse of the GPPRB is achieved when environmental measures are mandatory through normative provisions. To this effect, it is necessary that measures are linked to the subject of procurements, and clearly defined in procedures.

In the current frame of economic recession, SMEs are the driving force for the economy [113]. So as to prevent SMEs from being left out of larger procurement, and consequently limit the number of lots awarded to the same company, procurements can be divided into batches. Likewise, to contribute to economic development and job creation in regions during procurement execution phase, it is advisable to invite regional SMEs involved in the innovation sector linked to sustainability [113]. In this sense, to promote the public procurement of innovative solutions, various Administrations should provide support to companies and research institutions, able to provide technological solutions or advice on innovation.

4.4. Do Public Administrations Meet Their Own Challenges?

Public procurement opens up a great opportunity to renovate buildings according to principles of energy efficiency and circular economy. Additionally, this task is complex, time passes and not all administrations seem to keep up with the ambitious goals set. Generally, an administration efficient enough to speed up long procurement processes is needed.

Regarding the improvement of the GPPBR, the Administration can undertake actions in nearly all areas: information, coordination and monitoring, training, awareness, dissemination, as well as financing (Figure 13).

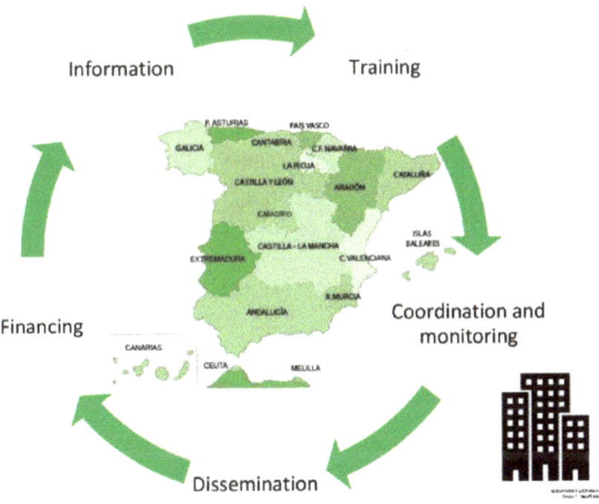

Figure 13. Actions of the Administration to improve the GPPBR in Spanish regions and autonomous cities (Map of Spain obtained from: Ministerio de Agricultura, Pesca y Alimentación; icon attribution: buildings by Untashable from the Noun Project).

4.4.1. Is There Quality Technical Information to Guide the GPPBR?

Generally, the information source found is scattered in different documents, and there are few specialized guides in the GPPBR. Due to this, it is vital to draft specific guides for an easy and feasible incorporation of environmental measures in the GPPRB, according to the type of building (teaching, administrative, health, etc.).

The proposed measures should be well-defined to clearly understand how they can be met and verified in the different phases of the GPPBR. With regard to the latter, some guides mentioned above [67,69] already include more precise and detailed information. Furthermore, with a better definition, consultations or revisions that might lengthen procedures could be avoided [114].

It would be very useful if measures had associated indicators, objectives or guidelines, such as value for money, cost-effectiveness, impact on the total cost of investment, savings generated, technologies involved, implementation period, number of citizens affected, etc. In this way, there would be criteria to bring in measures according to the type of work. Additionally, product performance suggested should be linked to its final functionality, for example photocatalytic effect, bactericidal or bacteriostatic surfaces, etc.

In order to become more functional, measures should include references to web sites or other guides for further information, and also a provision of good practices with real cases of application. It would also be desirable to connect measures with supplying companies that could evaluate practice and generate an "e-marketplace". In addition, to facilitate the implementation of GPPBR measures in small towns, regional administrations should draft sheets with practical, useful and schematic information to help suppliers make decisions, all based on a Think Tank for each municipality. If the contracting body did not have sufficient resources, it could hire a consulting service to better define measures according to the type of building, needs, environmental commitment or relationship between environmental benefits and economic costs.

In general, with regard to viability of environmental measures related to products, it is essential to know in advance the regional market in terms of construction products, especially the innovative ones. Furthermore, innovative measures need to enable better solutions with optimized costs [115]. For example, in the PPI strategy of the Junta de Andalucía, public tender processes are highlighted to promote start-up markets [116]. Likewise, progress should be made in research to propose the most effective measures considering the life cycle approach, and thus find optimal solutions in building renovation. For example, Level (s) is the EU tool recommended for integrating life cycle assessment into GPP [3]. Moreover, given the growing number of certification systems for building sustainability, they could be regarded as a reference.

Finally, it would be practical to have digital guides for updating and improving measures on a regular basis. In addition, it would be very convenient to add monitoring and inspection of environmental measures to procurements. In this connection, effectiveness and implementation could be easily verified, as well as identifying shortcomings that can improve environmental performance.

4.4.2. From Theory to Practice: Is There Sufficient Coordination among Public Administrations? How to Detect if Tools Are Being Applied?

To increase effectiveness of actions, joining forces is essential, along with coordination by different administration departments, and the different administrations, as well as implementing the monitoring of actions realized.

Because of the political organization of Spain at a territorial level, a large part of the definition in the programs for GPP, implementation and monitoring is performed by the regions themselves. This decentralization enables regional administrations to draw up laws, regulations and guides within the state regulatory framework, resulting in a large amount of documentation, as mentioned above. This requires well-trained professionals and a high level of organization. For this reason, and due to the different resources of regions, greater collaboration and coordination between these and the State should be desirable. Certain aspects of the GPPBR could be addressed centrally: for example, by setting up a national agency in which all regions were represented to support procurement authorities through guidance, training, and documentation as a common reference, especially guides. Based on them, regional administrations could draw up additional papers to address specific aspects of each area, if necessary. Thus, these tools could enhance the different regional economic bases, support specialization of various productive sectors and favor local products.

On the other hand, within the scope of the General State Administration, an interministerial commission was set up to bring criteria into the GPP [35]. In the same vein, it is necessary to create an interdepartmental committee for each region, in order to coordinate the onboarding of ecological criteria, development of strategies for GPP and monitoring.

With the aim of detecting the degree of implementation of the GPP goals and execution of actions, an annual monitoring concerning the application of environmental measures in the public procurement awarded should be carried out [89,117]. For this purpose, the interdepartmental committee should add results provided by departments for the GPP, with results provided in a standardized format or use platforms and tools for integration [118,119].

Regarding specific measures, for environmental enhancement check-up during the use phase of the building, actions such as checking consumption or conducting a survey for building users to analyze the perception of comfort should be undertaken [70]. A key factor in the environmental impact of buildings is energy use during occupation. The monitoring of CDW management would be required, since waste traceability is lost due to proliferation of illegal dumping [120].

4.4.3. Is There Enough Training for Professionals to Be Qualified?

One of the main problems detected for measure implementation is a lack of training for all professionals involved in the GPPBR process. Given the novelty and complexity of the subject, particularly in aspects related to the circular economy, it is needed to boost actions for an exchange of technical knowledge. Staff training in environmental criteria is one of the first steps to increase their incorporation into contracts.

For all this, training plans on environmental measures should be proposed, as well as tools developed, as part of procurement documents, including cases of good practice. In particular, emphasis is needed on the relevance of specific energy training plans and eco-labelling product presentations, given that both construction companies and designers should familiarize with these [41,69]. Training administration staff in the "Building Information Modeling" (BIM) work methodology is essential. This methodology is a requirement for public funding projects. At the State level, an Interministerial Commission was created for bringing BIM methodology into public procurement [121].

4.4.4. Are We All Sufficiently Involved?

A solid communication plan of actions carried out in the GPPBR can maximize impacts, by setting up web platforms to ensure visibility of hiring. Thus, the EU GPP Good Practice [122] platform presents success stories, in which improvements are emphasized, which is a feature inspirational for their replicability.

Likewise, in order to gain complicity of citizens, tools should be created to communicate the beneficial effects of the GPP. For example, an information panel could be displayed on energy efficiency actions, renewable production and other circular economy measures applied [37]—the same as in certain public buildings, where it is mandatory to display energy efficiency labelling [123]. Occupants in buildings could also receive information on carbon footprint over their lifetime, by installing automated systems to obtain data on electricity use, air quality and building waste production. All this leads to awareness raising on environmental impacts generated by buildings and behavior-correction, where appropriate. Additionally, user participation tools can be implemented in the phases of planning, execution and evaluation of the projects, as contributors to needs and ideas [109]. This direct involving encourages a greater acceptance of the results of the GPPBR.

4.4.5. Is Sufficient Funding Planned?

Faced with the serious social and economic crisis caused by the COVID-19 pandemic, in order to access the EU reconstruction funds, we come across both the constraint and opportunity that "the recovery is green". This implies that part of the EU aid must be invested in actions of the Green Deal to combat climate change and achieve the EU decarbonization objective from the present to 2050.

5. Conclusions

Scientific literature indicates that the GPPBR represents an opportunity to implement environmental policies in a prominent activity such as public procurement in construction—a key economic sector in Spain. However, this analysis concludes that the inclusion of environmental measures in public procurement for building rehabilitation is low, and the total costs continue to be weighted much higher than environmental criteria. Among the possible causes of this scenario, various studies note a lack of knowledge of parties involved, because of the absence of practical information, among other reasons.

Taking the State regulatory framework as a reference, Spanish regional administrations within respective competencies draw up laws, regulations and guides on GPP, climate change, environment, energy efficiency, circular economy, waste and innovation. By analyzing this documentation, it was possible to collect a large amount of information in the form of environmental measures to bring in the GPPBR specifications. It is worth noting the difficulty in finding specific and quality information on environmental measures to be incorporated into the GPPBR. Most instruments refer to GPP generically.

The study shows the need to develop advanced technical procedures. It is evident that these should compile ecological measures to put into practice in the GPPBR with a required level of definition. This way, the agents involved will understand how to comply with them and verify in the different phases of tender procedures. From the information consulted, the great support found in references to real cases of good practice, and even specific guides for further information is remarkable. It is also noted that if measures were associated with indicators, contracting authorities could add to them with better criteria. In addition, the information on the feasibility of applying measures is a key factor, since there might be barriers that should be detected before being proposed in the tender specifications.

It is found that the Administration plays a vital role in promoting the GPPBR. There can be no doubt that environmental measures are incorporated, and monitoring is carried out when there are regulatory provisions that comply with them. Additionally, with greater collaboration and coordination between regional administrations and the State, there would be sufficient resources to develop the mentioned technical procedures, especially guides. These could serve as a reference for regions to adapt to their specific features. Once information on environmental measures has been gathered in the GPPBR, it is essential to make this widely known to professionals and provide awareness and dissemination campaigns for everyone.

Based on this approach to environmental measures includible in the GPPBR, the necessity of future research is significant. A detailed study could be made on environmental measures in the procurement specifications of regional public administrations, as well as their assessment. It is also considered worthwhile to identify what the most effective strategies are after application, with an analysis of the results. Delving into research on good practice examples is always encouraging to promote further initiatives. Likewise, it is possible to deepen the analysis of life cycle costs of existing products on the market regarding building renovation, as well as the development of innovative products. Eventually, an in-depth investigation should be carried out on the real impact of GPPBR environmental measures on the reduction in environmental consequences caused by the constructing sector in Spain.

Author Contributions: All authors have contributed equally to the investigation and final manuscript. All authors have read and agreed to the published version of the manuscript.

Funding: This research received no external funding.

Conflicts of Interest: The authors declared no conflict of interest.

Appendix A

Table A1. Papers relating to the GPPBR developed by regional Public Administrations.

Regions		Papers: Laws (L), Regulations (R) and Guides (G)
Andalucía	L	• Law 8/8 October 2018, on measures to combat climate change, for transition to a new energy model in Andalucía. • Draft on Circular Economy Law in Andalucía. 2020.
	R	• Decree 169/31 May 2011, approving the Regulation for the Promotion of Renewable Energies, Savings and Energy Efficiency in Andalucía. • Andalucía Circular Bioeconomy Strategy (approved by Agreement of the Governing Council on 18 September 2018). • Public Procurement Strategy for Innovation in the Public Administration of the Junta de Andalucía (approved by Agreement on 4 September 2018). • Andalucía 2020 energy strategy (approved by Agreement on 27 October 2015). • Action Plan 2018–2020 for the Andalucia Energy Strategy (approved by the Evaluation Body of the Ministry of Employment, Business and Commerce on 17 December 2018). • Agreement 18 October 2016, by the Governing Council, promoting the incorporation of social and environmental clauses in the contracts of Andalucía. • Draft Decree approving the Comprehensive Waste Plan for Andalucía. Towards a Circular Economy in the Horizon 2030 (PIRec 2030).
	G	• Basic Project Guides (Greens). Horizon 2020. 2018. • Guide for good practice in green public procurement (final result of GPP4 Growth Activity 1.2 within the Interreg Europe program). 2017. • Guide for the inclusion of social and environmental clauses in the Junta de Andalucía procurement 2016.
Aragón	R	• Decree 46/1 April 2014, by the Government of Aragón, regulating actions in the field of energy efficiency certification in buildings and creating registration within the Region of Aragón. • Order HAP/522/7 April 2017, publicizing the Agreement on 28 March 2017, by the Government of Aragón, adopting measures for the strategic use of public contracts in support of common social objectives and deficit reduction in the Autonomous Community of Aragón. • Circular Aragón strategy. 2020. • Aragón Climate Change Strategy. Horizon 2030. 2019. • Aragón Energy Plan 2013–2020. 2014.
	G	• Green purchasing. Green Public Procurement and Contracting. 2nd Catalog of criteria, products and suppliers. 2009. • Guide for good environmental practice in contracting by public administrations. 2004.

Table A1. *Cont.*

Regions		Papers: Laws (L), Regulations (R) and Guides (G)
Islas Canarias	**L**	• Draft for the Islas Canarias Climate Change Law. 2020. • Prior public consultation on the Circular Economy Law of the Islas Canarias. 2020.
	R	• Islas Canarias Energy Strategy 2015–2025. 2017. • Draft for the Integral Waste Plan in the Islas Canarias.
	G	• Sustainable public procurement and procurement guide by the Islas Canarias Government.
Cantabria	**R**	• Decree 75/23 May 2019, establishing general policy guidelines on the incorporation of social criteria and clauses in contracting the public sector in Cantabria. • Decree 32/12 April 2018, approving the Action Strategy against Climate Change in Cantabria 2018–2030. • Decree 14/23 March 2017, approving the Waste Plan in Cantabria 2017–2023. • Decree 35/10 July 2014, approving the Energy Sustainability Plan of Cantabria 2014–2020.
	G	• Guidelines for energy efficiency: Public services. 2013.
Castilla La Mancha	**L**	• Law 7/29 November 2019, on Circular Economy in Castilla-La Mancha. • Law 1/2007, 15 February 2007, on the promotion of Renewable Energies and Incentive Plan for Energy Saving and Efficiency in Castilla-La Mancha.
	R	• Decree 78/2016, 20 December 2016, approving the Integrated Waste Management Plan for Castilla-La Mancha. • Decree 29/2014, 8 May 2014, regulating actions regarding certification of energy efficiency in buildings in Castilla-La Mancha and creating the Regional Registry for Energy Efficiency Certificates in Buildings of Castilla-La Mancha. • Order 4/18 January 2019, by the Ministry of Agriculture, Environment and Rural Development, approving the Climate Change Strategy in Castilla-La Mancha, Horizons 2020 and 2030. • Resolution 19 October 2016, by the General Secretariat for Finance and Public Administrations, which provides for the publication of the Instruction of the Governing Council 18 October 2016, on social clauses, gender and environmental perspective in contracting the regional public sector.
	G	• Relevant contributions to energy saving and efficiency in buildings by local administration. 2018. • Social and environmental criteria in public procurement (Federation of Municipalities and Provinces Castilla-La Mancha). 2010.

Table A1. *Cont.*

Regions		Papers: Laws (L), Regulations (R) and Guides (G)
Castilla y León	I	• Decree 11/20 March 2014, approving the Regional Plan of Sectorial Scope called "Comprehensive Waste Plan of Castilla y León" • Agreement 26/4 June 2020, by the Junta de Castilla y León, approving measures against climate change within Castilla y León. • Agreement 2/18 January 2018, by Junta de Castilla y León, approving the Energy Efficiency Strategy of Castilla y León 2020. • Agreement 128/26 November 2009, by the Junta de Castilla y León, approving the Regional Strategy against Climate Change 2009–2020. • Proposal for Circular Economy Strategy in Castilla y León 2020–2030 (processing public information).
	G	• Practical guide on bioclimatic construction solutions for present architecture. Castilla y Leon meeting. 2015. • Photovoltaic solar energy guides. EREN. 2013. • Green buying guide. University of Valladolid. 2007.
Catalunya	L	• Law 16/2017, on climate change.
	R	• Decree Law 16/2019 on urgent measures to face climate emergency and the promotion of renewable energies. • GOV/55/2020 Agreement, 31 March, approving the objectives and minimum content of the Catalunya Public Procurement Strategy. • Agreement GOV/84/11 June 2019, to promote strategic public procurement for innovation in the Administration of the Generalitat de Catalunya and public sector. • According to the Government of Catalunya, approvement of the Energy Efficiency and Saving Plan in buildings and facilities, Generalitat de Catalunya (GENERCAT), in the energy transitional framework in Catalunya, 2018–2022, 4 December 2018, period 2018–2022, 4 December 2018. • Strategy to Promote Green Economy and Circular Economy by the Government of Catalunya (approved by Government Agreement GOV/73/26 May 2015). • Catalunya strategy for energy renewal in buildings (ECREE) (approved by the Government's cord on 25 February 2014). • Catalunya ecodesign strategy for a circular and eco-innovative economy (2014). • Catalunya 2020 general waste and resource prevention and management program (PRECAT20).
	G	• Guide for the drafting of provisions of particular administrative specifications and technical requirements on energy performance contracts under guarantee provisions, subject to harmonized regulation (service contracts). 2018. • Col·lecció Quadern Pràctic (technical reference guides for professionals in energy sector). • GPP 2020 project. Procurement for a low-carbon economy. Final Results. 2016.
Ceuta	R	• Public cleaning and waste management ordinance. 2020. • Ecological action plan to improve environmental quality (under preparation).
	G	• Works Inspection Guide.

Table A1. *Cont.*

Regions			Papers: Laws (L), Regulations (R) and Guides (G)
Comunidad de Madrid	L	•	Draft Law on Climate Change and Energy Transition.
	R	•	Resolution 5 March 2020, by the President of the Administrative Contracting Advisory Board, providing adjustments in the models of specifications, in particular administrative clauses of general application informed by the Administrative Contracting Advisory Board. • Sustainable Waste Management Strategy for the Comunidad de Madrid 2017–2024 (Approved by Agreement 27 November 2018). • Agreement 3 May 2018, by the Governing Council, establishing the reservation of public contracts in favor of entities of social economy and promoting the use of social and environmental clauses in public procurement within the Comunidad de Madrid. • Air Quality and Climate Change Strategy of the Comunidad de Madrid (2013–2020). Blue Plan + (approved by Order 665/3 April 2014). • Energy Saving and Efficiency Plan in Public Buildings in the Comunidad de Madrid. 2017. • Energy Plan of the Comunidad de Madrid Horizon 2020.
	G	•	General Guide on Environmental Aspects for companies. Comunidad de Madrid. 2020 review. • MADRID7R Circular Economy Campaign. Comunidad de Madrid. 2017. • Frequently asked questions and quick answers for a Responsible Public Procurement. 2017.
Comunitat Valenciana	L	•	Draft of the Valencia Law on Climate Change and Ecological Transition. • Decree-Law 14/7 August 2020, by the Consell de la Generalitat, on measures to stimulate the establishment of facilities for the use of renewable energy due to climate emergency and the need of urgent economic recovery.
	R	•	Decree 55/5 April 2019, by the Consell, approving the revision of the Integral Waste Plan in the Comunitat Valenciana. • Decree 39/2 April 2015, by the Consell, regulating certification in energy efficiency in buildings. • Decree 200/1 October 2004, by the Consell de la Generalitat, regulating the use of suitable inert waste in restoration, conditioning and filling works, or construction purposes. • Order 19 October 2004, by the Ministry of Territory and Housing, on environmental requirements and criteria to be introduced in administrative clause specifications governing contracts of the Ministry of Territory and Housing, autonomous entities and public law entities—linked or dependent. • Agreement 16 December 2016, by the Consell, approving the Plan for energy saving and efficiency, promotion of renewable energies and self-consumption in buildings, infrastructures and equipment in public sector of the Generalitat Valenciana. • Valencia Strategy for Climate Change and Energy 2030. 2020.

Table A1. *Cont.*

Regions		Papers: Laws (L), Regulations (R) and Guides (G)
	G	• Green Guide, environmental measures in public procurement in the field of Constructing. Generalitat Valenciana; IVE, 2020. • Guide on circular economy. Making the Circular Economy Work. Guidance for regulators on enabling innovations for circular economy (prevention and waste recycling). 2019. • Passive design strategy guide for buildings. Forum for sustainable building. Comunitat. • Valenciana, IVE 2014. • Guide for the incorporation of renewable energies in buildings. Forum for sustainable building Comunitat Valenciana, IVE 2012. • Procedure to Promote Public Procurement of Innovative solutions (PPI).
Extremadura	L	• Law 12/26 December 2018, regarding socially responsible public procurement in Extremadura.
	R	• Decree 115/24 July 2018, regulating actions on energy efficiency certification of buildings Extremadura, creating the Registry for Certifications of Energy Efficiency in Buildings. • Resolution 25 February 2016, by the Counselor, providing publication of the Agreement by the Governing Council of the Extremadura Regional Government, 23 February 2016, approving the Instruction for incorporating social and environmental criteria, promotion of SMEs and promotion of sustainability in public procurement of the Junta de Extremadura and public sector entities. • Instruction 01/2020 on processing self-consumption facilities in Extremadura. • Energy efficiency strategy in public buildings of Extremadura Administration 2018–2030 (approved by the Governing Council on 28 November 2018). • Extremadura Climate Change Strategy 2013–2020 (approved by the Governing Council on 4 January 2014). • Green and circular economy strategy Extremadura 2030. 2018. • Extremadura Integrated Waste Plan 2016–2022. (approved by Agreement of the Governing Council, 28 December 2016). • Draft for the Extremadura Integrated Energy and Climate Plan 2021–2030. 2018.
	G	• Models of specifications for public works, supply and service contracts. Junta de Extremadura. 2019 and 2020.
Galicia	L	• Law 14/2013, December 26, for rationalization of autonomous public sector.
	R	• Decree 130/3 October 2019, approving the Interdepartmental Commission for Promotion and Coordination of the Galicia Strategy for Climate Change and Energy 2050. • Decree 128/2016, regulating energy certification of buildings in Galicia. • Galicia Strategy for Climate Change and Energy 2050 (spproved by the Council of the Xunta de Galicia on 3 October 2019). • Integrated Regional Plan for Energy and Climate 2019–2023 • Energy Saving and Efficiency Strategy in the Autonomous Public Sector of Galicia 2015–2020. • Galicia circular economy strategy 2019–2030.

Table A1. *Cont.*

Regions		Papers: Laws (L), Regulations (R) and Guides (G)
	G	• Guide for good practice to favor Public Procurement of Innovative solutions in Galicia. 2016. • Guide for a socially responsible public procurement in the Galicia public sector.
Illes Balears	L	• Law 10/22 February 2019, on climate change and energy transition. • Law 8/19 February 2019, on waste and contaminated land of the Illes Balears.
	R	• Instruction for socially responsible and environmentally sustainable contracting in the Insular Council of Menorca. 27 October 2018.
La Rioja	R	• Agreement on the integration for environmental aspects in public procurement of La Rioja Administration, Delegate Commission of the Government of Acquisitions and Investments, 28 February 2003. • La Rioja Energy Plan 2015–2020. • Strategic approach for energy policy 2015–2025. • Local Plan 21 Action Plan in La Rioja. • Bases for Sustainable Development Strategy 2020–2025.La Rioja. • La Rioja Waste Master Plan 2016–2026.
	G	• Public Procurement of Innovative solutions. Government of La Rioja.
Melilla	R	• Agreement by Honorable Assembly, 29 October 2019, regarding climate emergency in the Autonomous City of Melilla. • Ordinance for the Protection of Public Spaces in Relation to Cleaning and Waste Removal. • Regional operational program ERDF of the Autonomous City of Melilla 2014–2020.
Navarra	L	• Regional Law 2/13 April 2018, on Public Contracts.
	R	• Navarra climate change roadmap. 2017–2030–2050. • Navarra Energy Plan. 2030 Horizon. • Developing Plan for Circular Economy in Navarra 2030—Actions. • Navarra Waste Plan 2017–2027.
	G	• Responsible purchase. Region of Navarra. Environmental Clauses

Table A1. *Cont.*

Regions		Papers: Laws (L), Regulations (R) and Guides (G)
Pais Vasco	L	• Law 4/21 February 2019, on Energy Sustainability of the Pais Vasco.
	R	• Decree 25/26 February 2019, concerning energy efficiency certification in buildings in the Pais Vasco, control and registration procedure. • Recommendation 2/21 June 2018, on public procurement advisory board. Government of Pais the Vasco. Purpose: environmental clauses in public procurement. • Climate Change Strategy 2050 of Pais Vasco. • Euskadi Energy Strategy 2030. • Euskadi Circular Economy Strategy 2030. • Waste prevention and management plan for the CAPV 2020.
	G	• Environmental Criteria. 4.17. New construction of administrative and office buildings. Ihobe, Government of País Vasco. • Environmentally sustainable building guides, Ihobe, Government of País Vasco. 2010. • Sustainable building and renovation guide for housing in PaisVasco. 2015. • Public Procurement of Innovative solutions. Proposal paper. Ihobe. Government of País Vasco. 2016.
Principado de Asturias	R	• Agreement 16 November 2017, by the Governing Council, declaring Public Procurement of Innovative solutions as strategic objective for the Principado de Asturias Administration, creating the Commission for promoting public purchase of innovation. • Sectoral measures to combat climate change in the Principado de Asturias. • Setting-up the Joint Commission to evaluate impact of energy transition in the Principado de Asturias. Executive paper. Current situation, forecast and recommendations. October 2019. • Asturias Paradise Hub 4 Circularit. Economic Development Agency of the Principado de Asturias. • Strategic waste plan for the Principado de Asturias 2017–2024. • Agreement on 16 November 2017, by the Governing Council, declaring Public Procurement of Innovative solutions as strategic objective for the Administration of the Principado de Asturias and creating the Commission for promoting Public Procurement of Innovative solutions. • Public Procurement of Innovative solutions. Economic Development Agency of the Principado de Asturias (IDEPA).
	G	• Preliminary Need File for the project on Public Procurement of Innovative solutions. Government of the Principado de Asturias. • Practical guide for including social and environmental responsibility clauses in administrative contracting of the Principado de Asturias Administration and public sector. Governing Council Agreement, 3 May 2018.
Región de Murcia	R	• Strategy for mitigation and adaptation to climate change in the Región de Murcia. • Table of contents on circular economy strategy in the Región of Murcia (ESECIRM) 2017–2030. Draft. • Energy Plan. Región of Murcia 2016–2020. • Waste Plan for the Región of Murcia 2016–2020. • Research and Innovation Strategy for Smart Specialization 2014–2020.

References

1. European Parliament. Resolution of the European Parliament on Plenary Session of 28 November 2019: European Parliament Resolution on the Climate and Environment Emergency. 2019/2930(RSP). Available online: https://www.europarl.europa.eu/doceo/document/RC-9-2019-0209_EN.html (accessed on 25 June 2020).
2. European Commission. Commission Recommendation (EU) 2019/786 of 8 May 2019 on Building Renovation (Notified under Document C (2019) 3352). Official Journal of the European Union L127. 2019. Available online: https://eur-lex.europa.eu/eli/reco/2019/786/oj (accessed on 25 June 2020).
3. European Commission. Communication from the commission to the European Parliament, the Council, the European Economic and Social Committee and the Committee of the Regions A new Circular Economy Action Plan For a Cleaner and More Competitive Europe. COM/2020/98 Final. 2020. Available online: https://eur-lex.europa.eu/legal-content/EN/TXT/?qid=1583933814386&uri=COM:2020:98:FIN (accessed on 27 June 2020).
4. Pombo, O.; Rivela, B.; Neila, J. Life cycle thinking toward sustainable development policy-making: The case of energy retrofits. *J. Clean. Prod.* **2019**, *206*, 267–281. [CrossRef]
5. European Commission. Communication from the Commission to the European Parliament, the European Council, the Council, the European Economic and Social Committee and the Committee of the Regions. The European Green Deal. Brussels, 11.12.2019. COM (2019) 640 Final. 2019. Available online: https://eur-lex.europa.eu/legal-content/EN/TXT/?uri=COM%3A2019%3A640%3AFIN (accessed on 27 June 2020).
6. The European Parliament and the Council of the European Union. Directive 2012/27/EU of the European parliament and of the Council of 25 October 2012 on Energy Efficiency, Amending Directives 2009/125/EC and 2010/30/EU and Repealing Directives 2004/8/EC and 2006/32/EC. Official Journal of the European Union, L315. 2012. Available online: https://eur-lex.europa.eu/eli/dir/2012/27/oj (accessed on 27 June 2020).
7. The European Parliament and the Council of the European Union. Directive 2010/31/EU of the European Parliament and of the Council of 19 May 2010 on the Energy Performance of Buildings. Official Journal of the European Union, 2010, L153. Available online: https://eur-lex.europa.eu/legal-content/ES/TXT/?uri=celex%3A32010L0031 (accessed on 27 June 2020).
8. Sönnichsen, S.D.; Clement, J. Review of green and sustainable public procurement: Towards circular public procurement. *J. Clean. Prod.* **2020**, *245*, 118901. [CrossRef]
9. The European Parliament and the Council of the European Union. Directive 2008/98/EC of the European Parliament and of the Council of 19 November 2008 on Waste and Repealing Certain Directives. Official Journal of the European Union, L312. 2008. Available online: https://eur-lex.europa.eu/eli/dir/2008/98/oj (accessed on 27 June 2020).
10. European Commission. Europe 2020: A Strategy for Smart, Sustainable and Inclusive Growth. COM/2010/2020 Final. 2020. Available online: https://eur-lex.europa.eu/legal-content/EN/TXT/HTML/?uri=CELEX:52010DC2020&from=ES (accessed on 27 June 2020).
11. The European Parliament and the Council of the European Union. Directive 2014/24/EU of the European Parliament and of the Council of 26 February 2014 on Public Procurement and Repealing Directive 2004/18/EC Text with EEA Relevance. Official Journal of the European Union, L94. 2014. Available online: https://eur-lex.europa.eu/eli/dir/2014/24/oj (accessed on 24 June 2020).
12. European Commission. Communication from the Commission to the European Parliament, the Council, the European Economic and Social Committee and the Committee of the Regions Public Procurement for a Better Environment. COM (2008) 400 Final. 2008. Available online: https://eur-lex.europa.eu/legal-content/EN/TXT/HTML/?uri=CELEX:52008DC0400&from=ES (accessed on 24 June 2020).
13. Bouwer, M.; de Jong, K.; Jonk, M.; Berman, T.; Bersani, R.; Lusser, H.; Nissinen, A.; Parikka, K.; Szuppinger, P. Green Public Procurement in Europe 2005—Status overview. *Virage Milieu Manag. bv, Korte Spaarne* **2005**, *31*, 2011, AJ Haarlem, The Netherlands. Available online: https://ec.europa.eu/environment/gpp/pdf/Stateofplaysurvey2005_en.pdf (accessed on 26 August 2020).
14. Braulio-Gonzalo, M.; Bovea, M.D. Relationship between green public procurement criteria and sustainability assessment tools applied to office buildings. *Environ. Impact Assess. Rev.* **2020**, *81*, 106310. [CrossRef]
15. Cheng, W.; Appolloni, A.; D'Amato, A.; Zhu, Q. Green public procurement, missing concepts and future trends—A critical review. *J. Clean. Prod.* **2018**, *176*, 770–784. [CrossRef]

16. Sparrevik, M.; Førsund Wangen, H.; Magerholm Fet, A.; De Boer, L. Green public procurement—A case study of an innovative building project in Norway. *J. Clean. Prod.* **2018**, *188*, 879–887. [CrossRef]

17. Ministerio de la Presidencia, Relaciones con las Cortes e Igualdad. 2019. Orden PCI/86/2019, de 31 de enero, por la que se publica el Acuerdo del Consejo de Ministros de 7 de diciembre de 2018, por el que se aprueba el Plan de Contratación Pública Ecológica de la Administración General del Estado, sus organismos autónomos y las entidades gestoras de la Seguridad Social (2018–2025) (Order PCI/86/2019, 31 January, which Publishes the Agreement from the Council of Ministries of 7 December 2018, Approving the Green Public Procurement Plan of the General State Administration, Its Autonomous Organisations and Management Authorities of Social Security). Boletín Oficial del Estado, n 30. Available online: https://www.boe.es/diario_boe/txt.php?id=BOE-A-2019-1394 (accessed on 25 June 2020).

18. Ministerio de Fomento. Observatorio de Vivienda y Suelo. Boletín Especial Censo 2011 Parque edificatorio (Special Bulletin Census 2011 Building Stock). 2014. Available online: http://www.fomento.gob.es/MFOM.CP.Web/handlers/pdfhandler.ashx?idpub=BAW021 (accessed on 22 June 2020).

19. Ministerio de Transportes, Movilidad y Agenda Urbana. Actualización 2020 de la estrategia a largo plazo para la rehabilitación energética en el sector de la edificación en España (2020 update of the long-term strategy for energy renovation in the building sector in Spain). 2020. Available online: https://ec.europa.eu/energy/sites/ener/files/documents/es_ltrs_2020.pdf (accessed on 22 June 2020).

20. Ellen MacArthur Foundation. Available online: https://www.ellenmacarthurfoundation.org/ (accessed on 15 July 2020).

21. Ministerio de Energía, Turismo y Agenda Digital. Plan nacional de acción de eficiencia energética 2017–2020 (Energy Efficiency National Action Plan 2017–2020). 2017. Available online: https://ec.europa.eu/energy/sites/ener/files/documents/es_neeap_2017_es.pdf (accessed on 27 June 2020).

22. Comisión Nacional de los Mercados y la Competencia. E/CNMC/004/18 Overview of Public Procurement Procedures in Spain. 2019. Available online: https://www.cnmc.es/sites/default/files/2797776.pdf (accessed on 27 June 2020).

23. Jefatura del Estado. Ley 9/2017, de 8 de noviembre, de Contratos del Sector Público, por la que se transponen al ordenamiento jurídico español las Directivas del Parlamento Europeo y del Consejo 2014/23/UE y 2014/24/UE, de 26 de febrero de 2014 (Law 9/2017, 8 November, Contracts of the Public Sector, transposing to the Spanish legislation the Directives from the European Parliament and the Council 2014/23/EU and 2014/24/EU, from 26 February 2014). Boletín Oficial del Estado, n 272. 2017. Available online: https://www.boe.es/buscar/act.php?id=BOE-A-2017-12902 (accessed on 25 June 2020).

24. Congreso de los Diputados. Proyecto de Ley de cambio climático y transición energética. (Climate change and energy transition law project). Boletín Oficial de las Cortes Generales, 2020, 19. Available online: http://www.congreso.es/public_oficiales/L14/CONG/BOCG/A/BOCG-14-A-19-1.PDF (accessed on 27 June 2020).

25. Ministerio para la Transición Ecológica y el Reto Demográfico. Borrador actualizado del plan nacional integrado de energía y clima 2021–2030 (Updated draft of the energy and climate integrated national plan 2021–2030). 2020. Available online: https://www.miteco.gob.es/images/es/pniec_2021--2030_borradoractualizado_tcm30-506491.pdf (accessed on 25 June 2020).

26. Ministerio para la Transición Ecológica y el Reto Demográfico. Borrador de Estrategia de descarbonización a largo plazo 2050 (Draft for the Long-Term Decarbonisation Strategy 2050). 2020. Available online: https://energia.gob.es/es-es/Participacion/Paginas/DetalleParticipacionPublica.aspx?k=336 (accessed on 22 July 2020).

27. Ministerio para la Transición Ecológica y el Reto Demográfico. Estrategia Española de Economía Circular 'España Circular 2030' (Spanish Strategy of Circular Economy 'España Circular 2030'). 2020. Available online: https://www.miteco.gob.es/es/calidad-y-evaluacion-ambiental/temas/economia-circular/estrategia/ (accessed on 27 July 2020).

28. Ministerio de Fomento. Real Decreto 732/2019, de 20 de diciembre, por el que se modifica el Código Técnico de la Edificación, aprobado por el Real Decreto 314/2006, de 17 de marzo (Royal Decree 732/2019, 20 December, modifying the Technical Building Code, approved by the Royal Decree 314/2006, 17 March). Boletín Oficial del Estado, n 311. 2019. Available online: https://www.boe.es/diario_boe/txt.php?id=BOE-A-2019-18528 (accessed on 15 July 2020).

29. Ministerio para la Transición Ecológica y el Reto Demográfico. Actuaciones para reducir las emisiones de gases de efecto invernadero (Interventions to Reduce Greenhouse Gas Emissions). Available online: https://www.miteco.gob.es/es/cambio-climatico/temas/mitigacion-politicas-y-medidas/edificacion.aspx (accessed on 13 July 2020).

30. European Commission. European Semester Thematic Factsheet. Public Procurement. 22.11.2017. Available online: https://ec.europa.eu/info/sites/info/files/file_import/european-semester_thematic-factsheet_public-procurement_es.pdf (accessed on 27 July 2020).

31. Fuentes-Bargues, J.L.; González-Cruz, C.; González-Gaya, C. Environmental criteria in the Spanish public works procurement process. *Int. J. Environ. Res. Public Health* **2017**, *14*, 204. Available online: https://www.ncbi.nlm.nih.gov/pmc/articles/PMC5334758/ (accessed on 23 June 2020). [CrossRef] [PubMed]

32. De Giacomoa, M.R.; Testab, F.; Fabio Iraldob, F.; Formentinid, M. Does green public procurement lead to life cycle costing (LCC) adoption? *J. Purch. Supply Manag.* **2019**, *25*, 100500. [CrossRef]

33. Jolien, J.; Grandia, P.M.; Kruyen, P. Assessing the implementation of sustainable public procurement using quantitative text-analysis tools: A large-scale analysis of Belgian public procurement notices. *J. Purch. Supply Manag.* **2020**, 100627. [CrossRef]

34. Bertone, E.; Sahin, O.; Stewart, R.; Zou, P.; Alam, M.; Blaird, E. State-of-the-art review revealing a roadmap for public building water and energy efficiency retrofit projects. *Int. J. Sustain. Built Environ.* **2016**, *5*, 526–548. [CrossRef]

35. Ministerio de Hacienda y Función Pública. Informe relativo a la contratación pública en España—2017 (Report on public procurement in Spain—2017). 2018. Available online: https://contrataciondelestado.es/wps/wcm/connect/b73beca9-843f-43e4-bc89-09de10471717/2017+PUBLIC+PROCUREMENT+REPORT-SPAIN+.pdf?MOD=AJPERES (accessed on 27 June 2020).

36. Tang, Z.W.; Ng, S.T.; Skitmore, M. Influence of procurement systems to the success of sustainable buildings. *J. Clean. Prod.* **2019**, *2018*, 1007–1030. [CrossRef]

37. Presidencia de las Illes Balears. Ley 10/2019, de 22 de febrero, de cambio climático y transición energética. Comunidad Autónoma de las Illes Balears (Law 10/2019, 22 February, of climate change and energy transition). Butlletí Oficial de les Illes Balears, 2019, n 27. Available online: http://www.caib.es/sites/canviclimatic2/es/la_ley_de_ccyte/ (accessed on 30 June 2020).

38. Presidencia de la Junta de Andalucía. Ley 8/2018, de 8 de octubre, de medidas frente al cambio climático y para la transición hacia un nuevo modelo energético en Andalucía (Law 8/2018, 8 October, of measures against climate change and to ensure the transition towards a new energy model in Andalucía). Boletín Oficial de la Junta de Andalucía, 218, n 199. 2018. Available online: https://www.juntadeandalucia.es/boja/2018/199/1 (accessed on 29 June 2020).

39. Departamento de la Presidencia de la Generalidad de Cataluña. Ley 16/2017, del cambio climático. Law 16/2017, of climate change). Diari Oficial de la Generalitat de Catalunya, 2017, n 7426. Available online: https://canviclimatic.gencat.cat/web/.content/03_AMBITS/Llei_cc/docs/LEY_16_2017_CC_CAST.pdf (accessed on 27 June 2020).

40. Presidencia. Ley 4/2019, de 21 de febrero, de Sostenibilidad Energética de la Comunidad Autónoma Vasca. (Law 4/2019, February 21, on Energy Sustainability of the País Vasco). Boletín oficial del País Vasco, n 42. 2019. Available online: https://www.euskadi.eus/bopv2/datos/2019/02/1901087a.pdf (accessed on 30 August 2020).

41. Generalitat de Catalunya. Departament de la Presidència. Secretaría del Govern: Acord del govern per a l'aprovació del pla d'estalvi i eficiència energètica als edificis i equipaments de la Generalitat de Catalunya (2018-2022) (Government Agreement for the Approval of a Plan for Energy Efficiency and Savings in Buildings and Facilities of Generalitat de Cataluña 2018–2022). 2018. Available online: https://contractacio.gencat.cat/web/.content/gestionar/regulacio-supervisio/acords-govern/ag20181204.pdf (accessed on 27 June 2020).

42. Consejería de Economía, Empleo y Hacienda. Plan de Ahorro y Eficiencia Energética en Edificios Públicos (Energy Efficiency and Savings Plan in Public Buildings). Comunidad de Madrid. 2017. Available online: https://www.comunidad.madrid/transparencia/sites/default/files/plan/document/plan_de_ahorro_y_eficiencia_energetica_en_edificios_publicos_de_la_comunidad_de_madrid.pdf (accessed on 20 June 2020).

43. Conselleria de Economía Sostenible, Sectores Productivos, Comercio y Trabajo. Acuerdo de 16 de diciembre de 2016, del Consell, por el que se aprueba el Plan de ahorro y eficiencia energética, fomento de las energías renovables y el autoconsumo en los edificios, infraestructuras y equipamientos del sector público de la Generalitat (Agreement of 16 December 2016, of the Council, approving the plan for energy efficiency and savings, development of renewable energies and buildings, infrastructures and facilities self-consumption in the public sector of the Generalitat). Diari Oficial de la Generalitat Valenciana, 2017, n 7957. 2016. Available online: http://www.dogv.gva.es/datos/2017/01/13/pdf/2017_214.pdf (accessed on 15 July 2020).

44. Gobierno de Navarra. Plan energético de Navarra Horizonte 3030 (Energy Plan of Navarra Horizon 2030). 2018. Available online: https://www.navarra.es/NR/rdonlyres/9F32A10F-A290-4F13-842D-90E799CA5ABA/403478/PEN2030DEFINITIVO20171226comprimido.pdf (accessed on 8 July 2020).

45. Gobierno Vasco. Estrategia Energética de Euskadi 2030 (Energy Strategy of Euskadi 2030). 2017. Available online: https://www.eve.eus/EveWeb/media/EVE/pdf/3E2030/EVE-3E2030-castellano.pdf (accessed on 8 July 2020).

46. Junta de Extremadura. Estrategia de eficiencia energética en edificios públicos de la administración regional de Extremadura 2018–2030 (Strategy for energy efficiency in public buildings of the regional administration of Extremadura 2018–2030). 2017. Available online: https://www.agenex.net/files/E4PAREX_2018.pdf (accessed on 19 July 2020).

47. Junta de Extremadura. Estrategia de economía verde y circular Extremadura 2030 (Strategy of Circular and Green Economy Extremadura 2030). 2018. Available online: https://extremadura2030.com/ (accessed on 19 July 2020).

48. Xunta de Galicia. Estrategia Gallega de economía Circular 2019–2030 (Circular Economy Strategy of Galicia 2019–2030). 2018. Available online: https://ficheiros-web.xunta.gal/transparencia/informacion-publica/EGEC_cas.pdf (accessed on 25 July 2020).

49. Gobierno Vasco. Estrategia de Economía Circular de Euskadi 2030 (Circular Economy Strategy of Euskadi 2030). 2018. Available online: https://www.euskadi.eus/contenidos/documentacion/economia_circular/es_def/adjuntos/EstrategiaEconomiaCircular2030.pdf (accessed on 14 July 2020).

50. Gobierno de Aragón. Estrategia Aragón Circular (Strategy for circular Aragón). 2020. Available online: https://aragoncircular.es/aragon-circular-2030/ (accessed on 14 July 2020).

51. Gobierno de Navarra. Agenda para el desarrollo de la Economía Circular en Navarra 2030 (Circular economy development plan Navarra 2030). 2019. Available online: https://gobiernoabierto.navarra.es/sites/default/files/3291_anexo_agenda_para_el_desarrollo_de_la_economia_circular.pdf (accessed on 30 June 2020).

52. Presidencia de la Junta de Castilla-La Mancha. Ley 7/2019, de 29 de noviembre, de Economía Circular de Castilla-La Mancha (Law 7/2019, November 29, for Circular Economy of Castilla-La Mancha). Diario Oficial de Castilla La-Mancha, 2019, n 244. Available online: https://docm.castillalamancha.es/portaldocm/detalleDocumento.do?idDisposicion=1575527306132180219 (accessed on 29 August 2020).

53. COTEC Foundation for Innovation. Situation and Evolution of Circular Economy in Spain. Report 2019. Available online: https://cotec.es/media/informe-cotec-economia-circular-2019.pdf (accessed on 29 August 2020).

54. Consejería de Medio Ambiente y Ordenación del Territorio. Comunidad de Madrid. Estrategia de Gestión Sostenible de los residuos de la Comunidad de Madrid 2017–2024 (Strategy of Sustainable Waste Management of Comunidad de Madrid 2017–2024). 2018. Available online: https://www.comunidad.madrid/sites/default/files/doc/medio-ambiente/residuos_cm_tdc_interactiva_mod_2_1.pdf (accessed on 18 July 2020).

55. Gobierno de la Comunidad de Madrid. MADRID7R Economía Circular (MADRID7R Circular Economy). Comunidad de Madrid. Available online: http://www.madrid7r.es/ (accessed on 18 July 2020).

56. Gobierno de Cantabria. Plan de Residuos de la Comunidad Autónoma de Cantabria 2017–2023 (Waste Plan of the Region of Cantabria 2017–2023). 2017. Available online: https://boc.cantabria.es/boces/verAnuncioAction.do?idAnuBlob=311012 (accessed on 29 August 2020).

57. Comunidad de Madrid. Modelos de pliegos de cláusulas administrativas particulares recomendados por la Junta Consultiva de Contratación Administrativa de la Comunidad de Madrid (Specification Models of Particular Administrative Clauses Recommended by the Junta Consultiva de Contratación Administrativa de la Comunidad de Madrid). 2020. Available online: http://www.madrid.org/cs/Satellite?cid=1354385753025&language=es&pagename=PortalContratacion%2FPage%2FPCON_listado (accessed on 15 July 2020).

58. Ministerio para la Transición Ecológica y el Reto Demográfico. Instituto para la Diversificación y Ahorro de la Energía. Modelo de pliego de condiciones técnicas para la contratación de los servicios energéticos de los edificios de las administraciones públicas. Modelo de pliego de condiciones administrativas para la contratación de los servicios energéticos de los edificios de las administraciones públicas (Technical specification model for the procurement of energy services in public administration buildings. Administrative specification model for the procurement of energy services in public administration buildings). Available online: https://www.idae.es/tecnologias/eficiencia-energetica/edificacion/edificios-publicosconsulta-publica-sobre-los-borradores-de-modelos-de-pliegos-de-clausulas-administrativas-y-tecnicas (accessed on 14 July 2020).

59. Generalitat de Catalunya. Guia per a la redacció dels plecs de clàusules administratives particulars i de prescripcions tècniques dels contractes de rendiment energètic amb estalvis garantits, subjectes a regulació harmonitzada (contractes de serveis) (Guide for Drafting Specification Models of Administrative Clauses and Technical Prescriptions for Energy Efficiency and Savings Contracts under a Harmonised Regulation). 2018. Available online: http://icaen.gencat.cat/web/.content/10_ICAEN/17_publicacions_informes/04_coleccio_QuadernPractic/quadern_practic/arxius/12_contractesRendimentEnergetics_pdf (accessed on 25 July 2020).

60. Comunidad de Madrid. Guía general de aspectos relativos a la protección del medio ambiente (General Guide of Environmental Protection Aspects). 2020. Available online: https://www.comunidad.madrid/sites/default/files/doc/medio-ambiente/guia_general_aspectos_ambientales_febrero_2020.pdf (accessed on 25 July 2020).

61. Comunidad de Madrid. Preguntas frecuentes y respuestas rápidas para una Contratación Pública Responsable (Frequently Asked Questions and Speedy Answers for a Responsible Public Procurement). 2017. Available online: https://www.economiasolidaria.org/wp-content/uploads/2020/06/GUIA_CONTRATACION_PUBLICA_RESPONSABLE.pdf.pdf (accessed on 25 July 2020).

62. Junta de Andalucía. Guía para la inclusión de cláusulas sociales y medioambientales en la contratación de la Junta de Andalucía (Guide for the Inclusion of Social and Environmental Clauses in the Procurement of Junta de Andalucía). 2016. Available online: https://www.juntadeandalucia.es/export/drupaljda/GUIA_CSM.pdf (accessed on 25 July 2020).

63. Gobierno del Principado de Asturias. Guía práctica para la inclusión de cláusulas de responsabilidad social y medioambiental en la contratación administrativa de la administración del Principado de Asturias y su sector público (Practical Guide for the Inclusion of Social and Environmental Responsibility in Administrative Procurement in the Public Sector of Asturias). 2018. Available online: https://sede.asturias.es/Proveedores/FICHEROS/ESTRUCTURA%202011/NORMATIVA%20PERFIL/INSTRUCCIONES%20LEY%209_2017/guia_practica_clausulas_social_y_m_amb_07-05-2018.pdf (accessed on 25 July 2020).

64. Portal de Contratación de Navarra. Fichas de cláusulas medioambientales de Navarra (Factsheets of Environmental Clauses of Navarra). 2017. Available online: https://portalcontratacion.navarra.es/documents/880958/1690345/FICHASCLAUSULASMEDIOAMBIENTALES.pdf/29ac17b9-557f-32c5-9c2e-8000ffc2c539?t=1549874773602 (accessed on 15 June 2020).

65. Consejo de Gobierno de Cantabria. Decreto 32/2018, de 12 de abril, por el que se aprueba la Estrategia de Acción frente al Cambio Climático de Cantabria 2018–2030 (Decree 32/2018, 12 April, approving the Strategy Action against Climate Change in Cantabria 2018–2030). Boletín Oficial de Cantabria, 2018 n 78. 2018. Available online: https://boc.cantabria.es/boces/verAnuncioAction.do?idAnuBlob=325226 (accessed on 13 July 2020).

66. Greensproject. Guías básicas para la aplicación de criterios ambientales a la compra de productos/servicios en la administración pública. (Green Public Procurement Basic Guide for the Application of Environmental Criteria in Public Administration). 2020. Available online: http://greensproject.eu/es/contratadores-publicos/ (accessed on 25 July 2020).

67. Generalitat Valenciana. Instituto Valenciano de la Edificación. Guía Verde, medidas medioambientales en la contratación pública en el ámbito de la edificación de la Generalitat (Green Guide, Public Procurement Environmental Measures in the Building Sector of the Generalitat). Generalitat Valenciana. 2020. Available online: http://www.habitatge.gva.es/es/web/arquitectura/guia-verda (accessed on 7 July 2020).

68. Gobierno de Cantabria. Guías para la eficiencia energética. Servicios públicos (Guides for Energy Efficiency. Public Services). Gobierno de Cantabria. 2013. Available online: https://www.cambioclimaticocantabria.es/documents/3528731/5184577/7-126+serviciospublicos.pdf/20d1e1d3-44b7-7672-3cef-355d21a00042 (accessed on 7 July 2020).

69. Sociedad Pública de Gestión Ambiental del Gobierno Vasco. Criterios Ambientales. Categorías de productos y servicios. Edificación. Construcción de edificios administrativos o de oficinas (Environmental Criteria. Products and Services Categories. Construction. Building of Administrative or Office Buildings). 2020. Available online: https://www.Ihobe.eus/criterios-ambientales (accessed on 10 July 2020).

70. Gobierno Vasco. Guía de edificación y rehabilitación sostenible para la vivienda en la comunidad autónoma del País Vasco (Guide to Sustainable Residential Building and Renovation in the Commity of the Basque Country). 2015. Available online: https://www.euskadi.eus/contenidos/documentacion/guia_edificacion/es_pub/adjuntos/GUIA_EDIFICACION_SOSTENIBLE_castellano_2015.pdf (accessed on 25 July 2020).

71. Generalitat de Catalunya. Institut Català d'energia. Estalvi i eficiència energètica en edificis públics Col·lecció Quadern Pràctic Número 2. Guia pràctica per informar i formar els professionals mostrant tecnologies, aparells, hàbits i exemples reals d'aplicació de l'estalvi i l'eficiència als edificis (Energy efficiency and savings in public buildings Col·lecció Quadern Pràctic Número 2. Practical guide for practitioners showcasing technologies, systems, habits and real examples for savings and efficiency in buildings). 2009. Available online: http://icaen.gencat.cat/web/.content/10_ICAEN/17_publicacions_informes/04_coleccio_QuadernPractic/quadern_practic/arxius/02_edificis_publics.pdf (accessed on 27 July 2020).

72. Institute for Energy in Catalunya. Saving and Energy Efficiency Plan in Building and Facilites of the Generalitat de Cataluña 2018–2022. Results. 2020. Available online: http://icaen.gencat.cat/ca/plans_programes/eficiencia_generalitat/ (accessed on 29 August 2020).

73. González Martínez, J.A. Saving and Energy Efficiency Plan in Building and Facilites of the Comunidad de Madrid. 2019. Available online: http://www.amiasociacion.es/portalsSrvcs/news/files/254_PLAN_AHORRO_ENERGETICO_EDIFICIOS_CM.pdf (accessed on 30 August 2020).

74. Ceice. gva. Available online: http://www.ceice.gva.es/es/inicio/area_de_prensa/not_detalle_area_prensa?id=731479 (accessed on 30 August 2020).

75. Asociation for Universities of Extremadura. Green Circular Economy for municipalities (2019). Available online: http://redextremaduraverde.org/web/wp-content/uploads/2019/03/Econom%C3%ADa-circular-desde-y-para-el-municipio.pdf (accessed on 30 August 2020).

76. Concello de Lugo. Building Green Boost, awarded in the Junta de Gobierno local del Concello de Lugo. 2020. Available online: https://www.lugobiodinamico.eu/el-edificio-impulso-verde-adjudicado-en-la-junta-de-gobierno-local-del-concello-de-lugo/ (accessed on 30 August 2020).

77. País Vasco Government. Ihobe. Section I Th estate of the art y and good practice for circular economy and waste management in Euskadi. 2019. Available online: https://www.ihobe.eus/publicaciones/economia-circular-y-gestion-residuos-en-euskadi-2 (accessed on 30 August 2020).

78. Government of The País Vasco. Ihobe. Purchase and Green public procurement in the País Vasco. Report on activity 2019. 2020. Available online: https://www.ihobe.eus/publicaciones/programa-compra-y-contratacion-publica-verde-pais-vasco-informe-seguimiento-2019-2 (accessed on 29 August 2020).

79. Generalitat de Catalunya. Institut Català d'energia. Edificis de consum d'energia gairebé zero (Buildings with nearly zero-energy consumption), Col·lecció Quadern Pràctic Número 11. 2017. Available online: http://icaen.gencat.cat/web/.content/10_ICAEN/17_publicacions_informes/04_coleccio_QuadernPractic/quadern_practic/arxius/11_Edificis_energia_zero.pdf (accessed on 22 July 2020).

80. Junta de Castilla León. Manual práctico de soluciones constructivas bioclimáticas para la arquitectura contemporánea (Toolkit of bioclimatic solutions for contemporary architecture). Junta de Castilla y León. 2015. Available online: http://www.iccl.es/certificacion/manual-practico-de-soluciones-constructivas-bioclimaticas (accessed on 20 July 2020).

81. Comunidad de Madrid. Plan de Ahorro y Eficiencia Energética en Edificios Públicos de la Comunidad de Madrid (Plan of Energy Efficiency and Savings in Public Buildings). 2015. Available online: https://www.comunidad.madrid/transparencia/informacion-institucional/planes-programas/plan-ahorro-y-eficiencia-energetica-edificios-publicos (accessed on 14 July 2020).

82. Gobierno de Cantabria. Decreto 35/2014, de 10 de julio, por el que se aprueba el Plan de Sostenibilidad Energética de Cantabria 2014-2020 (Decree 35/2014, 10 July, approving the Energy Sustainability Plan of Cantabria 2014-2020). Boletín Oficial de Cantabria, 2014, n 137. Available online: https://boc.cantabria.es/boces/verAnuncioAction.do?idAnuBlob=271320;https://parlamento-cantabria.es/sites/default/files/dossieres-legislativos/PSEC_2014-2020_1.pdf (accessed on 16 July 2020).

83. Ciudad autónoma de Melilla. Programa operativo regional feder de la ciudad autónoma de Melilla 2014-2020 (Regional programme feder of Melilla 2014–2020). 2014. Available online: https://www2.melilla.es/melillaPortal/RecursosWeb/DOCUMENTOS/1/1_14428_1.pdf (accessed on 15 July 2020).

84. Junta de Castilla y León. Estrategia de Eficiencia Energética de Castilla y León 2020 (Strategy for Energy Efficiency in Castilla y León 2020). 2018. Available online: https://energia.jcyl.es/web/es/ahorro-eficiencia-energetica/estrategia-eficiencia-energetica-2020.html (accessed on 27 July 2020).

85. Consejería de Economía y Hacienda. Castilla y León. Acuerdo 26/2020, de 4 de junio, de la Junta de Castilla y León, por el que se aprueban medidas contra el cambio climático en el ámbito de la Comunidad de Castilla y León (Agreement 26/2020, 4 June, Junta de Castilla y León, approving measures against climate change in the area of Castilla y León). Boletín Oficial de Castilla y León, 2020, n 111. Available online: http://bocyl.jcyl.es/boletines/2020/06/05/pdf/BOCYL-D-05062020-6.pdf (accessed on 15 July 2020).

86. Ministerio de Ciencia y Tecnología. Real Decreto 842/2002, de 2 de agosto, por el que se aprueba el Reglamento electrotécnico para baja tensión (Royal Decree 842/2002, August 2, approving Electrotechnical Regulation for Low Voltage). Boletín Oficial del Estado, n 224. 2002. Available online: https://www.boe.es/buscar/pdf/2002/BOE-A-2002-18099-consolidado.pdf (accessed on 7 July 2020).

87. Generalitat de Catalunya. Institut Català d'energia. Instal·lació d'infraestructura de recàrrega del vehicle elèctric (Installing infrastructures for charging electric vehicles) Col·lecció Quadern Pràctic Número 9. 2016. Available online: http://icaen.gencat.cat/web/.content/10_ICAEN/17_publicacions_informes/04_coleccio_QuadernPractic/quadern_practic/arxius/09_3aEdi_infraestructura_vehicle_electric.pdf (accessed on 15 July 2020).

88. Gobierno de Aragón. Manual de buenas prácticas ambientales en la contratación por administraciones públicas (Toolkit of good environmental practices in procurement of public administrations). 2004. Available online: https://www.aragon.es/documents/20127/674325/manual_practicas_admon.pdf/bfcb1933-b65e-5059-0737-bff39c50acd4 (accessed on 27 July 2020).

89. Comunidad de Madrid. Estrategia de Gestión Sostenible de los residuos de la Comunidad de Madrid 2017–2024 (apartados contratación pública) (Strategy for a Sustainable Waste Management in Comunidad de Madrid 2017-2024, public procurement sections). 2016. Available online: https://www.comunidad.madrid/servicios/urbanismo-medio-ambiente/estrategia-residuos-comunidad-madrid (accessed on 27 July 2020).

90. Presidencia de la Junta de Extremadura. Ley 12/2018, de 26 de diciembre, de contratación pública socialmente responsable de Extremadura (Law 12/2018, 26 December, of socially responsible public procurement of Extremadura). Diario Oficial de Extremadura, 2018, n 251. Available online: http://doe.gobex.es/pdfs/doe/2018/2510o/18010013.pdf (accessed on 28 July 2020).

91. Gobierno de Navarra. Ley Foral 2/2018, de 13 de abril, de Contratos Públicos (Regional Law 2/2018, 13 April, of Public Contracts). Boletín Oficial de Navarra, 2018, n 73. Available online: http://www.navarra.es/NR/rdonlyres/6253BAA2-031B-45D6-AF41-710B8B321839/418876/LeyForal.pdf (accessed on 27 July 2020).

92. Consejería de Agricultura, Ganadería, Pesca y Desarrollo Sostenible. Acuerdo de 9 de enero de 2020, del Consejo de Gobierno, por el que se aprueba la formulación del Plan Andaluz de Acción por el Clima (PAAC) (Agreement, 9 January 2020, of the Government Council, approving the Climate Action Plan in Andalucía, PAAC). Boletín Oficial de la Junta de Andalucía, 2020, n 8. Available online: https://www.juntadeandalucia.es/boja/2020/8/2 (accessed on 28 July 2020).

93. Consejería de Agua, Agricultura, Ganadería, Pesca y Medio Ambiente. Estrategia de mitigación y adaptación al cambio climático de la Región de Murcia (Strategy against climate change in the Region of Murcia). 2019. Available online: http://cambioclimaticomurcia.carm.es/index.php?option=com_k2&view=item&id=390:estrategia-de-mitigacion-y-adaptacion-al-cambio-climatico-de-la-region-de-murcia&Itemid=303 (accessed on 8 June 2020).

94. Xunta de Galicia. Estrategia gallega de economía circular 2019–2030 (Strategy for a circular economy in Galicia 2019-2030). Available online: https://ficheiros-web.xunta.gal/transparencia/informacion-publica/EGEC_cas.pdf (accessed on 8 June 2020).

95. Portal for public procurement of the Comunidad de Madrid. Available online: http://www.madrid.org/cs/Satellite?language=es&pagename=PortalContratacion%2FPage%2FPCON_home (accessed on 30 August 2020).

96. Presidencia de la Junta. Ley 7/2019, de 29 de noviembre, de Economía Circular de Castilla-La Mancha (Law 7/2019, 29 de November, of Circular Economy in Castilla-La Mancha). Diario Oficial de Castilla-La Mancha, 2019, n 244. Available online: https://docm.castillalamancha.es/portaldocm/descargarArchivo.do?ruta=2019/12/12/pdf/2019_11104.pdf&tipo=rutaDocm (accessed on 8 June 2020).

97. Comunidad de Madrid. Estrategia de Gestión Sostenible de los residuos de la Comunidad de Madrid 2017–2024 (Strategy for a Sustainable Waste Managemente in Comunidad de Madrid 2017-2024). 2018. Available online: https://www.comunidad.madrid/sites/default/files/doc/medio-ambiente/residuos_cm_tdc_interactiva_mod_2_1.pdf (accessed on 10 June 2020).

98. Conselleria de Territorio y Vivienda. Orden de 19 de octubre de 2004, de la Conselleria de Territorio y Vivienda, sobre requisitos y criterios medioambientales a introducir en los pliegos de cláusulas administrativas que rijan en los contratos de la Conselleria de Territorio y Vivienda, las entidades autónomas y entidades de derecho público vinculadas o dependientes de la misma (Order of 19 October 2004, about environmental requirements and criteria to introduce in specifications of administrative clauses of the contracts of the Conselleria de Territorio y Vivienda, the autonomous organizations and related public entities). Diario Oficial de la Comunitat Valenciana, 2004, n 4.872. Available online: http://www.dogv.gva.es/datos/2004/10/28/pdf/2004_X10844.pdf (accessed on 10 June 2020).

99. Federación de Municipios y Provincias Castilla-La Mancha. Criterios sociales y medioambientales en la contratación pública (Social and environmental criteria in public procurement). 2010. Available online: https://www.oneplanetnetwork.org/sites/default/files/criterios_sociales_y_medio_ambientales_en_la_contratacion_publica_espanha_16.pdf (accessed on 10 June 2020).

100. Consejería de Hacienda y Administraciones Públicas. Resolución de 19/10/2016, de la Secretaría General de Hacienda y Administraciones Públicas, por la que se dispone la publicación de la Instrucción del Consejo de Gobierno de 18/10/2016, sobre la inclusión de cláusulas sociales, de perspectiva de género y medioambientales en la contratación del sector público regional (Resolution 19/10/2016 of the Secretaría General de Hacienda y Aministraciones Públicas, which publishes the Instruction of the Government Council of 18/10/2016, about inclusion of social, gender perspective and environmental clauses in regional public procurement). Diario Oficial de Castilla-La Mancha, 2016, n 209. Available online: https://contratacion.castillalamancha.es/sites/contratacion.castillalamancha.es/files/documentos/paginas/archivos/instruccion_del_consejo_de_gobierno_de_18102016_sobre_la_inclusion_de_clausulas_sociales.pdf (accessed on 15 June 2020).

101. Portal de Contratación de Navarra. Compra responsable (Responsible purchase). Comunidad Foral de Navarra. Cláusulas medioambientales. 2010. Available online: https://portalcontratacion.navarra.es/documents/880958/1690345/FICHASCLAUSULASMEDIOAMBIENTALES.pdf/29ac17b9-557f-32c5-9c2e-8000ffc2c539?t=1549874773602 (accessed on 15 June 2020).

102. Sociedad Pública de Gestión Ambiental del Gobierno Vasco. Etiquetado ambiental de producto. Guía de criterios ambientales para la mejora de producto de Ihobe (Environmental product labelling. Guide of environmental criteria for improving Ihobe's product). 2011. Available online: https://www.Ihobe.eus/publicaciones/etiquetado-ambiental-producto-guia-criterios-ambientales-para-mejora-producto (accessed on 15 June 2020).

103. Estrategia Aragón circular (Circular Strategy of Aragón). Available online: https://aragoncircular.es/aragon-circular-2030/ (accessed on 16 June 2020).

104. Sociedad Pública de Gestión Ambiental del Gobierno Vasco. Manual de directrices para el uso de áridos reciclados en obras públicas de la Comunidad Autónoma del País Vasco de Ihobe (Toolkit for the use of recycled aggregates in public construction works of Ihobe). 2009. Available online: https://www.Ihobe.eus/publicaciones/manual-directrices-para-uso-aridos-reciclados-en-obras-publicas-comunidad-autonoma-pais-vasco (accessed on 28 July 2020).

105. Gobierno de la Rioja. Plan Director de Residuos de La Rioja 2016-2026 (Waste Plan of La Rioja 2016-2026). Available online: https://www.larioja.org/larioja-client/cm/medio-ambiente/images?idMmedia=838694 (accessed on 16 June 2020).

106. Conselleria de Territorio y Vivienda. Decreto 200/2004, de 1 de octubre, del Consell de la Generalitat, por el que se regula la utilización de residuos inertes adecuados en obras de restauración, acondicionamiento y relleno, o con fines de construcción (Decree 200/2004, 1 October, regulating the use of inert waste appropriate for restoration and filling works, or for construction purposes). Diari Oficial de la Generalitat Valenciana, 2004, 4860. Available online: https://www.dogv.gva.es/datos/2004/10/11/pdf/2004_10263.pdf (accessed on 16 June 2020).

107. Ministerio de la Presidencia. Real Decreto 105/2008, de 1 de febrero, por el que se regula la producción y gestión de los residuos de construcción y demolición (Royal Decree 105/2008, 1 february, por regulating construction and demolition waste production and management). Boletín Oficial del Estado, 2008, n 38. Available online: https://www.boe.es/buscar/act.php?id=BOE-A-2008-2486 (accessed on 16 June 2020).

108. European Commission. GPP In practice. Issue no. 92. November 2019. Available online: https://ec.europa.eu/environment/gpp/pdf/news_alert/Issue_92_Case_Study_175_Ribera.pdf (accessed on 28 August 2020).

109. European Commission. GPP In practice. Issue no. 57. December 2017. Available online: https://ec.europa.eu/environment/gpp/pdf/news_alert/Issue57_Case_Study116_Galicia.pdf (accessed on 28 August 2020).

110. Ihobe. Services and resources. Good practice. Available online: https://www.ihobe.eus/buenas-practicas/reforma-y-ampliacion-edificio-consistorial-amorebieta-etxano-bizkaia-con-criterios-edificacion-ambientalmente-sostenible (accessed on 28 August 2020).

111. Generalitat de Cataluña. Decreto-ley 16/2019, de 26 de noviembre, de medidas urgentes para la emergencia climática y el impulso a las energías renovables (Decree-law 16/2019, 26 November, of urgent measures for climate emergency and use of renewable energies). Comunidad Autónoma de Cataluña. Diario Oficial de la Generalitat de Cataluña, 2019, n 8012. Available online: https://www.boe.es/buscar/pdf/2020/BOE-A-2020-445-consolidado.pdf (accessed on 25 July 2020).

112. Observatorio de la contratación pública. Contratación de energía eléctrica 100% procedente de energías renovables (Procurement of renewable electric energy). 2020. Available online: http://www.obcp.es/opiniones/contratacion-de-energia-electrica-100-procedente-de-energias-renovables (accessed on 15 July 2020).

113. Economía solidaria. Preguntas frecuentes y respuestas rápidas para una Contratación Pública Responsable (Frequently asked questions and speedy answers for a responsable public procurement). 2017. Available online: http://www.economiasolidaria.org/files/GUIA_CONTRATACION_PUBLICA_RESPONSABLE.pdf (accessed on 29 July 2020).

114. Instituto Valenciano de Investigaciones Económicas. Evaluación de la contratación pública de la Generalitat Valenciana (Evaluation of public procurement) (1) Elaboración de la base de datos y análisis de la duración de los procedimientos de contratación (Elaboration of a database and analysis of the procurement procedures). 2018. Available online: http://www.argos.gva.es/fileadmin/argos/Documentos/Bibliografica/Estudios_analisis_politicas_publicas/Informe_10_Contratacion.pdf (accessed on 29 July 2020).

115. European Commission. Commission announcement. Guidance on public procurement in the field of innovation. 2018. Available online: https://ec.europa.eu/transparency/regdoc/rep/3/2018/ES/C-2018-3051-F1-ES-MAIN-PART-1.PDF (accessed on 31 August 2020).

116. Consejería de Economía, Conocimiento, Empresas y Universidad. Estrategia de Compra Pública de Innovación en la Administración Pública de la Junta de Andalucía (Strategy for Public Purchase of Innovation in the Public Administration of Junta de Andalucía). 2019. Available online: https://www.juntadeandalucia.es/export/drupaljda/planes/18/09/ECPI_vdef.pdf (accessed on 12 June 2020).

117. Consejería de Hacienda y Administración Pública. Acuerdo de 18 de octubre de 2016, del Consejo de Gobierno, por el que se impulsa la incorporación de cláusulas sociales y ambientales en los contratos de la Comunidad Autónoma de Andalucía (Agreement of 18 October 2016, driving the incorporation of environmental and social clauses in contracts). Boletín Oficial de la Junta de Andalucía, 2016, n 203. Available online: https://www.juntadeandalucia.es/boja/2016/203/2 (accessed on 15 June 2020).

118. Gobierno Vasco. Green public procurement and procurement program of the Pais Vasco. 2020. 2016. Available online: https://www.ihobe.eus/Publicaciones/ficha.aspx?IdMenu=97801056-cd1f-4503-bafa-f54fa80d9a44&Cod=b34b0641-45b2-4b5a-9870-469079626014&Idioma=es-ES&Tipo= (accessed on 27 August 2020).

119. Government of the País Vasco; Ihobe. Green procurement and public procurement in the Pais Vasco. Activity Report. Fiscal year 2019. April 2020. Available online: https://www.ihobe.eus/publicaciones/programa-compra-y-contratacion-publica-verde-pais-vasco-informe-seguimiento-2019-2 (accessed on 27 August 2020).

120. Fundación Conama. Economía circular en el sector de la construcción. Grupo de trabajo GT-6 Congreso Nacional del Medio Ambiente 2018 (Circular economy in the construction sector. Work group GT-6 National Conference of the Environment 2018. Available online: http://www.conama.org/conama/download/files/conama2018/GTs%202018/6_final.pdf (accessed on 25 June 2020).

121. Ministerio de la Presidencia, Relaciones con las Cortes e Igualdad. Real Decreto 1515/2018, de 28 de diciembre, por el que se crea la Comisión Interministerial para la incorporación de la metodología BIM en la contratación pública (Royal Decree 2015/2018, 28 December, creating the Interministerial Commission to incorporate BIM methodologies in public procurement). Boletín Oficial del Estado, 2018, n 29. Available online: https://www.boe.es/buscar/act.php?id=BOE-A-2019-1368 (accessed on 25 July 2020).

122. European Commission. Environment: GPP Good Practice. Available online: https://ec.europa.eu/environment/gpp/case_group_en.htm. (accessed on 28 August 2020).

123. Ministerio de la Presidencia. Real Decreto 235/2013, de 5 de abril, por el que se aprueba el procedimiento básico para la certificación de la eficiencia energética de los edificios (Royal Decree 235/2013, 5 April, approving a basic procedure for certifying the energy efficiency in buildings). Boletín Oficial del Estado, 2013, n 89. Available online: https://www.boe.es/buscar/pdf/2013/BOE-A-2013-3904-consolidado.pdf (accessed on 25 July 2020).

© 2020 by the authors. Licensee MDPI, Basel, Switzerland. This article is an open access article distributed under the terms and conditions of the Creative Commons Attribution (CC BY) license (http://creativecommons.org/licenses/by/4.0/).

Article

The Church Tower of Santiago Apóstol in Montilla: An Eco-Sustainable Rehabilitation Proposal

M. Araceli Calvo-Serrano [1,*], **Isabel L. Castillejo-González** [1], **Francisco Montes-Tubío** [1] and **Pilar Mercader-Moyano** [2,*]

[1]   Department of Graphic Engineering and Geomatics, Campus de Rabanales, University of Cordoba, 14071 Córdoba, Spain; ma2cagoi@uco.es (I.L.C.-G.); ir1motuf@uco.es (F.M.-T.)
[2]   Department of Building Construction I, University of Seville, Reina Mercedes Avenue 2, 41012 Seville, Spain
*   Correspondence: acalvoserrano@me.com (M.A.C.-S.); pmm@us.es (P.M.-M.);
    Tel.: +34-660-503-494 (M.A.C.-S.); +34-618-305-559 (P.M.-M.)

Received: 29 July 2020; Accepted: 28 August 2020; Published: 31 August 2020

Abstract: Is it possible to carry out eco-sustainable rehabilitations on specially protected buildings? This is the main question and starting point for this research. We will use the tower of the Church of "Santiago Apóstol" in Montilla as a case study; with its most remote antecedents in the 15th century, it is an emblematic building of one of the most important cities in the Cordovan countryside and is listed as an Asset of Cultural Interest (*Bien de Interés Cultural* or BIC) as of 2001. The application of eco-efficiency criteria in the rehabilitation of this type of building might stimulate the reactivation of the construction sector in the rural area, positively impacting the promotion of a circular economy. To this end, a general methodology has been established for carrying out eco-sustainable renovations on this type of building, which defines indicators for evaluating the eco-sustainability of such interventions. This methodology is applied to the case study of this important building in Montilla to ensure that a feasible intervention has been proposed, aligned with three basic pillars of sustainability that considers its environmental, economic, and social impact.

Keywords: patrimony; tower; buildings rehabilitation; eco-sustainable; sustainable; town planning; resource efficiency; eco-efficient construction solutions

---

1. Introduction

The restoration of historical buildings aims to preserve their architectural and artistic elements so that they faithfully illustrate their evolution through history, allowing the history of a building to be read through its architecture. Special attention must be paid when buildings are listed as Assets of Cultural Interest (BIC in Spanish) [1], regulated by the Spanish and Andalusian Historical Heritage laws [2,3].

In order to preserve the beauty of this heritage, it is necessary to indicate the pertinent works of repair, conservation, reconstruction, and restoration. For the intervention in a building classified as Historical Heritage two processes must be carried out: A historical study and a safety evaluation. We wonder if it is possible to carry out a rehabilitation of this type of building while meeting eco-efficiency criteria. In the case of the Church Tower of Santiago Apóstol, multiple interventions have been carried out over the years. The last one was executed in 1989, and an additional conservation intervention would be necessary shortly after.

We think rehabilitation and eco-sustainable interventions on specially protected historical buildings can be key to reactivating the construction sector and general economy in rural areas. For decades, these regions have suffered from depopulation. As a result of the health and economic crisis in which we find ourselves today, a return of the population to these areas is highly plausible, with all that this may entail. The application of strategies and methods in the field of eco-sustainability, while respecting the environment and the impact on buildings, will also favor the flow of human, material, and financial resources in the area, generating employment and promoting tourism in the region and bolstering its economic development.

In the enhancement of these architecturally and historically relevant buildings, we understand that two essential aspects must be considered: Materials and eco-sustainability. The physical qualities of the materials used (texture, shine, surface, appearance, etc.) must be in line with the space, awakening the senses of its viewer, integrating the renovated portions with the history of the building and the environment [4,5]. The materials used will play a fundamental role in the eco-sustainability of the rehabilitation. Carrying out the intervention from an eco-sustainable point of view will reduce not only energy consumption, but also the impact on the environment and the generation of waste.

In the context of the current economic crisis, it is also necessary to search for strategies that contribute to reactivating the construction sector, while achieving the preservation and maintenance of these specially protected historical buildings.

The legal concept of sustainability arises from the United Nations Framework Convention on Climate Change [6], when society and governments begin to gain awareness of the danger of greenhouse gas emissions. After that, the Kyoto Protocol [7] continues along the same lines at the international level, and Directives 2010/31/EU, 2012/27/UE, and 2018/844/UE establish a series of minimum requirements and objectives at a European level for the energy efficiency of buildings [8], including a 20% reduction in minimum consumption compared to 1990 levels by 2020. There is an exemption in this regard for religious and officially protected buildings. Also, the Spanish and Andalusian Historical Heritage Laws [2,3] establish a principle of minimum intervention in these buildings, requiring that the contributions of all existing historical periods be respected in the restorations.

It is becoming common to carry out a rehabilitation of these buildings while applying eco-sustainable criteria and taking into account the challenges and limitations that may present themselves in the process. Some of these problems include space, the conservation requirements for protected elements, and the economic investment that this requires, which in many cases is not easily assumed by the property holder.

This article aims to address and analyze the rehabilitation of the tower of the Church of Santiago Apóstol in Montilla from an eco-sustainable point of view, while giving the necessary importance to its context.

2. Materials and Methods

The eco-sustainable rehabilitation of specially protected buildings is a topic of great interest today. For this reason, in this article, we will establish a general method for such an undertaking, analyzing its applicability through specific study case, the tower of the Church of Santiago Apóstol in Montilla (Córdoba). We will also detail how to carry out a future intervention by applying eco-sustainable rehabilitation parameters, detecting the challenges that may arise, and how to solve them.

2.1. Methods for Evaluating Rehabilitation with Eco-Sustainable Criteria

The absence of a uniform method for evaluating the rehabilitation of a historic building with eco-sustainable criteria [9,10] makes it possible to propose a model, which establishes key eco-sustainable evaluation performance indicators. Our main objective is to establish an eco-sustainable model for the rehabilitation of protected buildings like the tower of the Church of Santiago.

Using the methodology of the Integrative Model of Values for Sustainable Evaluations (MIVES) [11] as a starting point, we have determined some quality indicators that are representative, quantifiable, and precise for evaluating the rehabilitation of historical and protected buildings such as this tower; we have done so while considering the three pillars of eco-sustainability (environmental, social, and economic). This method has been developed by the Polytechnic University of Catalonia. It is a methodology of multicriteria decision-making that establishes value indicators for evaluating alternatives in order to solve problems in favor of sustainable rehabilitations. It establishes a series of phases, starting from the delimitation of the decision, and organizes the aspects to be considered in a branching manner in the form of a decision-making tree. It takes into account who takes the decision, what the limits are, and its conditions, including, for example, the requirements, components and life cycle of materials, among others. It then creates value functions and assigns a weight to each of the elements to then define possible alternatives to be considered when making decisions. Finally, these alternatives will be evaluated and analyzed to contrast the medium and long-term results in a control phase.

We think that starting from the MIVES methodology is convenient, given that one of its most important characteristics, as Viñolas explains, is that decisions are made prior to the intervention, defining both the aspects to be taken into consideration and how they are to be evaluated.

Though it is difficult to reduce the assessment of the sustainability of a rehabilitation to a series of indicators, establishing acceptance rates is the most objective way to do so. For this reason, some indicators have been identified in this regard and grouped as they relate to the three focuses of sustainability:

1. Environmental Focus:

    -   Energy consumed in the transport of material or components (ECTM) needed for the intervention. The fact that they must come from companies close to the place, as will be discussed below within the social focus, will have a positive impact on this indicator.
    -   Energy consumed in the installation and application process of the materials used (ECIP).
    -   Assessment of the degree of eco-sustainability of the material and natural resources used (ALEQ). In this case, measuring whether the resources used count as eco-sustainable or are in the process of obtaining an official quality certification in this regard.
    -   Level of use of reusable and recycled materials (LURRM), establishing a ratio between the total quantity of materials used and the portion that is reusable and recycled. A high rate of recycled and reusable materials will help indicate an eco-sustainable rehabilitation, and support the circular economy model, which in turn supports the reduction of raw materials through the application of the 3R concept (reduce, reuse, and recycle).
    -   Management of generated waste, measuring the level of waste generated and the treatment given. The aim is to minimize polluting waste in favor of biodegradable or reusable waste. We will make a distinction here between the level of reusable (MRW) and recycled waste (MRcW)

2. Social Focus:

    -   Ecological Footprint of Service and Resource Providers (EFSRP), as an indicator of the environmental impact of companies providing services and resources. Here, possessing an accreditation or being in the process of becoming accredited with some reputable certification will be considered [12].
    -   Distance from providers to the intervention site (DPIS). Reduced distance will bring benefits in terms of the impact on employment and generation of wealth in the local environment. It will also have positive consequences for the level of pollution generated by transportation.
    -   Hiring of labor from the local environment (LE). This will promote employment and generate wealth within the geographical environment where the intervention must be carried out. Especially when considering the economic crisis in which we find ourselves today as a consequence of the coronavirus pandemic, this can have direct positive effects.

3. Economic Focus: In the framework of a circular economy, it is not only necessary to reduce the cost for the company in its activities, but to reduce the entry of virgin materials and the production of waste. This criterion is directly related to the ecological footprint indicator previously explained, as well as the use of renewable energy. We must analyze the resulting logistics costs, which include the managing cost, storage cost, transport cost, administrative cost, and financial cost.

Table 1 shows how to calculate each of these indicators and what their measurement units are, in addition to establishing acceptance rates for each of them. From ECTM and ECIP indicators, the acceptance rates should be established once suppliers have been selected, the distance at which they are located is known, and the electrical tools to be used have been determined.

Table 1. Calculation of the proposed quality indicators to eco-sustainable rehabilitation.

INDICATOR	CALCULATION	UD	ACCEPTANCE INDEX
ECTM	fuel consumption per km traveled $\times$ km traveled	l	
ECIP	Electric power of the tool used (Kw) $\times$ Usage time (h)	Kwh	
ALEQ	$\frac{\text{ecosustainable resources used}}{\text{total resources used}} \times 100$	%	90%
LURRM	$\frac{\text{reusable and recycled materials used}}{\text{total resources used}} \times 100$	%	80%
MRW	$\frac{\text{total reusable waste}}{\text{total waste generated}} \times 100$	%	80%
MRcW	$\frac{\text{total non recycled waste}}{\text{total waste generated}} \times 100$	%	80%
EFSRP	$\frac{\text{providers with quality certification}}{\text{total provider}} \times 100$	%	90%
DPIS	distance between provider and the workplace (km)	km	20 km
LE	$\frac{\text{total local workers}}{\text{total hired workers}} \times 100$	%	80%

3. Results

3.1. Descriptive and Historical Analysis

To demonstrate the methodology outlined above, we will use the case study of the Santiago Apóstol Church of Montilla, a little town in Cordova, located in the center of Andalusia. For centuries it has been the most important church in the city, as part of the inherited cultural landscape that is integrated in the current urban reality. Various transformations have been carried out over the centuries, and it was declared an Asset of Cultural Interest in 2001.

Its exact origins are unknown. The first vestiges date from the 13th century [13–19]. In 1237, the territory was conquered under the command of King Ferdinand III, who in 1257 entrusted them to Gonzalo Yáñez, also known as Gonçalo Eanes Do Vinhal. The population settled and the medieval town was built around the hilltop castle and its fortress. It reached maximum splendor under Fernández de Córdoba and the lineage of the Casa de Aguilar [13], who resided there until 1508. Around it, as with most medieval towns, there were buildings and constructions of various kinds, among which the Church of Santiago stood out.

Within religious architecture, churches have great relevance. They satisfy the need of the people to have a space for prayer, which over time is transformed along with their environment. Since their origins, the conception of these buildings has followed some general guidelines. From the pre-Romanesque era, they began to be equipped with towers as a way of bringing humans closer to God. During the Middle Ages the *alarifes* (architects) increased their height. A weathervane with a cross or a rooster, which is a symbol of the Resurrection, was often placed on the roof of the towers.

The evolution reflected in the construction of the Church of Santiago is proof of the inseparable relationship between society and architecture. Although the building has undergone continuous transformations, its underlying structure has hardly been altered. According to Ramirez de Arellano, at the beginning of the 16th century, the Marques of Priego paid for some important work to be done on the building, which is now considered the first great reform, probably in response to the great demographic growth of the town, then the second most populated in the state of Priego [14]. Hernán Ruiz's reconstruction of the tower happened at this time.

Characteristics of the Mudejar style are observed, although some have disappeared with the passage of time, such as the coffered ceiling or the original main doorway [19]. This style prevails in the churches built in the current region of Cordova that come from the time of the Reconquest by Fernando III, as it is the case of the *fernandine* churches in Cordova city.

We will focus our attention on two of the most important exterior elements that have survived until our times and about which very little information exists: the façade and the tower.

We must go back to the era of the Reconquest, where Christian temples were either built inside of old Muslim mosques, as was the case with the Church of San Lorenzo and the Cathedral of Cordova, or new churches were built to stand alone. The conversion of mosques into Christian churches has been addressed throughout history. García and Ortega [20] explore the topic of Christianized mosques in Toledo and Cordova, identifying three types of processes: Expansion, emptying, and general demolition. The objective of the conversion of the mosques into Christian churches during the Reconquest was to change the identity of the property, marking it for Christian worship. For this reason, the first action taken in this process was to add Christian furniture and liturgical elements before subsequently carrying out the architectural changes. The most frequent intervention was the transformation of minarets into church towers, as seen in the Cordovan churches of Santiago and San Lorenzo. The use of other architectural elements such as columns or ashlars and even walls was also frequent.

In Cordova there are very important *fernandine* churches, with a mix of Romanesque, Gothic, and Mudejar elements, with their signature wooden coffered ceilings and additional influence from the French Gothic and Cistercian styles [21].

Some studies [14,16,17,22,23] establish that, after conquering the territory, the Church of Santiago was built upon an old mosque and consecrated to Christian worship, although no evidence has been found so far in the work carried out on the church or the surrounding area. Lorenzo Muñoz attributes the name of its main church to the day of the city's conquest on 25 July 1240 [15]. Antonio Jurado de Aguilar refers to the testament of Alonso Fernández de Córdoba, dated on 25 October 1235, in which he leaves ten thousand maravedis to the church to use for masses for his soul and for the construction of its main chapel. He also states that his grandson, Gonzalo Fernández de Córdoba, who was very devoted to the Apostle Santiago, contributed to the construction of a new altarpiece, presided over by a sculpture of Santiago [16]. There is not another written record on the church's architecture until the beginning of the 15th century. On 15 July 1437, a letter from the Bishop of Cordova dated 3 March 1422, about the obligation to pay tithes to the church, is read and published [19].

Almost no records about the original façade of the church have been found. There are just a few brief references from the beginning of the 17th century, in which Lucas Jurado y Aguilar points out that the façade and main doorway were constructed in 1624 [17]. The façade appears in a drawing of the Montilla barn (*alhorí*) made by the Cordovan architect Juan Antonio Camacho, master builder of the Duke of Medinaceli and the Bishopric of Cordova [24]. In the upper right section of this drawing dated 1723, we can see the façade and the tower before its destruction by the 1755 earthquake (Figure 1). Typical details of a Renaissance doorway are present including the great attention paid to the harmony of proportions, following the guidelines established in Vitruvio's treatise *De Architectura*, a reference book for the architects of the time. The façades of this style consist of two main parts and a pediment, with classicist structural and decorative elements. Regarding the main bodies, the semicircular arch, columns, entablatures, and roof with coffered ceilings stand out, with values such as symmetry, simplicity, and structural clarity prevailing. As for the decorations, the pilasters, scrolls, keys, medallions, and the pediment are typical. Usually, a niche was included in the second section over the doorway with a sculpture of the saint to whom the church was dedicated [25]. We understand that it is not the original, since its style, as we have explained, is one that postdates the original construction of the church. Furthermore, documentation exists stating that it was rebuilt in 1624 [17]. If we compare this doorway with the *alhorí* built by Camacho we see how the architect was inspired by the doorway of the church (Figure 2).

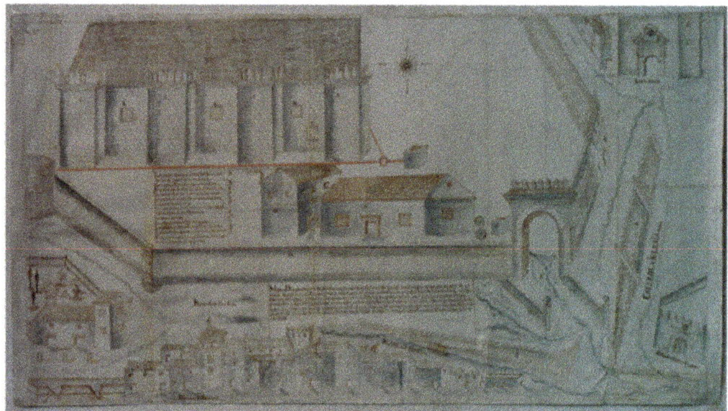

Figure 1. Plan for Montilla's barn built by Juan Antonio Camacho in 1723. (Archivo Ducal de Medinaceli, Sección Priego, Original. Leg 113, Num I).

Figure 2. Facade of Montilla's barn built by Juan Antonio Camacho. (Author: A. Calvo).

To gain a clearer understanding of how the original façade of the church might have looked (Figure 3), and having consulted some documented references about its existence from as early as the 14th century, we must compare the façades of the *fernandine* churches in Cordova. They are simple, with buttresses and a rose window, and a main doorway at the foot of the central nave with pointed arches adorned with diamond tips (*"punta de diamante"*) or dog teeth (*"diente de perro"*) and crowned by a roof tile on modillions (Figure 4).

Figure 3. Outline of the possible doorway of the church of Santiago Apostol in Montilla in the 15th Century. (Author: A. Calvo).

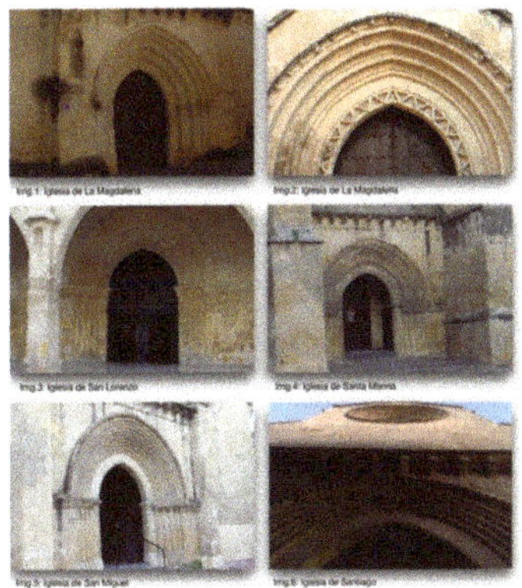

Figure 4. Details of the doorways of the most important fernandine churches in Cordova (Author: A. Calvo).

Taking Camachos's drawing (Figure 1) and the documentation consulted in various sources as a reference, it has been possible to approximate the main exterior of the church lost after the reform of the doorway in the early 17th century, when the tower designed by Hernán Ruiz was still standing (Figure 5). The imagined reconstruction of Hernan Ruiz's tower has been put together based on the description preserved in the Montilla Historical Archive [26] and drawings made by Enrique Garramiola, located in the Montilla Municipal Archive (Figure 6). The digital reconstruction of the 15th century doorway in (Figure 4) is based on a study of the *fernandine churches* of Cordova, built during this period.

Figure 5. Elevation of the facade after the reconstruction of 1624 and the original tower of the Church of Santiago Apóstol of Montilla (Author: A. Calvo).

Figure 6. The tower of the Church of Santiago built by Hernán Ruiz, by Enrique Garramiola. (Municipal Archive of Montilla. Church of Santiago Folder).

On the outside, its tower—the main subject of this article—stands out prominently and can be seen from almost anywhere in the city.

It is well known that the current tower is not the original. Ramirez de Arellano said that, at the beginning of the 16th century, the Marquis of Priego paid for what can be understood as the first important reform [13]. This included a reconstruction of the tower by Hernán Ruiz, that crowned the church from the end of the 16th century to the 18th century. No documentation has been found about the architecture of the previous tower, although it is known that in between 1576 and 1577 Hernan Ruiz, senior teacher of the Bishopric of Cordova presented himself in Montilla for the rebuilding of this tower. According to the notes of Enrique Garramiola conserved in the Municipal Archive of Montilla, information on the status of the tower's construction and the church bells was collected in the Chapter Records on 21 April 1574. Some official writings on 30 March 1577 signed by the scribe Juan Martínez de Córdoba provide a wealth of information about it. This information includes the decision to demolish the primitive tower and use the demolition material for reconstruction, as well as to lower the bells of the tower and leave them in on the walkway of the church in the meantime [21]. They establish a period of three months from the 15th of April for the execution of the project and describe how the construction is to be executed, alluding to a drawing that accompanies the writing and that must have been lost at some point.

Garramiola refers to it in an article in which he describes how the tower should be built according to the contract they signed [18], alluding to a drawing by Hernán Ruiz that would serve as a guide for the construction project. The contract indicates that it should have a first stone section from the Aguilar quarry that is 20 feet long and a cornice. The second section would be of brick plastered with lime and sand and capitals of the same stone as the first section that framed the windows, with blue tiles embedded in the stone adorning the windows. There would be stone pyramids on the corners of the roof and a spiral staircase ascending to the bell of the tower.

Hernán Ruiz II explained in his architectural treatise how he organizes the construction of this tower, widespread throughout Andalusia. The absence of drawn plans of the Santiago tower leads us to think that it was similar to what Hernán Ruiz explains in the last pages of his manuscript, when he talks about the square floorplan and the development of the staircase of the church tower of San Lorenzo in Cordova, also his work. In the contract for the construction of the Santiago tower [26], the existence of a spiral staircase that goes up to the last section is mentioned. We can extrapolate that it was built in a way that is very similar to the way described by Hernán Ruiz in his treatise. Here, as

Professor Navascués indicates [27], the sections are vaulted, sometimes with split voissoirs, others on half-barrel or sloping barrel vaults, or even with flat and sloping roofs. In addition, a comparison of the stairs of some other towers built by Hernan Ruiz and his son and grandson by the same name, such as the tower of the churches of Nuestra Señora de Soterraño (Aguilar de la Frontera), El Salvador (Pedroche), and San Mateo (Lucena), can help to clarify how the one in Santiago might have been. The one in Pedroche is very special, because it has a double handrail, which also functions as a drain.

With the help of these detailed descriptions, Garramiola draws a picture of the tower (Figure 6), which is kept in the Municipal Archive of Montilla. Both the written account and this drawing help to make an approximation of the tower, and lead to a comparison with towers of other churches from the time built by the Hernán Ruiz I-III in the kingdom of Cordova. Also, in the image of Montilla prepared by Pier Maria Baldi in 1668 for the book "Cosme de Medicis's journey though Spain and Italy", plate XXXVII shows a small sketch of this tower, which closely resembles those presented by Garramiola.

On November 1, 1755, one of the most virulent earthquakes remembered took place and would go down in history as the Lisbon earthquake. The devastating effects spread beyond Portugal. In Montilla, it was felt between 09:45 until 10:00 in the morning, affecting almost all the buildings in the city. Its effects on the Church of Santiago were devastating. A large portion of its tower collapsed, also affecting the façade, the roof, the choir section, and some of its chapels. The people of Montilla turned to the Duke of Medinaceli, patron of the church, for help. Fray Francisco Álvarez and the architect José Vela carried out a survey of the property in 1769 and declared the damage of the tower to be severe, arguing for the need of its demolition, carried out by Vela in 1771.

Six years later, two projects for the construction of the new tower were presented to the Duke, one by Vela himself and the other by Fernando Moradillo, a famous Spanish architect of the time based in Madrid, who carried out important projects in the capital city (Figure 7). Finally, the Duke, in line with the wishes of the Montillans, chose Vela's proposal, which underwent some changes. The Moradillo tower (Figure 8) consisted of three sections. The first one with rustic pilasters preserved the style of the Hernan Ruizes. The second one was simple without any adornments, and the bells covered by a double capital with a weathervane. In the center, there was a staircase with a square floor plan, which was supported by the tower walls and illuminated by windows on its sides [28].

Figure 7. Tower of the Church of Santiago by José Vela. Torre de la iglesia parroquial. Planta y alzado. s.f. (siglo XVIII). (Archivo Ducal de Medinaceli., Mapas y Planos, cajón M. Ref. 1988025218).

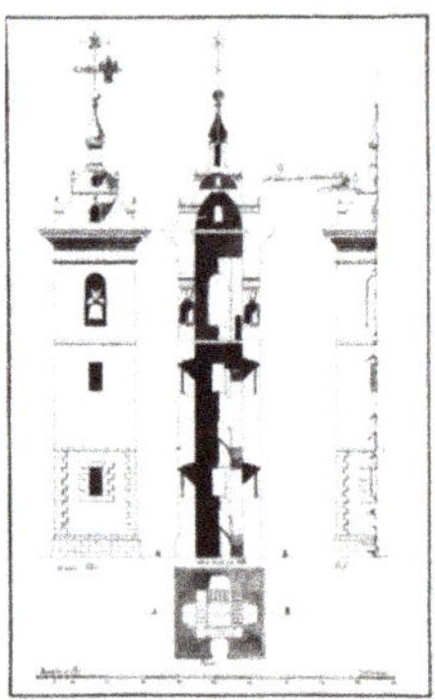

Figure 8. Tower of the Church of Santiago by Fernando Moradillo. Torre de la iglesia parroquial mayor del Señor Santiago. Fernando Moradillo, arquitecto (1777). (Fuente: A.D.M., Mapas y Planos, cajón M. Ref. 1988025219).

In 1989, the architects Juan Carlos Cobos Morillo and Felipe de la Fuente Darder presented a project to rehabilitate the church to the Andalusian Department of Culture [29]. With regards to the tower, they emphasize the structural damage to the three lower bodies, the bell tower, and its roof. Due to the lack of maintenance and the quality of the materials used for the masonry, part of the floor and a bell were in dire condition. The vaults and steps of the ascending staircase also suffered significant damage, with the stair nosing completely ruined, a partial detachment of the paint and mortar, missing pieces and breakage of some windows sills, detachment of the flooring, and heavily deteriorated plaster. The architects also drew attention to the poor design of the last section of the staircase, which made it difficult to evacuate rainwater. The top of the bell tower's cornice, ceramic brick pieces with mortar, was also in a state of deterioration, drowning in the vegetation that grew at the edges of the roof and at the intersection with the wall, which was also causing detachment of brick, the façade, and the breakage of the enamel of the ceramic pieces.

Faced with this situation, the architects proposed an intervention with three focuses: Restoration, restitution, and reconstruction. In the project, in accordance with the time, the principles of eco-sustainable rehabilitation were not yet considered. They began by conducting research of the original construction materials for their replacement and they centered the project around four improvements:

1. Restoration of the windows, with sealing and replacement of defective or worn parts.
2. Cleaning and waterproofing of the roof and its main channels for the evacuation of rainwater.
3. Restoration of the cracks in the dome and the disfigured plaster of the window.
4. General review of plaster.

3.2. Intervention Proposal Method Applied to the Case Study

In recent years, various interventions have been carried out in the church, but not in the tower, except for small projects in the bell tower. In a first phase, in 2012, part of the roof and the central nave were reformed, changing the lighting. In 2019, the second phase was concluded with work done on sections including the chapels of the Nacimiento and the Baptistery, the left side of the church, the area of archpriesthood and the primitive sacristy, and the top of the right side.

Although the general state of conservation of the tower after the rehabilitation carried out at the end of the 20th century is acceptable, we understand that it is necessary to carry out interventions that favor its conservation. For this reason, it is necessary to identify the risks that can affect, in the medium and long term, the conservation of the property, defining their impact on the deterioration process and identifying the most vulnerable parts and the degree of protection [30,31]. The windows that line the stairwell are open, as well as the ascent to the bell tower, giving way to two main problems: The entry of birds, which deposit highly corrosive organic waste there and nest along the interior of the tower, and the rainwater tower. This can lead to negative effects on the materials and conservation of the tower in the medium to long term. For this reason, we understand that three main interventions can be carried out:

1. Cleaning the roof.
2. Review of the wire mesh window coverings.
3. Cleaning of the inside of the tower.

3.3. Execution of an Eco-Sustainable Rehabilitation within the Case Study

As we have just stated, the proposal for an eco-sustainable rehabilitation on the tower of the Church of Santiago Apóstol must focus on three aspects, which we will explain in detail below, while paying special attention to the eco-sustainability of the materials used.

Figures 9–11 show images of the current state of the tower, where the damage caused by the birds that nest inside it and other factors are observed, which help to understand the need for this intervention.

Figure 9. Detail of the current state of the dome.

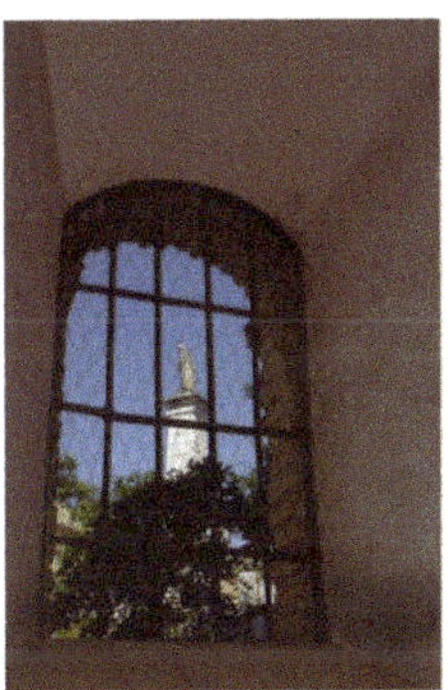

Figure 10. Detail of the current enclosure of one of the windows.

Figure 11. Stairs access to the bell tower.

As it is a specially protected building, the decision-making regarding interventions depends of the Public Administration, so they must be declared to the administration and their established procedure must be followed. As illustrated in Section 2.1, indicators have been determined, as value functions when we extrapolate them to the MIVES method. This way, acceptance rates can be assigned that help when evaluating each indicator in order to evaluate the sustainability of the rehabilitation. These rates will help in the selection of alternatives when one of the aspects gets in the way of an eco-sustainable evaluation of the intervention.

As for the first intervention, cleaning the roof, it is advisable to use soft brushes and small spatulas and to then continue with a deeper cleaning by applying eco-sustainable products to remove plant and organic debris. These products should have a low contamination index and should not negatively affect the glazing of the tiles. A general review of the roof should be carried out to, if needed, replace the defective tiles and check the waterproofing of the roof.

For the second intervention, it is necessary to review the wire mesh covering the windows (Figure 10), as well as the covering made with the same material on the spiral staircase that ascends to the bell tower (Figure 11) in order to avoid the entry of birds by this route. The wire mesh currently in place is in very poor condition. Most of the covering "screens" or nets have either been lost or have large tears that facilitate the entry of birds. For this reason, and to avoid the negative effects of the nesting and the highly corrosive organic residues coming from them on the monument, the revision of the wire mesh screens should be undertaken, and anti-bird netting should be installed.

Finally, with regards to the interior cleaning of the tower, it is necessary to start with an in-depth cleaning of the entire ascending staircase before proceeding to repaint the interior walls.

3.4. Problems Detected for Eco-Sustainable Rehabilitation in the Case Study

As we said previously, the rehabilitation of buildings classified with some type of historical-artistic protection such as the one in question is complex, although essential, since it is the responsibility of the towns to safeguard them as part of the cultural heritage of the community [32,33]. Still, we consider the eco-sustainable rehabilitation of protected historical assets as a great challenge.

It must comply not only with the requirements established by the aforementioned legislation, but with various other requirements that are in continuous flux and that have their particularities according to their regulatory framework. A sustainable rehabilitation is one whose interventions on a building are in line with the definition of sustainability understood in a broad sense. In other words, it must consider environmental, economic, and social aspects [34,35].

It should be noted that as it is a specially protected property, interventions on it are limited and those limitations must be respected. The previsions of the Spanish Technical Building Code (CTE) [36] and its Basic Documents (DB), which are understood to be all applicable, except that of Energy Saving (HE), which can be exempted, cannot be ignored here.

In the case of the tower studied here, due to the characteristics presented by the study, focus should be placed mainly on the DB of Health (HS), specifically in two of its five requirements: HS1 (Protection against moisture) and HS5 (Water evacuation). Within the concept of sanitation established by the DB-HS, the risk of deterioration of the building due to different factors is included, as we understand happens to this case.

The execution of the three interventions that we consider necessary to undertake in the tower (cleaning the roof, enclosing the windows and the ascent to the bell tower, and cleaning its interior) can pose some problems from an eco-sustainable point of view.

As established by Al Khatib, Soon Poh, and El-Shafie [37], the identification of problems with sustainable materials in interventions in historical settings is an urgent need, since it will help to save on the cost of execution, also reducing the time each intervention takes to complete, contributing to sustainable construction.

Considering their study, we understand that the main obstacles for the eco-sustainable execution of these three interventions in the tower would be:

1. Using eco-sustainable materials and resources.
2. Finding local providers of eco-sustainable resources.
3. Completion time.
4. Management of the waste.

Next we will discuss possible solutions to these obstacles.

The use of eco-sustainable materials such as disinfectants, cleaning products and eco-sustainable paints must be given fundamental importance, as well as the use of brushes, and other tools made with mostly natural elements, with a long durability and high quality. They should also have a low impact on the environment and preserve the historical and aesthetic value of the building.

Regarding the sustainability of the cleaning products used, containers should be made of fully recycled material whenever possible and should be highly effective and efficient. They should also be safe for human health and free of carcinogenic or mutagenic components, heavy metals, ammonia, petroleum, distillates, etc. Its acute oral toxicity must be greater than 5000 mg/kg and the acute cutaneous toxicity greater than 2000 mg/kg. It is recommended that the pH be between 2 and 11.5 with a flash point above 93 °C. The cleaning products should also be highly concentrated, in order to reduce water consumption, the actual product to be used per m^2 should be calculated, and products should not have been tested on animals. After being diluted aquatically, they should be biodegradable [38].

Traditional paints are synthetic and contain highly polluting elements, such as lead, cadmium, acetone, etc., which release their chemicals over the course of years. We can find four types of ecological paints whose main components are of vegetable or mineral origin. Among them, it is worth highlighting the lime paints traditionally used in Andalusia, which have a high impermeability and fungicidal and antiseptic capacity, preventing the proliferation of fungi, mold, and bacteria. They are very resistant both for indoor and outdoor use. The second kind worth mentioning are clay paints, which have several different shades and are very delicate. They have been used and continue to be used traditionally for plastering walls and their interior layers, such as mortar. Plant-based paints, of mainly organic and vegetable origin, are increasingly used for interiors. The last type, silicate paints, have antibacterial, disinfecting, and fungicidal properties but require specialized professionals for their application and have a higher cost [39].

Regarding the covering of the windows with anti-bird screens, there are several possibilities, depending on the providers. This type of mesh, with its low visual impact, is usually made of two materials: Nylon and polyethylene, both recyclable and highly resistant. The rolls of the highest quality mesh are made by thermoformed knots, with a thickness of between 1.2 and 1.5 mm, diamonds of 4 to 5 cm and a resistance of between 200 to 250 kg/m^2. This type of mesh lasts between 5 and 15 years. Its durability varies depending on whether the mesh is subjected to UV filters, which delay deterioration over time. Another possibility would be the use of metal hunting mesh, which is low cost, has quick installation, and ensures respect for the environment. It is made of galvanized wire with a galvanized finish. Hunting mesh has a low visual impact and high resistance, with the thickness of the wires ranging from 1.9 mm to 3 mm. The second obstacle centers around having enough environmentally sustainable materials provided. Due to the geographical location of Montilla, in the center of Andalusia, we understand that if it is not possible to find suppliers in Montilla itself that meet these requirements, they can be found in other parts of the region.

The duration of the intervention should also be considered, as it will influence the cost of the project. That is why it is necessary to leave a certain margin when determining the execution times, considering weather problems that prevent work or can cause delays in the delivery of materials.

The last of the problems raised is the management of the waste generated, for which it is necessary to abide by the current legislation, mainly R.D. 105/2008 [40] and Decreto 73/2012 [41] and other European and Spanish legislation applicable.

First of all, it must be taken into consideration that construction and demolition waste (RDC) is "any substance or object generated in construction or demolition and that its owner disposes of or has the intention or obligation to dispose of" [42]. Those residues pose a problem. Their environmental impact can affect soil and water contamination, the non-use of resources, and the deterioration of the landscape, contributing, in turn, to savings in raw materials and the conservation of natural resources. Their treatment is necessary so that they can be correctly disposed of and, where appropriate, recycled and reused, depending on the material in question. As we can see of the general characteristics of the materials recommended for the intervention, they are eco-sustainable materials that do not needing special treatment in terms of general waste management. They can therefore be disposed of at either of the two waste management plants for construction and demolition materials located near Montilla.

As stated above, although the tower is currently in good condition overall, interventions that favor its conservation must be carried out, and we think that, despite the problems of starting work on a protected building, it is possible to do so by applying eco-efficiency criteria. To this end, we have defined the medium- and long-term risks that may affect the conservation of the property in addition to the problems that must be dealt with for an eco-efficient rehabilitation, identifying the most vulnerable parts of this risk and the level of protection [31].

The geographical location of Montilla and the development of the businesses in the area make it possible to acquire materials and contract suppliers and labor locally, causing a positive impact on the local economy.

The roof with four sloping sides facilitates water evacuation, thanks to its four natural drainage points. Its cleaning and the revision of its tiles and waterproofing do not entail great difficulties.

The covering of gaps in the windows will also protect the structure from harsh weather and from the damage caused by entering birds. Given the materials that are proposed for this project, it is understood that the visual impact will be minimal while the changes will be high durable and easy to maintain.

Lastly, the use of eco-sustainable products will be possible in sanitization the of the tower's interior.

4. Conclusions

The conservation of specially protected historical buildings, such as the tower of the Church of Santiago Apóstol, is essential for the preservation of the historical, artistic, and cultural values of towns and it facilitates their renovation and renewal over time [43,44]. Now, the question we should be asking ourselves is whether it is possible to rehabilitate or carry out interventions on these buildings using eco-efficient criteria.

Eco-sustainability applied to renovations is a recent practice and is usually applied in new buildings or in the rehabilitation of urban periphery and industrial buildings being turned into housing [34,45–48]. Eco-sustainable interventions present important challenges when carried out in historical buildings as we must consider the materials used, their impact on the building, the period of execution, cost, economic impact, generation of resources, etc. [37]. Today and the future's society is responsible for the conservation of historical buildings and preservation of the beauty of the heritage left by our ancestors. The approach from an eco-sustainable point of view will generate benefits in the short and long term, also redounding in the surrounding economy [32].

The current health and economic crises make it necessary to find ways to reactivate the economy and, specifically, the Construction Sector. For decades, the depopulation of the rural environment has continued. This crisis is leading part of the population to reconsider returning to rural environments, which are the keepers of a large part of the historical assets and monuments that must be preserved. The eco-sustainable rehabilitation of these buildings is, without a doubt, a process that helps support a circular economy, the enhancement of the environment, the generation of employment in these places, and in some cases, an increase in income derived from tourism.

To determine if an eco-sustainable rehabilitation or intervention is feasible in a specially protected historic building, we have developed a methodology based on the MIVES model and three approaches to sustainability (environmental, social, and economic). In our methodology, the impact on the environment will be measured through indicators that include: Energy consumed, ratio of recycled and reusable materials, management of the waste generated, and number of local service and resource providers and laborers to generate economic benefits in the area. For this assessment, a quality index had been established that will help to determine the execution of an eco-sustainable intervention.

At the end of the 20th century, the tower underwent an important rehabilitation, so today it is not necessary to carry out a major intervention on its structure, although it does require conservation work to ensure its preservation. We specifically recommend an intervention in three parts: (1) Cleaning the roof, (2) covering the windows and access to the bell tower to prevent the entry of birds, and (3) cleaning the interior of the tower. We have determined that there are four main considerations when evaluating an eco-sustainable intervention (the use of eco-sustainable materials and resources, sourcing of eco-sustainable materials and laborers that come from a geographical area near the monument, reasonable duration of the intervention, and safe and effective management of the generated waste), all of which are achievable in our case study.

For these reasons, we think that eco-efficiency criteria can and should be applied, not only as in this case study, but also to other interventions on specially protected historic buildings in rural areas. It will bring benefits not only to their structure and conservation, but also to the surrounding area.

Author Contributions: Introduction, M.A.C.-S.; materials and methods, M.A.C.-S.; results, M.A.C.-S.; conclusions, M.A.C.-S.; writing-original draft preparation and editing, M.A.C.-S.; writing-review, P.M.-M., I.L.C.-G. and F.M.-T. All authors have read and agreed to the published version of the manuscript.

Funding: This research received no external funding.

Conflicts of Interest: The authors declare no conflict of interest.

References

1. Decreto 111/2001, de 30 de Abril, por el que se Declara Bien de Interés Cultural, Con la Categoría de Monumento, la Iglesia de Santiago, en Montilla (Córdoba). Available online: https://www.juntadeandalucia.es/boja/2001/81/37 (accessed on 28 July 2020).
2. Ley 16/1985, de 25 de Junio, del Patrimonio Histórico Español. Available online: http://noticias.juridicas.com/base_datos/Admin/l16-1985.html (accessed on 28 July 2020).
3. Ley 14/2007, de Patrimonio Histórico de Andalucía. Available online: http://noticias.juridicas.com/base_datos/CCAA/an-l14-2007.html (accessed on 28 July 2020).
4. Barbero-Barrera, M.M.; Fernández-Rodríguez, M.A.; Flores-Medina, N.; Pinilla-Melo, J. La materialidad de la arquitectura. In *V Encuentro sobre Experiencias Innovadoras en la Docencia*; Universidad Complutense de Madrid: Madrid, Spain, 2015; Available online: https://www.edificacion.upm.es/innovacion/2015/18.La%20Materialidad_BARBERO_FERNANDEZ_FLORES_PINILLA.pdf (accessed on 28 July 2020).

5. García-Alvarado, R. Los límites de la materialidad. Tres cualidades arquitectónicas para el próximo siglo. In *Arquitectura del Sur*; Universidad del Bío-Bío: Santiago, Chile, 1998; Volume 14, p. 27. Available online: https://dialnet.unirioja.es/servlet/articulo?codigo=5231434 (accessed on 28 July 2020).

6. Convención Marco de las Naciones Unidas sobre el Cambio Climático (CMNUCC). Available online: https://www.miteco.gob.es/es/cambio-climatico/temas/el-proceso-internacional-de-lucha-contra-el-cambio-climatico/naciones-unidas/CMNUCC.aspx (accessed on 28 July 2020).

7. Protocolo de Kioto. Available online: http://unfccc.int/resource/docs/convkp/kpspan.pdf (accessed on 28 July 2020).

8. Directiva 2010/31/EU del Parlamento Europeo y del Consejo, de 19 de Mayo de 2010, Relativa a la Eficiencia Energética de los Edificios. Available online: https://eur-lex.europa.eu/legal-content/ES/TXT/?uri=celex%3A32010L0031 (accessed on 28 July 2020).

9. Bell, S.; Morse, S. Sustainability Indicators Past and Present: What next? *Sustainability* **2018**, *10*, 1688. [CrossRef]

10. Viñolas, B.; Cortés, F.; Marques, A.; Josa, A.; Aguado, A. MIVES: Modelo Integrado de Valor para Evaluaciones de Sostenibilidad. In *II Congrés Internacional de Mesura i Modelització de la Sostenibilitat*; Centro Internacional de Métodos Numéricos en Ingeniería (CIMNE): Barcelona, Spain, 2009; Available online: https://upcommons.upc.edu/handle/2117/9704?locale-attribute=es (accessed on 28 July 2020).

11. Rüdisser, J.; Leitinger, G.; Schirpke, U. Application of the Ecosystem Service Concept in Social-Ecological Systems-from Theory to Practice. *Sustainability* **2020**, *17*, 2960.

12. Wüest, T.; Grobe, L.O.; Luible, A. An innovative Gaçade Element with Controlled Solar-Thermal Collector and Storage. *Sustainability* **2020**, *12*, 5281.

13. Quintanilla Raso, M.C. *Nobleza y Señoríos en el Reino de Córdoba: La casa de Aguilar (s. XIV-XV)*, 1st ed.; Monte Piedad y Caja de Ahorros de Córdoba: Córdoba, Spain, 1979; Available online: https://dialnet.unirioja.es/servlet/libro?codigo=609190 (accessed on 28 July 2020).

14. Ramírez de Arellano, R. *Inventario-Catálogo Histórico Artístico de Córdoba*, 1st ed.; Monte Piedad y Caja de Ahorros de Córdoba: Córdoba, Spain, 1982; p. 431.

15. Lorenzo Muñoz, F.B. *Historia de la M. N.L. Ciudad de Montilla*, Montilla, Spain, Unpublished work. 1779.

16. Jurado y Aguilar, A. *Historia de Montilla*, Montilla, Spain, Unpublished work. 1777.

17. Jurado y Aguilar, L. *Historia de Montilla*, Montilla, Spain, Unpublished work. 1763.

18. Garramiola Prieto, E. Las torres de Santiago (I). *Munda-Montulia*, Excmo Ayuntamiento de Montilla: Montilla, Spain, Unpublished work. 1979.

19. Nieto Cumplido, M. Aproximación histórica a la historia de Montilla en los siglos XIC y XV. In *Montilla, Aportaciones Para su Historia (I Ciclo de Conferencias Sobre la Historia de Montilla)*; Delegación de Cultura del Excmo. Ayuntamiento de Montilla: Montilla, Spain, 1982; p. 299.

20. García Ortega, A.J. De Mezquitas a Iglesias. Formalización y Trazado en los Procesos de Reconversión de Toledo y Córdoba. *EGA Expr. Gráfica Arquit.* **2015**, *26*, 202–211. Available online: https://polipapers.upv.es/index.php/EGA/article/view/4053 (accessed on 28 July 2020).

21. Suárez Medina, R. El Sonido del Espacio Eclesial en Córdoba. El Proyecto Arquitectónico Como Procedimiento Acústico. Ph.D. Thesis, Universidad de Sevilla, Sevilla, Spain, 2002; pp. 24–31. Available online: https://idus.us.es/handle/11441/15659 (accessed on 28 July 2020).

22. Delgado López, D. Historia de Montilla. *Breve Resumen General de España, Tomo I*, Montilla, Spain, Unpublished work. 1895.

23. Morte Molina, J. *Montilla: Apuntes Históricos de Esta Ciudad*; Imprenta, Papelería y Encuadernación de M. de Sola Torices: Montilla, Spain, 1888; Available online: http://www.bibliotecavirtualdeandalucia.es/catalogo/es/consulta/registro.cmd?id=1001022 (accessed on 28 July 2020).

24. Baena Sánchez, A. Estudio Histórico-Técnico y Reconstrucción Virtual del Alhorí de los Duques de Medinaceli de Montilla a Través de la Obra del Arquitecto Juan Antonio Camacho de Saavedra. Ph.D. Thesis, Universidad de Córdoba, Córdoba, Spain, 2018. Available online: https://helvia.uco.es/xmlui/handle/10396/16876 (accessed on 28 July 2020).

25. Martín González, J.J. *Historia del Arte. Tomo II*; Editorial Gredos: Madrid, Spain, 1978; pp. 38–56.

26. Archivo Histórico de Protocolos Notariales de Montilla (A.H.P.N.M.), Escribanos siglo XVI, Legajo 143. Fundación Biblioteca Manuel Ruiz Luque: Montilla (Córdoba), Spain, Unpublished work. 1577.

27. Navascués Palacio, P. *El Libro de la Arquitectura de Hernán Ruiz, el Joven. Estudio y edición crítica*, 1st ed.; Escuela Técnica Superior de Arquitectura de Madrid: Madrid, Spain, 1979; p. 16.

28. Sánchez González, A. *El arte en la representación del espacio. Mapas y planos de la colección Medinaceli*, 1st ed.; Servicio de Publicaciones de la Universidad de Huelva: Huelva, Spain, 2017; pp. 340–341.

29. Archivo Central de la Consejería de Cultura y Patrimonio Histórico de Andalucía (A.C.C.C.P.H.A.) A9.004.14. Consejería de Cultura y Patrimonio Histórico: Sevilla, Spain, 1577; Unpublished work.

30. Suárez, M.; Gómez Baggethun, E.; Benayas, J.; Tilbury, D. Towards an Urban Resilience Index: A Case Study in 50 Spanish Cities. *Sustainability* **2016**, *8*, 774.

31. Herráez, J.A.; Pastor, M.J.; Durán, D. *Guía para la Elaboración e Implantación de Planes de Conservación Preventiva*; Ministerio de Cultura y Deporte: Madrid, Spain, 2019. Available online: https://es.calameo.com/read/00007533531922e9ec8e8 (accessed on 28 July 2020).

32. Go-Eum, K.; Jeon-Ran, L. The impact of Historic Building Preservation in Urban Economics: Focusing on Accommodation Prices in Jeonju Hanok Village, South Korea. *Sustainability* **2020**, *12*, 5005.

33. Quintana López, T. *Urbanismo Sostenible. Rehabilitación, Regeneración y Renovación Urbanas*, 1st ed.; Tirant-LoBlanch: Valencia, Spain, 2016; pp. 373–381.

34. Ruiz Palomeque, G. Sustainable renovation management of Large Urban Ensembles by local public administration. *Inf. Constr.* **2015**, *67*. [CrossRef]

35. Mercader Moyano, P.; Roldán Porras, J. Evaluating Environmental Impact in Foundations and Structures through Disaggregated Models: Towards the Decarbonisation of the Construction Sector. *Sustainability* **2020**, *12*, 5150. [CrossRef]

36. Código Técnico de Edificación Español. Available online: https://www.codigotecnico.org (accessed on 28 July 2020).

37. Al Khatib, B.; Soon Poh, Y.; El-Shafie, A. Materials Challenges in Reconstruction of Historical Projects: A Case Study of the Old Riwaq Project. *Sustainability* **2019**, *11*, 4533. [CrossRef]

38. Logic Clean. ¿Cómo es y Cómo se Hace una Limpieza Sostenible? Available online: https://www.logicclean.es/una-limpieza-ecologica-sostenible/ (accessed on 18 July 2020).

39. Arquitectura sostenible. Pinturas ecológicas, 100% Sostenibles. Available online: https://arquitectura-sostenible.es/pinturas-ecologicas-100-sostenibles/ (accessed on 10 July 2020).

40. Real Decreto 105/2008, de 1 de Febrero, por el que se Regula la Producción y Gestión de Residuos de Construcción y Demolición. Available online: http://www.juntadeandalucia.es/medioambiente/portal_web/web/temas_ambientales/residuos_2/marco_normativo/boe_rcd.pdf (accessed on 28 July 2020).

41. Decreto 73/2012, de 22 de Marzo, por el que se Aprueba el Reglamento de Residuos de Andalucía. Available online: https://www.juntadeandalucia.es/boja/2012/81/4 (accessed on 28 July 2020).

42. Agencia de Obra Pública de la Junta de Andalucía; Consejería de Fomento y Vivienda de la Junta de Andalucía. *Gestión y Tratamiento de Residuos de Construcción y Demolición (RCD). Guía de buenas prácticas*; Agencia de Obra Pública de la Junta de Andalucía: Córdoba, Spain; Consejería de Fomento y Vivienda de la Junta de Andalucía: Córdoba, Spain, 2015. Available online: http://www.juntadeandalucia.es/medioambiente/portal_web/servicios_generales/doc_tecnicos/2015/gestion_tratamiento_residuos_construc_demolic/gestion_tratamiento_residuos_RCD_buenas_practicas.pdf (accessed on 28 July 2020).

43. Sanz Cabrera, J. Restauración de Sta. María la Mayor (SMLM). Baena. Córdoba. Spain. La austeridad y eficiencia en la recuperación de los espacios perdidos. In *Libro de Actas del 3er Congreso Internacional de Construcción Sostenible y Soluciones Ecoeficientes*; University of Seville: Sevilla, Spain, 2017; Available online: https://idus.us.es/bitstream/handle/11441/59609/Sanz%20cabrera%2C%20jerónimo%20%28espa%29.pdf?sequence=2&isAllowed=y (accessed on 28 July 2020).

44. Corgnati, S.; Filipi, M.; Perino, M. A new approach for the IEQ assessment, research in building physics and building engineering. In *Proceeding of the 3rd International Conference on Research in Building Physics IBPC 2006*; Taylor & Francis Group: Montreal, QC, Canada, 2006; Available online: https://www.researchgate.net/publication/279878849_A_new_approach_for_the_IEQ_Indoor_Environment_Quality_assessment (accessed on 28 July 2020).

45. Blandón, B.; Palmero, L.; di Ruocco, G. The Revaluation of Uninhabited Popular Patrimony under Environmenthal and Sustainability Parameters. *Sustainability* **2020**, *12*, 5629.

46. Diulio, M.P.; Mercader-Moyano, P.; Gómez, A.F. The influence of the envelope in the preventive consevation of books and paper records. Case study: Libraries and archives in La Plata. *Elsevier Energy Build.* **2019**, *183*, 727–738. [CrossRef]
47. Candelas-Gutiérrez, A. The Power of Geometrix Relationships in Mudéjar Timber Roof Frames. *Nexus Netw. J.* **2017**, *19*, 521–545. [CrossRef]
48. Candelas-Gutiérrez, A.; Borrallo-Jiménez, M. Methology of Restoration of Historical Timber Roof Frames. Application to Traditional Spanish Structural Carpenty. *Int. J. Archit. Herit.* **2020**, *14*, 51–74. [CrossRef]

© 2020 by the authors. Licensee MDPI, Basel, Switzerland. This article is an open access article distributed under the terms and conditions of the Creative Commons Attribution (CC BY) license (http://creativecommons.org/licenses/by/4.0/).

MDPI

St. Alban-Anlage 66

4052 Basel

Switzerland

Tel. +41 61 683 77 34

Fax +41 61 302 89 18

www.mdpi.com

Sustainability Editorial Office

E-mail: sustainability@mdpi.com

www.mdpi.com/journal/sustainability

www.ingramcontent.com/pod-product-compliance
Lightning Source LLC
LaVergne TN
LVHW070202170726
843515LV00007B/1981